Total Construction
Project Management

About the Authors

George J. Ritz was a leading expert in construction and project management, with 40 years' experience in executing projects in the United States and abroad. He worked on a broad spectrum of construction projects, ranging from schools for students who are physically challenged to world-class petrochemical plants. Mr. Ritz was a registered professional engineer and a frequent lecturer on project management and related topics. He wrote *Total Engineering Project Management*, also published by McGraw-Hill Education.

Sidney M. Levy has 35 years' experience as a senior executive in a major general-contracting firm in New England. He is the author of 30 books on construction means, methods, and operations in the United States and abroad. He has lectured to industry groups in the United States, Mexico, Europe, Japan, and Korea. His book *Project Management in Construction* was awarded the British Chartered Institute of Building Silver Medal. Mr. Levy currently resides in Baltimore, Maryland, where he is an independent construction consultant.

Total Construction Project Management

George J. Ritz

Sidney M. Levy

Second Edition

New York   Chicago   San Francisco
Lisbon   London   Madrid   Mexico City
Milan   New Delhi   San Juan
Seoul   Singapore   Sydney   Toronto

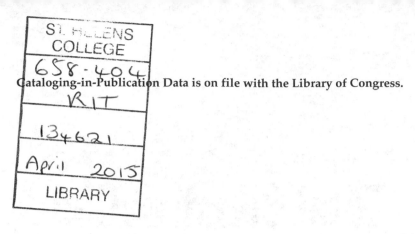
Cataloging-in-Publication Data is on file with the Library of Congress.

McGraw-Hill Education books are available at special quantity discounts to use as premiums and sales promotions, or for use in corporate training programs. To contact a representative please e-mail us at bulksales@mcgraw-hill.com.

Total Construction Project Management, Second Edition

Copyright © 2013, 1994 by McGraw-Hill Education, LLC. All rights reserved. Printed in the United States of America. Except as permitted under the United States Copyright Act of 1976, no part of this publication may be reproduced or distributed in any form or by any means, or stored in a data base or retrieval system, without the prior written permission of the publisher.

1 2 3 4 5 6 7 8 9 0   DOC/DOC   1 9 8 7 6 5 4 3

ISBN 978-0-07-180137-9
MHID 0-07-180137-5

The pages within this book were printed on acid-free paper.

Sponsoring Editor	**Proofreader**
Larry S. Hager	Cenveo® Publisher Services
Acquisitions Coordinator	**Production Supervisor**
Bridget Thoreson	Pamela A. Pelton
Editorial Supervisor	**Composition**
David E. Fogarty	Cenveo® Publisher Services
Project Manager	**Art Director, Cover**
Yashmita Hota, Cenveo® Publisher Services	Jeff Weeks
Copy Editor	
Cenveo® Publisher Services	

Contents

Preface

Mr. George J. Ritz, author of *Total Construction Project Management*, passed away before a second edition of this popular book could be produced. The editors at McGraw-Hill Education selected Sidney M. Levy to assume this task.

Mr. Ritz was a registered professional engineer with decades of experience in engineering and construction projects, and he had that rare insight that a construction project consists of more than the plans and specifications. The managers of the project, along with the other members of the team—the owner, design consultants, vendors, and subcontractors—must form a relationship of mutual trust and respect in order to create a successful project. The author of this second edition shares these same views.

Technological advances over the past several decades have manifested themselves in both construction and design, and the information technology industry has had a profound impact on the way we design and build today's projects. Nowadays a project superintendent or manager can take a photograph of a questionable detail with a smartphone, transmit it immediately to the architect or engineer, and possibly get a response before his or her coffee gets cold.

Computer-assisted design has advanced with the development of software permitting three-dimensional drawings to be simultaneously transmitted to each team member for review and comment. This software for building information modeling has transformed the industry and reduced many constructability and design conflicts, thereby speeding up the entire process and lowering the potential for disagreements and disputes.

But as Mr. Ritz stated throughout his book, the human element and the relationships people form and build on will continue to define a successful project.

As the term *construction manager* is used throughout the book, it is intended to mean the person responsible for managing the construction process at the jobsite, whether stationed in the field, as a project superintendent, site supervisor, construction engineer, or the project manager assigned to a single or multiple projects, stationed in the

· company's corporate office. And of course, the term can connote the constructor who is acting as the owner's *agent* rather than the third party between owner and design consultants on the one hand and builder on the other.

But anyone responsible for field activities and operations would be wise to embrace the concept of *total construction project management* to meet his or her project and personal goals.

George J. Ritz wrote in the preface of his initial edition:

> I hope the information passed along to you in this book will help you enjoy managing in the construction industry as much as I have. Remember, none of us will ever know all there is to know about construction management, so keep an open mind.
>
> Good luck on your projects!

As the author selected to update and revise this book, I plan to follow Mr. Ritz's advice.

SIDNEY M. LEVY

Total Construction Project Management

CHAPTER 1

The Construction Management Environment

If one is going to practice management within an industry, it is a good idea to define the arena in which the management techniques will be applied. We really need to know just what business we are in to evaluate our present goals, find out where we have been, and see where we hope to go from here.

The theme of this book is the practice of management across all facets of project execution in construction. Therefore, construction technology will be introduced only as it bears on total construction project management. It is important for many reasons to master the basic technology applicable to your specialized field of construction. Knowledge and understanding of construction operations and technology are key to running a successful project; mastery is an essential ingredient in progressing up the management ladder and improving your professional standing.

The "total" part of this book's title means that we are addressing the application of construction management practices in an integrated systems context. One must apply all the techniques presented here over the life of the construction project to be successful. The construction project manager's genuine commitment to total construction project management must be absolute. Superficial utilization will not meet the goal of effective construction project management.

Some History of the Construction Process

Construction has had a long history, closely paralleling the development of human civilization. The Egyptian pyramids, constructed in 2600 BCE (before the Common Era—often used in lieu of BC,

1

before Christ), reveal a remarkable knowledge about geometry and mechanics: the use of the lever. And the mystery of Stonehenge and the massive stone figures on Easter Island also convey a sense of the same types of knowledge.

The temple and tomb builders of the Mediterranean civilizations left many high-profile projects of their era. Construction technology had moved along with the development of the wedge, lever, sledge, rollers, and the incline plane to make those monuments possible. The Egyptians, in the construction of the pyramids, used these levers, wedges, rollers, and inclined planes.

Ancient Greece developed architectural systems around the fifth century BCE; many of the products of these systems stand today as testimony to their beauty and structural integrity. The Greek word for *architect* is *architektoniki*, which literally translated means "master builder." The Roman Empire, which existed from 27 BCE to 476 CE, picked up many of these engineering concepts—witness the Colosseum, completed in 80 CE. We see these two disciplines of construction and architecture merging and taking on new meanings as we approach the Renaissance (meaning "rebirth"), which spanned the period from the fourteenth to the seventeenth century, saw the beginning of the Scientific Revolution, and produced some of the world's greatest architects and builders.

And then there was Marcus Vitruvius Pollio, referred to simply as Vitruvius, an architect and military engineer who developed the *Ten Books on Architecture*. Each volume is devoted to a different topic—site work, materials, temple construction (details, doors, and altars), etc.—making it perhaps the first book of architectural specifications. Vitruvius is perhaps better known for his sketch of Vitruvian Man, immortalized by Leonardo da Vinci (Fig. 1.1), which takes the measurements of a man—fingers, palms, height, length of the arm—and relates them to proportions to be used in the design of a building. One of his statements in Book VI may not please designers: he says, "Poor but honest makes a good architect." Filippo Brunelleschi (1377–1446) was one of the foremost architects and engineers during the Renaissance; he created the dome of the Florence Cathedral, a marvel of engineering for its day.

Today's designers using computer-assisted design create three-dimensional drawings that can actually permit a building to be built without creating one paper drawing, as General Motors did for a plant in Michigan several years ago. Adding the fourth dimension, time, creates a virtual schedule; software engineers consider the fifth dimension one that creates material-quantity takeoffs as design progresses. But in the field, we still seek tests to determine proper soil bearing capacity for our foundations, Erector-set structures of steel or concrete, and mechanical, electrical, and plumbing systems installed pipe by pipe, duct by duct.

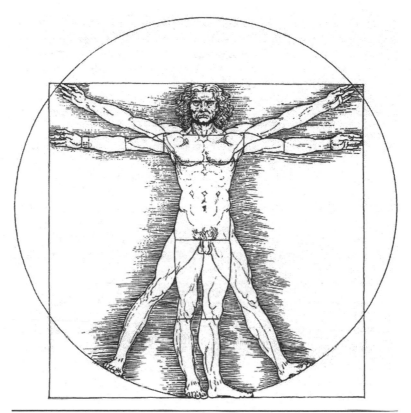

FIGURE 1.1   Da Vinci's Vitruvian Man.

Where Do Managers of Construction Projects Come From?

Today, there are any number of colleges, universities, and local community colleges offering degree or certificate courses in construction and project management. But many project managers took another route. An astute foreperson or project superintendent, recognized by management as a comer, will be given more authority and training and will climb the ladder of improvement within the field; this process of advancement continues today.

Although many universities offer four-year degree programs in project or construction management, there are colleges that offer certificate programs requiring attendance for only one or two years, and there is even online instruction. There are also bachelor-of-science programs offering degrees in information technology that are applicable to a career in construction.

Many community colleges around the country provide courses in project management for either day or evening students, thereby allowing someone to learn the basics of the profession while pursuing a day job, possibly in another field.

The University of Texas at Austin's Cockrell School of Engineering is the home of the Construction Industry Institute, which offers continuing-education courses, online-education courses, and numerous workshops that also offer the ability to hone the development of one's skills in the construction industry.

How Large Is the Construction Industry?

We all know that the depressed economic conditions—not only in this country, but worldwide—beginning in the first decade of the twenty-first century had an impact on the construction industry.

The following chart tracks some of these ups and downs. Rather than display complete decades, we selected individual years: we began with 1993, which was prior to the beginning of the world economic meltdown, and proceeded through the boom and bust years of 2000, 2007, and 2010. The figures for 2011 were the latest available at the time of this writing.

Year	Total	Nonresidential*	Residential	Public
1993	458,808	150,025	194,150	113,906
2000	784,954	247,213	353,065	184,676
2007	1,429,997	593,821	556,078	280,098
2010	1,107,559	550,565	262,843	294,151
2011	1,103,029	567,835	248,548	286,646

*Nonresidential includes commercial, healthcare, educational, religious, transportation, communication, manufacturing, etc.
Source: U.S. Census Bureau.

Value of Construction Put in Place per Year (Millions of Dollars)

The following provides a snapshot of the construction industry based upon the latest information available as of this writing:

Figure 1.2: Employment in the largest occupations in the building industry as of May 2010, including not only trade workers but also supervisors, managers, and administrative staff.

Figure 1.3: Mean hourly wages for the largest occupations in the building industry as of May 2010, including supervisors, managers, and administrative staff.

Table 1.1: 2010 Employment by occupation, including construction managers.

Table 1.2: Average earnings and hours for all employees in the industry, average earnings and hours for production and non-supervisory employees, and earnings by occupations, all as of December 2011.

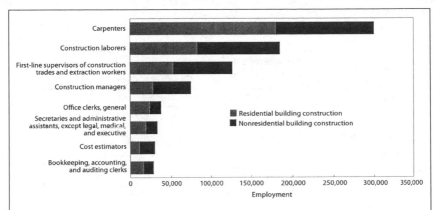

- Construction occupations accounted for 64 percent of employment in residential building construction and 62 percent of employment in nonresidential building construction.

- Carpenters made up almost half of the construction occupation employment in residential building construction but accounted for less than a third of the nonresidential building construction employment.

- First-line supervisors of construction trades and extraction workers and construction managers were more prevalent in nonresidential building construction than in residential building construction, accounting for about 4 percentage points more of the overall construction occupations employment.

FIGURE 1.2   Employment in the building industry. (*Source: U.S. Bureau of Labor Statistics*)

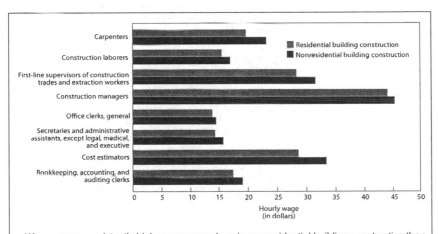

- Wages were consistently higher among workers in nonresidential building construction than those in residential building construction.

- The mean hourly wage for construction occupations was $19.55 in residential building construction, compared with $22.64 in nonresidential building construction.

- Cost estimators had the largest nominal difference in mean hourly wages with a $4.74 spread between nonresidential and residential building construction.

- Carpenters had the largest percent difference in mean hourly wages—17.6 percent—between nonresidential and residential building construction.

FIGURE 1.3   Mean hourly wages for the industry's largest occupations as of May 2010. (*Source: U.S. Bureau of Labor Statistics*)

Employment by Occupation	
Data Series	**Employment, 2010**
Carpenters	509,700
Construction laborers	608,810
Construction managers	153,100
Electrician	384,100
Operating engineers and other construction equipment operators	203,020

Source: Occupational Employment Statistics

TABLE 1.1 Employment by Occupations by Trade as of 2010

Earnings and Hours of All Employees					
Data Series	**Back Data**	**Sep. 2011**	**Oct. 2011**	**Nov. 2011**	**Dec. 2011**
Average hourly earnings		$26.20	$26.23	$26.26	$26.36*
Average weekly hours		38.0	38.1	37.5	37.4*

Earnings and Hours of Production and Nonsupervisory Employees					
Data Series	**Back Data**	**Sep. 2011**	**Oct. 2011**	**Nov. 2011**	**Dec. 2011**
Average hourly earnings		$23.28	$23.16	$23.04	$23.06*
Average weekly hours		38.8	38.9	38.1	38.0*

Earnings by Occupation				
	Wages, 2010			
	Hourly		**Annual**	
Data Series	**Median**	**Mean**	**Median**	**Mean**
Carpenters	$19.10	$21.10	$39,730	$43,890
Construction laborers	$14.49	$16.35	$30,140	$34,010
Construction managers	$40.08	$44.78	$83,370	$93,140
Cost estimators	$29.08	$31.54	$60,490	$65,600
First-line supervisors/managers of construction trades and extraction workers	$28.51	$30.21	$59,310	$62,830

*Preliminary

TABLE 1.2 Average Earnings and Hours for Production and Nonsupervisory Employees as of December 2011

Defining Construction Project Management

It is impossible to define a complex operation such as construction management in one simple sentence. We will have to dissect the term and define its many facets. Throughout this book we will use the term *construction manager* to apply interchangeably to a construction manager (CM) and a general contractor (GC) where we refer to that construction manager as simply a project manager. Although the CM acts as the owner's agent and the GC is a third-party participant in the construction process, both basically perform the same duties— management of the construction process.

What Is a Construction Project?

The term *construction project* means different things to different people. It can mean building a house, a high-rise, a dam, an industrial plant, or an airport, or even remodeling or upgrading a facility. Table 1.3 summarizes the various major categories of construction projects, delineating the diverse parts of the industry. It groups construction projects into various specialty areas based on the markets served by the industry. The intent is to make the list representative rather than exhaustive. You should be able to fit your type of construction projects into a suitable class.

The list of types of construction activities under each category provides an idea of the complexity of each type of project. The construction technology tends to become less complex as the categories move from left to right.

Although the types of projects shown in Table 1.3 differ, they do have at least four traits in common:

1. Each project is unique.

2. A project works against schedules and budgets to produce a specific result.

3. The construction team cuts across many organizations and functional lines that involve virtually every department in the company.

4. Projects come in various shapes, sizes, and complexities.

Looking at the matrix of construction projects depicted in Table 1.3, one can readily see that each type of project is unique and not repetitive. Rarely do we find that even two single-family homes are built exactly the same. The individuality of the owners assures that all construction projects will have some degree of uniqueness about them.

Process-Type Projects			
Liquid/Gas Processing Plants (1)	**Liquid/Solid Processing Plants (2)**	**Solids Processing Plants (3)**	**Power Plants (4)**
Refineries & petrochemical plants Organic & inorganic chemicals Monomers Basic chemicals LNG & industrial gases Nat. gas cleanup etc.	Pulp & paper Mineral & ore dressing plants Polymers & plastics Pigments & paints Synthetic fibers Food, beverage & pharmaceuticals Specialty chemicals Soaps & detergents Films & adhesives etc.	Cement, clays, & rock Mining & smelting Iron & steel Nonferrous metals Fertilizers & agricult. chemicals Glass & ceramics Rubber & polymer extrusion Activated carbon & carbon black etc.	Fossil fuel & hydroelectric plants Power transmission lines & substations Cogeneration plants Coal gasification & hot-gas clean-up Flue gas scrubbers Solid waste burning Nuclear power* etc.
% of TIC** 12%	10%	9%	7–8%
Types of Construction Activities			
Complex process piping systs., with much process equip. using exotic alloys Major computerized process contr. systs. Extensive power & control wiring systems Major utility equip. & distribution systems Piling & heavy equip. foundations Extens. light & heavy structural steel Thermal insulation Extens. safety systs. Major pollution & waste contr. systs. Minor bldgs. & enclosed spaces—min. architect. treatment Minor HVAC & plmbg.	Average process piping systems with mixture proc. & mech. equip. some in exotic alloys Major computerized proc. contr. systs. Extensive power & control wiring systs. Major utility equip. & distribution systems Piling & heavy equip. foundations Heavy & light structural steel Thermal insulation Exten. safety systs. Major pollution & waste control systs. Major industrial & support buildings Sanitary construction	Minor process piping systems with heavy mech. equip. Some exotic alloys & finishes Moderate computer proc. control systs. Extensive wiring & elect. systems Major utility equip. & distribution systems Piling & heavy equip. foundations Heavy & light structural steel Thermal insulation Exten. safety systs. Major pollution & waste-control systs. Major industrial & support buildings Sanitary construction	Minimal proc. piping systems, but large hi-temp. steam & gas piping systs. Heavy boiler & turbogenerator equip. Some alloys & special finishes Major boiler control & minor proc. contr. Major wiring & elect. substation work Major utility systs. Piling & heavy equip. foundations Heavy structural steel & stacks Thermal insulation Major industrial & support buildings Some architectural finished areas
Applicable codes: OSHA, MSHA, ASME, ANSI, EPA, ASTM, and local building, plumbing, and electrical codes.			
Key Craft Labor Used			
Major: Pipefitter-welders, ironworkers, riggers, boilermakers, millwrights, equipment operators, electricians, instrument technicians, and concrete workers. Minor: Insulators, painters, specialty brick masons, carpenters, laborers, and some architectural trades			

*Nuclear jobs are exponentially more complex due to NRC regulations.

TABLE 1.3 A Matrix of Construction Project Characteristics

Nonprocess-Type Projects				
Manufacturing Plants (5)	Civil Works Projects (6)	Support Facility Projects (7)	Commercial and A&E (8)	Miscellaneous Projects (9)
Automotive & heavy equip. assy. plants Light manufacturing Electronic component mfg. & assembly plants Aerospace manufacturing plants etc.	Dams & irrigation Highways & bridges Public transport systems Water & sewerage treatment systems Port & marine projects Airports Public works infra-structure projects Oil, gas, & water transmission lines etc.	Laboratories Test facilities Aerospace test facilities R&D facilities Pilot plants etc.	Office buildings High-rise buildings Shopping malls Healthcare facilities Institutions, schools, banks & prisons Multiple-family housing units Multiple-unit housing schemes Military facilities etc.	Plant turn-arounds Revamp projects Restoration projects Single-family housing units etc.
7%	6%	6%	4–6%	Variable
Types of Construction Activities				
Site devel. per size & local conditions Major indust. bldgs. with architect treatmt. Major HVAC, plmbg., utility & some minor process piping systems Light to medium assembly line & matl. handling systems Clean room assembly & laboratory areas Computer rooms Fire protection systs. Moderate electrical distrib. systems Heavy pollution control & waste treatmt.	Heavy earth-moving, blasting, & site preparation Large underground piping systems Major forming, rebar instl., & conc. pours Heavy concrete & asphalt paving Some industrial & passenger bldgs. Fuel, water, & waste treatment, & mech. equip. installation Railroad track installation Signalling & communications systs. Use of pre- & post-stressed concrete Sheet piling, coffer dams, & marine construction	Mixture of industrial & institutional bldgs. Some specialized equip. installation Moderate utility systems and minor process systems Heavy HVAC with fume-control systs. Some special finishes on bldgs. and equip. Moderate to heavy elect. wiring systems Average site dev. & concrete work per site conditions Heavy plumbing & waste treatment Interior & exterior architectural trades	Heavy architectural treatments & high-rise building construction technology Major HVAC & plumbing systems Major-use masonry & precast wall units Major use of window-wall installations Use of high productivity techniques for duplicate floors High-rise structural steel del. & erect. Installation of people-movers Heavy foundns. for high-rise bldgs. Smart bldg. control systems	Moderate to heavy demolition work Safety requirements for working in an operating facility Tight schedules Specialty craft requirements for restorations Other input is highly variable depending on type of project
Applicable codes: OSHA, DOT, ASME, ANSI, EPA, ASTM, and local building, plumbing, and electrical codes.				
Key Craft Labor Used				
Major: Ironworkers, equipment operators, riggers, electricians, plumbers, cement workers, laborers, sheetmetal workers, architectural trades, brick masons, welders, and carpenters Minor: Pipefitters, instrument techs, boilermakers, and insulators.				

**Design cost as a percentage of total installed cost (TIC)

TABLE 1.3 A Matrix of Construction Project Characteristics (*Continued*)

9

Figure 1.4 is a visual presentation of the organizational lines that construction managers may cross on any given day. Because not all the contacts have the same goals as the construction team, the construction manager must handle each contact with the utmost care—as we will later learn.

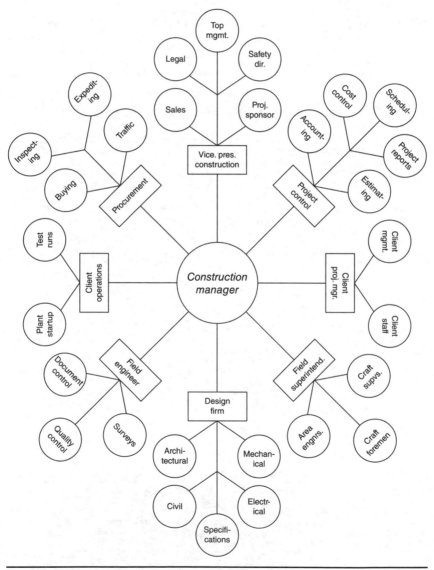

Figure 1.4   Construction manager coordination interfaces.

What Are the Project Variables?

Some unique project variables such as size, complexity, and life cycle also occur in the construction project environment. Usually large projects become complex just by reason of problems of the size and scope of work.

Small projects can also become complex by reason of new technologies, remote location, tight schedules, or other unusual factors. Small projects are often more difficult to execute than large ones, so do not take them lightly. One particular type of complex small project is a plant-maintenance turnaround project done on an extremely tight, hourly based schedule.

The project starts with a *conceptual* phase, then passes through a *definition* phase, CE-Add-and onto construction. And as construction nears completion, the work tapers off into a *turnover* phase, and the owner accepts the project. The project-execution phase accounts for most of the project resources, so it becomes the focal point of the life-cycle curve.

In many cases, a single project manager does not handle a project completely through the full cycle. He or she may start the project and develop the conceptual phase before handing the project over to other specialists who will follow through with the execution phase. The conceptual phase has a heavy accent on research and development, market analysis, licensing, financing, and economics. The early phase requires a set of skills different from those needed in the project-execution phase.

The life-cycle curve gains most of its vertical growth during the detail and procurement phases, and peaks during the construction phase. The major commitment of financial and human resources occurs during that part of the project. That is also the time when the project master plan must be in place, because pressure on budgets and schedules increase during that phase, and must be resolved prior to construction.

We have defined a project; now let's define a manager, The dictionary tells us that a manager is "a person charged with the direction of an institution, business or the like." Synonyms listed are *administrator*, *executive*, *supervisor*, and *boss*. At one time or another, the construction manager will perform all of these functions in executing the project.

Going one step further, we can define the construction management system. It is a centralized system of planning, organizing, and controlling the field work to meet the goals of schedule, cost, and quality.

Figure 1.5 graphically illustrates the definition by showing the key functions of the construction management system. The construction team uses the input of people, money, plans, specifications, and materials and organizes them to deliver the desired facility. The figure also shows the feedback-and-control loop that ensures that the construction project's goals are met.

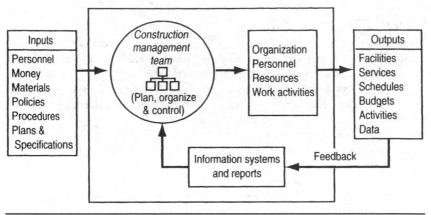

Figure 1.5   The construction management process.

Project Goals

The primary goal of the construction team is to *finish projects as specified*, on schedule, and within budget, and to attain the quality levels required by contract. This message has been dubbed the Construction Manager's Creed. It is so important a motto that it is recommended you enlarge it and hang it on your wall. It will serve as a reminder of your goals when the water in the swamp is rising and the alligators are snapping at you from every side.

The Construction Manager's Creed

Finish the project safely:

- As specified
- On schedule
- Within budget
- Meeting all specified quality levels

The whole system of construction management exists to ensure that construction teams meet those goals. Obviously, those goals have not been met every time; otherwise we would not continue to hear about projects that failed because they were poorly managed, late, over budget, or abandoned.

This book was designed to present the fundamental methods that construction management practitioners must know and use for effective total construction project management. Fortunately there is no particular mystique involved, just good common sense and more than a modicum of hard work. When construction projects fail, it is not from lack of a complex management system. More likely, it happens because one of the basic principles that we will be discussing in this book has been overlooked.

Goal-Oriented Project Groups

Let us look for a moment at the key goal-oriented groups involved in the execution of any capital construction project.

Group	Description
Owner/client	The one who commissions and owns the completed facility
Project team	The project manager(s) and key staff members
Architect/engineering firm	The entity responsible for the design of the facility
Construction team	The entity responsible for delivering the finished project.

The underlying thought for all capital-project and construction management people must be that they are creating facilities for a client who seeks to earn a profit or to provide a service at a reasonable cost. The first case applies to the business sector, the second to government and service agencies. Even a builder of a single-family home is providing living services to the home buyer or owner.

In addition to the functionality of the facility, we must consider aesthetics, safety, and the public welfare. The final trade off decisions on this group of project requirements will also involve the construction manager.

The Owner/Client's Goals

The owner creates the need for the facility and raises the necessary financial resources for its creation, so the owner certainly rates top billing in the goals department. Actually, the owner's needs are quite simple:

Need	Reason
1. The best facility for the money	To maximize profits or services at a reasonable cost
2. On-time completion	To meet production or service schedules and financial goals
3. Completion within budget	To meet financial plans for the facility and return capital
4. A good project safety record	To meet safety standards and avoid bad publicity

In their search for tools to maximize profits or services at a reasonable cost, owners want reliability, efficiency, safety, and good onstream time. We hope they do not want gold plating and overdesign, which add unnecessary costs.

On-time completion allows owners to meet productions quotas and schedules while avoiding the high cost of added interest on construction loans and start-up costs. On-time completion usually results in the added bonus of a smooth start-up and rapid acceptance of the facility. On-time completion also adds to the résumé of a successful builder.

Finishing within budget avoids nasty surprises that can upset the owner's financial plan. Project overruns lead to a slower payout and negate the chance of an early return on the owner's investment. On very large projects, these financial jolts are often felt right down to the owner's bottom line. For example, the city of Boston's Big Dig, officially known as the Central Artery/Tunnel Project, was scheduled to be completed in 1998 at a cost of $2.8 billion (in 1982 dollars; adjusted for 2006, the projected cost was $6 billion). The project was not completed until December 2007, at a cost of $14.6 billion. The *Boston Globe* newspaper estimated that the project would ultimately cost $22 billion and would not be paid off until 2038.

The Project Team's Goal

The project management team is responsible to the owner for the direction and coordination of all facets of the project. The team is often made up of representatives from the owner, the design firm, and the constructor, depending on the contractual arrangements. Their main responsibilities include overall project schedule, budget, quality, and performance to contract. Because this group has a composite of interests, their goals are the total goals covered in this section.

The Architect/Engineer's (A/E) Goals

The A/E group is the entity responsible for delivering design documents for the facility. It can be independent of the contracting group, such as an independent A/E firm or the owner's captive central engineering department. In the latter case, the ultimate client is the corporate division or service group owning the facility. It is desirable to have at least a quasicontractual understanding between the internal groups to ensure that the operating division's project goals are met.

The design team's goals are the following:

Goal	Reason
1. Make a profit	Remain financially healthy in order to attract and retain qualified employees
2. Finish on time	To satisfy the owner/client and meet contractual requirements
3. Design within budget	To ensure that goal 1 is met, satisfy the owner/client, and add to the company's reputation
4. Furnish quality per contract	To ensure that goal 1 is met, satisfy the owner/client, and meet contractual requirements
5. Get repeat business	To maintain the company's reputation and reduce marketing expense

The Construction Team's Goals

The construction team's assignment is to deliver the finished facility ready for acceptance by the owner. This may be as a separate third-party contractor, a construction manager acting as the owner's agent, or a member of a design-build team.

Goal	Reason
1. Make a profit on each project	This applies to GCs, CMs, and design-builders
2. Finish on time	To satisfy the owner/client and meet contractual requirements
3. Build within budget	To ensure that goal 1 is met and satisfy the owner/client
4. Furnish quality per contract	To ensure that goal 1 is met, satisfy the owner/client, and meet contractual requirements
5. Finish the job safely	To meet the company's and owner/client's safety goals
6. Get repeat business	To maintain the company's reputation and use as a sales tool using successful projects and their owners as references

Contracting firms must strive to make a profit on every project performed in order to remain in business. But there are several factors that, despite the best and sometimes extraordinary efforts of the construction team, may make the initial profit goal difficult to achieve. Severe weather may slow down construction without being severe enough to justify requesting a change order. Labor strikes, labor shortages, or strikes at a key supplier may be beyond the control of the team and may have an impact on schedule and hence costs.

Although profit is the lifeblood of a contractor, reputation may be the contractor's most important product. Even in the face of market forces—e.g., inflation, shortages of labor or material—the contractor must prevail and do the best job possible, even if it results in a decrease in profit in the short term.

The owner/client does not wish to see the contractor lose money—but a contract is a contract, and both parties must meet their obligations. Most owners are quick to recall how a diligent contractor pushed ahead in spite of adversity, and we have seen time and time again how owners value this attribute and willingly negotiate a contract for another project with such a contractor. Of course, this applies to private projects but not public projects, which must be competitively bid.

Completing the project on time and under budget is what every contractor strives for, and contractors who are consistent in doing so are also adding to their reputations and their quest to negotiate projects with those owner/clients and establish long-term relationships.

The Construction Team's Personal Goals

The last goal-oriented group to consider is the construction team. Here we are talking about people much like ourselves, so the matter becomes even more personal.

The project team's goals are the following:

Goal	Reason
1. Earn a financial reward	This can be in the form of a bonus or promotion
2. Identify with a successful project	To build a reputation and gain personal satisfaction
3. Finish on time and within budget	To attain goals 1 and 2 and satisfy the owner/client and company management
4. Maintain one's reputation in the business	To achieve personal satisfaction and demonstrate professionalism

The results of our labors are usually visible, and we can point with pride and say, "I built that building." The satisfaction of a job well done is something that we all strive for. Completing a difficult project on schedule and under budget gives an additional satisfaction, that of having met a challenge. Having both the client and management satisfied at the end of a project is indeed much more satisfying than the reverse would be.

Financial rewards can come in several forms: a bonus, an increase in salary, a promotion, or a combination of the three. Many contracting firms put their project teams on an incentive program, in which rewards are tied to a successful performance. Commercial and industrial clients also favor incentives because these ensure, at a relatively minor cost, that project goals will be met.

It is extremely important for a CM or GC to build and maintain a reputation in the field. Subcontractors prefer working with successful contractors, those that manage and operate their projects efficiently. And quite often these subcontractors, in a competitive-bid situation, will give these contractors a preferred price because they know that they will be able to work efficiently alongside them and end up achieving their profit goals.

News of successful projects spreads rapidly in this small community of contractors, subcontractors, and vendors—and, of course, potential clients. The reverse is also true, so this is another compelling reason to strive to create a successful project.

Basic Construction Project Management Philosophy

Our basic construction project management philosophy is simply stated in three words: *plan, organize, control*. We call this our Golden Rule of construction management. It should hang right alongside the

Construction Manager's Creed. Practicing the Golden Rule will deliver the goals in the Creed.

Many of you may recognize the Golden Rule as a distillation of a general management philosophy of plan, organize, execute, coordinate, and control. They ensure its easy application by busy construction managers and general contractors.

Total construction project management requires that you practice the Golden Rule on your projects in general and for each project activity. By that we mean that you must plan, organize, and control every activity on the project. A good example of applying this rule to a piece of the project is your approach to project meetings: Plan your meetings carefully and inclusively if you are the leader, organize their logistics, and control the meetings (we will discuss this in more detail later in the book).

As a corollary to the Golden Rule, add the KISS principle, to further stress the need for a simple, uncluttered approach to construction project management. *KISS* is an acronym for the phrase "Keep it simple, stupid," which turned up as a joke some years ago. Some people may balk at the term *stupid*, but it is not meant to be a derogatory term, merely to emphasize a term that most people can remember KISS.

The goal of introducing simplicity is to point out that overly detailed plans, budgets, schedules, and control systems do not automatically equate to more successful projects. In fact, the reverse is true: Overkill systems have hurt more small projects than underpowered ones have hurt large projects.

What Are We Going to Plan, Organize, and Control?

The key construction activities that we need to plan, organize, and control lie at the very heart of the total construction management system. They also establish the format of the project-execution portion of this book. They are the basic building blocks of the total construction project management approach.

Project-Planning Activities

The plans for a typical capital project are

- a construction-execution plan
- a time plan—field schedule
- money plans—a construction budget and cash flow, and
- a resources plan—people, materials, systems, and money

The *construction-execution plan* is the master plan for executing the field work, from bidding to completion to transfer to the owner. The master plan must operate within the constraints of overall project financing, strategic schedule dates, allocation of project resources, and contracting procedures.

The *time plan* is the itemized working plan for the project execution, which results in the detailed construction schedule. The construction team makes a work-breakdown structure by breaking the project scope into major work activities. It then assigns completion dates for each operation.

The *money plan*s consist of a detailed project budget, based on a sound construction-cost estimate, and a *cash flow* resulting from the budget and the schedule which forecasts how the budgeted funds will be spent.

The *resources plan* forecasts the human, material, and systems resources required to execute a construction project according to the master plan and the schedule.

Organization Activities

The activities in the organizing area cover deployment of the human resources and systems required to meet the master plan and the project schedule. In this area managers must

- Prepare organizational charts and personnel-loading curves
- Write key position descriptions
- Issue site operating procedures
- Mobilize and motivate field staff
- Arrange site facilities and systems
- Issue and start control procedures

Plan the field organization based on the scope of work and the personnel plan. Issue detailed construction procedures promptly to instruct and guide the new organization. Mobilize the staff and suitable facilities to kick off the project on schedule. Start the cost controls immediately; you are already spending money.

Control Activities

Controlling the construction plans and activities are essential construction project management responsibilities. Major areas of control are

- Quality—field engineering, materials, and construction
- Time—measured by the construction project schedule
- Money—measured by the construction budget and cash-flow plan
- Physical progress and productivity
- Project reporting

The controlling function must monitor the quality of all phases of the work to meet the universal goal of building the project as specified.

Time is monitored by checking physical progress (not necessarily labor hours expended) against the schedule. The project-reporting system regularly informs key project players as to the status of project activities and the results of the project-control systems in detail.

The lists just given constitute the skeleton of an effective method for the successful execution of capital projects. In later chapters we will fill in the flesh, sinew, and blood to give you a complete body of knowledge for total construction project management.

Construction Manager's Job Description

A discussion of the project environment would not be complete without addressing the duties of the construction manager or project manager. We have included a comprehensive construction manager's job description in the appendix; it is a hybrid of several versions that have been gathered over the years.

Writing an industry-wide job description suitable for the total CM or GC population would result in a glut of models to cover all cases. Therefore, we have selected a model for a contractor's environment, whether for a CM or a GC. Readers working in a noncontractor environment should substitute *contractor* for *client* in the model provided.

Section 1.0, Scope, defines the areas of construction activities, which extend to all aspects of construction execution in accordance with the terms and conditions of the contract with the owner or client.

Section 2.0, Concept, sets the stage for the granting of the CM or GC's charter and delegates responsibility to complete the project safely, as specified, on time, and within budget, fulfilling the quality standards required by the plans and specifications. This is also where top management makes its commitment to the construction management system. If management's charter statement is weak at the outset, strengthening it later will require a great deal of effort.

Section 3.0, Duties and Responsibilities, gets down to defining the on-the-job operating techniques required to execute a project. It gives a step-by-step approach to planning, organizing, and controlling a capital construction project. This section serves as a good checklist to follow as you proceed through the project life cycle, even if you do not use it as a job description.

Section 4.0, Authority, expands on the CM or GC's authority in relation to other company activities. Strength in the authority area is important to the creation of the manager's power base in the company hierarchy. Weakness here can seriously hobble the construction manager in staffing projects and marshaling the firm's construction resources to execute the projects. Authority must be earned; and what better way to earn it than to create a proven record of bringing in projects on time and within budget and developing good relationships with owners for which you have worked?

Section 5.0, Working Relationships, gives a few suggestions about the human relations involved in using the authority granted under section 4.0 when operating across departmental lines. You can expand on this list to include other interfaces in your particular environment.

Section 6.0, Leadership Qualities, lists some of the leadership activities that are most useful in motivating the construction team and others involved in the project. We will expand on these valuable qualities in later chapters.

The job description applies within the context of a contractor's environment, which usually grants a stronger construction management charter than that of a typical owner. This is primarily due to the competitive nature of the contracting business, in which the profit motive discussed earlier is of the essence. More than one highly successful construction company has gone broke because of one large failed project, so attention to meeting profit goals must be at the forefront of your leadership qualities.

If your construction management job is through employment by the project's owner, the main difference between the owner's and contractor's job descriptions is found in section 3.0, Duties and Responsibilities. The owner and contractor have the same duties and responsibilities, but delegate their execution to the contractor whenever a contractor is engaged to execute the work. The owner's responsibility is to see that the contractor is performing according to the contract and the project plan.

If you use the job description in the appendix as a model for your own position, you can adapt the parts that apply to your work and add any new material describing your particular situation. It makes good sense to write your own charter as broadly as possible and then let your management cut it back. You might as well see how far management is willing to go. But remember this: If you are given total responsibility to meet the construction project's goals, you must have the authority necessary to do the job!

We hope that you noticed the application of our Golden Rule of construction management in the layout of section 3.0. We divided the duties and responsibilities into subsections for the planning, organizing, and controlling functions that are essential to effective construction project management.

The job description in quite general, as it must be without a specific project assignment. Therefore, you should prepare a written list of project-specific goals to measure your performance and that of your supporting staff. You should incorporate those specific project goals into your staff's job descriptions. This will allow you to take an approach of management by objectives, which we will cover in more detail in later chapters.

Project Size

The one project variable that needs more detailed discussion is project size. Most people seem to have more difficulty with size than with any other aspect of construction management discussions.

Opinions of project participants vary somewhat among the different types of capital projects listed in Table 1.3. The common denominators for judging project size are scope of services, number of drawings, major equipment items, bulk-material quantities, and home-office and field labor hours. In some cases, the period of time required to complete the project may be extended by inflation of the cost of materials, increased costs of union labor when contracts are renegotiated, and commitment of the same team to the length of the project. Making a quick review of the project's major size indicators will give you a better idea of how best to approach your new assignment.

Since a high percentage of capital projects are small to midsize projects, some readers may feel overwhelmed when looking at the numbers for larger projects. However, larger projects do come along occasionally, so you should know how to deal with them. We have found that doing a single large project is sometimes easier than running several small ones simultaneously, in many cases because lead times are projected further out for large projects, allowing for a more orderly acquisition of major components of construction and long-lead items.

As a project becomes larger, it is broken down into smaller, more manageable units. An area engineer or manager is responsible for a specific unit and reports to the construction manager. On large projects, we may also sublet complete units to subcontractors who specialize in executing a particular type of work and hence are more familiar with its intricacies.

The most important differences between handling small and large projects are the systems and tools used for project control. However, with a variety of software programs—including for scheduling and spreadsheets for daily, weekly, and monthly cost control—hands-on control of even very large projects can be effected if qualified staff is available to accumulate and present this information for review and comment by the project manager.

Project-Size Ratios

Most contractors today prefer to subcontract all of their work for several reasons: firstly, because it allows control of costs, since a fixed or lump-sum contract will be negotiated with the subcontractor; secondly, because these specialized contractors perform the same type of work week after week, so their labor crews are generally more proficient; and thirdly, because it eliminates a layer of supervision for the contractor, who no longer has to monitor the productivity and cost of work performed by his or her own crews.

But for those contractors who do maintain some of their own crews—most typically laborers, masons, and carpenters—a method must be established to create a daily report to compare costs to budget, to determine whether unit costs are being met or exceeded, and, if the latter is the case, to determine what is required to get these costs on target. A method for reporting and analyzing these costs is discussed in a later chapter relating to project control.

CHAPTER 2

Bids, Proposals, and Contracts

The bidding, proposal, and contracting process plays the key role in total construction project management. This process is the lifeblood of the engineering and construction industries. Until we have reached an agreement or signed a contract, construction of the facility is on hold.

Our specific goal in this chapter is to discuss the construction project manager's role in the overall process. The process commences with an owner, whether private or public, deciding to construct a capital project. The steps required from concept to bidding to contract formatting to contractor selection are lengthy and complex.

The term *contractor-selection process* refers to a system by which the owner selects the contractor and negotiates the terms of the contract. The contract itself sets the ground rules and apportions the risks for executing the construction work. The overall process is shown schematically in Fig. 2.1.

The contracting process can be conveniently divided into two major phases: contract formation and contract administration. The contract formation can be established in several ways:

- The owner's design consultants prepare a detailed scope of work. This process is most often utilized for public projects where competitive bidding is required by statute, but it is widely used in the private sector as well. It is referred to as design-bid-build.

- The contractor works with the owner's design consultants to establish a scope of work commensurate with both the owner's requirements and project budget. The result of this process is often referred to as a negotiated contract.

- The designers and the builder form one entity, in which either the architect or the builder is the lead authority. This is called design-build, and is similar to a negotiated contract.

The contract itself sets out the scope of work and responsibilities for the execution of the work. It also assigns the risk sharing for the

Development phase	Contracting phase	Execution phase
Activities	Activities	Activities
Project planning	Contracting plan	Detailed engineering
Market development	Contractor screening	Procurement
Process planning	Selection of bidders	Construction
Cost estimating	Invitation for proposals	
Basic design	Contractor's proposals	
	Bid review	
	Contract award	
. . . By owner	. . . By owner and contractor	. . . By contractor

FIGURE **2.1**   The contractor selection process.

contracting parties, be they owner, design consultants, prime contractor, subcontractor, or material supplier.

Our drift toward a litigious society over the years has significantly affected the construction contracting and design environment. Sixty or seventy years ago, contracts were signed, filed, and forgotten until project completion. Handshake agreements were often used. Any minor differences of scope or what was actually intended but not specifically included in the contract were resolved between the parties. But lawyers turned this area into one of their more lucrative business opportunities, until owners, design consultants, and contractors finally realized there must be a better way.

As we will see in the section on contract formats, the desire to work together, realizing that some give-and-take can not only speed up the design/construct process but limit or eliminate the need to litigate.

The construction manager is responsible for the administration of the contract, whatever form it takes, and that makes it vital for him or her to be thoroughly knowledgeable about every facet of the contract.

Construction-Execution Approach

The construction bidding, proposal, and contracting process is initiated by the owner, who has a project to build. There are about as many approaches to contracting as there are owners. Each combination of owner and project has some unique features that need to be covered by the contract. The standard contract formats developed by the American Institute of Architects (AIA), the Associated General Contractors of America (AGC), the Construction Managers Association of America (CMAA), and the Design-Build Institute of America (DBIA) are frequently modified to fit the requirements of a specific building project. Engineering-driven contracts for water, wastewater treatment, petrochemical, and other industrial process projects often utilize contracts prepared by the Engineers Joint Contract Documents Committee (EJCDC).

Here we are concentrating on how the owner wishes to perform the work. Each one of these approaches is subject to further variations introduced by the type of contract used.

The owner can use the following contracting approaches:

1. Hiring a general contractor who will perform some of the work and/or subcontract the specialized portions of the work via a negotiated process

2. Utilizing a design-build arrangement—often referred to as turn-key—involving a single entity of designer-contractor: a designer and contractor working under a joint-venture-type contract

3. Awarding the design of the project to the owner's own design consultants and engaging a third-party builder through a competitive bidding process

4. Hiring a construction manager (CM) to act as the owner's agent, usually working throughout both the design and construction processes

These basic contracting strategies are shown graphically in Fig. 2.2, along with several alternatives. The traditional case (upper left) is the one used for projects involving public funds. The other five alternatives are used to varying degrees in the private sector. Even these basic plans can be cut and pasted to give virtually unlimited contracting possibilities. The owner's project team must consider the advantages and disadvantages of each approach and match the most favorable plan to the environment and goals of each project.

The single-contract, self-perform mode means that the contractor performs the bulk of the construction work with its own forces; this method has been used less and less in recent years. Certain trades, such as roofing, ceramic tile, insulation, HVAC, electrical, and possibly site work are subcontracted to others. Fewer contractors in today's market self-perform any significant amount of work, since they find that the more-or-less lump-sum price afforded by subcontractors reduces their risk and lowers the number of supervisors and forepersons on the jobsite.

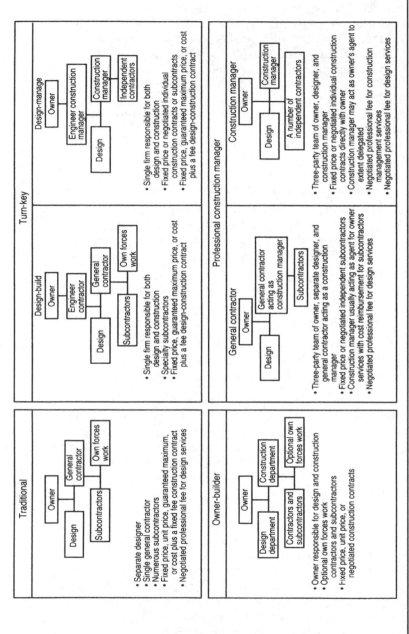

FIGURE 2.2 Basic contracting strategies—the proposal schedule.

In design-build mode, a single firm or joint venture provides the engineering, architecture, and construction services under a single contract. Because the construction manager or general contractor is part of the design-build team, his or her input as far as budgets, constructability issues, and schedule is jointly shared with the design consultants. As a result, the time frame from design to project completion is significantly reduced. A survey conducted in the year 1998 by the Pennsylvania State University College of Engineering has shown that completion of design-build projects is 33 percent faster than for CM-at-risk projects and 33 percent faster than for conventional design-bid-build projects. In some cases initial costs have exceeded those for the design-bid-build or CM approach, but the savings in construction interest loans and the ability of the owner to occupy the facility sooner more than makes up for this slight increase in cost. But more rapid construction is not the only benefit of the design-build mode. Because the owner looks to one entity for both design and construction, the potential for change orders is also reduced: The contractor cannot point to minor or even major design flaws, since he or she jointly shares this responsibility with the other team members.

The construction-management mode means that the construction work will be outsourced to other contractors and subcontractors, since the CM is acting as the owner's agent. In some instances the owner will allow the CM to bid on portions of the work, as long as the price is competitive. But generally the CM does not directly hire any craft labor; he or she advises the owner on subcontractor and vendor selection and these contracts and purchase orders are issued in the owner's name but administered by the CM. The CM uses a staff of project-control people, subcontract administrators, and craft specialists to attain the project goals of quality, schedule, budget, and compliance with requirements.

In recent years, two methods of construction-management contracting have evolved:

- CM for fee—The construction manager enters into a contract with the owner whereby the CM will provide the necessary personnel to manage the project and will receive his or her fee as a percentage of project costs, where no limit on final project cost is guaranteed.

- CM at risk—The construction manager will receive a fee expressed as a percentage of costs, but the CM will guarantee the costs. If these costs exceed the guaranteed maximum price, the overruns are paid for out of the CM's fee.

A variation on this CM option-AU is for the owner to act as his or her own general contractor by awarding and coordinating contracts. But only an owner with sufficient knowledge of the construction cycle and with staff personnel experienced in construction ought to pursue this variation.

The selection of the contracting mode is largely a function of the owner's project needs. A single contractor utilizing subcontracted services and possibly some self-perform trade workers is used by government owners who are required by law to operate on a fixed-price construction-contract mode where the bidder's list is open to all who are not blacklisted and can provide a bid bond. Generally, the design work is completed before the government agency prepares a bidding process; this may be delayed or awarded in stages due to annual budget and fiscal restrictions. CM approaches in the competitive-bid market are also an option for both private and public owners.

In recent years, public agencies have embraced the design-build process, and it is likely that this trend will continue.

Owner–Contractor Operating Mode

Once the owner decides to go outside for help on a project, another party is introduced into the project proceedings. We must now have an arm's-length agreement between the owner and the outside party, which brings a proposal and contractor-selection process into the picture.

The owner's decision to enter into a proposal phase brings a number of managers into the decision. The first one is the owner's project manager or representative, who heads up the owner's contractor-selection team. The owner's request for proposal (RFP) will, in turn, activate several contractors' project or construction managers, who are responsible for much of the proposal preparation effort. Remember, one of the precepts of good management is for the project construction managers to be responsible for their projects from initial proposal to project closeout. The project architect or engineer, usually employed by the owner, can also be involved as a noncontracting party.

The Contractor-Selection Process

The process for contractor selection that we are going to discuss here will apply across the board, from major prime contractors to minor subcontractors. Naturally, the process is less detailed when it is used for proposals on a smaller project with a smaller scope of work. We hope you will remember the KISS principle when you find yourself engaging in this activity!

Figure 2.1 shows the part of the contracting cycle in which we are now working. The contracting phase fits neatly between the project development and execution phases. The bottom line in the figure shows the party with the primary responsibility for the major activity taking place in the block above.

The Owner's Input

The owner must do a thorough job in the project development phase to allow the subsequent contracting and execution phases to proceed smoothly and efficiently. The design and construction phases are critical because they involve the largest commitment of human and financial resources. Poor definition during the project conception often brings delay and indecision, which are costly and goal wrenching in the phases that follow. The owner must have a firm project scope in mind in order to convey his or her needs to the design consultants.

Contractor Selection

The contractor-selection process must follow the contracting plan discussed earlier in the chapter. Changing the plan in midselection will likely cost money and valuable time in the strategic planning schedule.

Successful contractor selection can be best accomplished through an open and honest approach by all parties. Using a list of prequalified contractors can further improve the selection process for the owner. There is no point in working with a contractor with whom you are not willing to enter into a contract.

That point becomes more complicated when public bidding for government work is required by law. The selection team in public bidding must be especially careful to closely examine the financial and technical capabilities of all contractors who request bids. This must be done before bidding, because it is difficult to eliminate an unqualified low bidder after the bids have been opened. Public agencies require that contractors who are planning on submitting bids be able to present a bid bond. The bid bond is a measure of the contractor's financial health, since the bonding company issues the bond after a careful review of the contractor's financial statements.

If one bidder challenges another's bid and files a lawsuit contesting its validity, this could tie up the award process for quite a while. It is not uncommon for challenges to be made if a bidder inserts a qualification in the bid, taking exception to some aspect of the bid proposal, whether major or minor.

In private work it is not absolutely essential to award the work to the lowest bidder. In fact, negotiated contracts without any bidding process are prevalent in the private sector, especially when an owner has worked successfully with a contractor on a prior project. The negotiated-contract approach does not exclude the selection process, but it does simplify it.

Selecting a contractor for larger projects requires a competent team, one that represents a broad spectrum of technical and business skills. A project or construction manager usually heads up the selection team to coordinate the other departmental inputs to the selection process.

In addition to the technical people on the team, input from tax, risk-management, procurement, legal, and accounting specialists will be necessary. The selection team should participate for the duration of the selection process to minimize the adverse effects of repeated learning curves.

The owner's contracting plan is the foundation of the selection process. A well-conceived and management-approved plan is necessary at this point to avoid the need to revise it later. Management should tailor the contracting plan toward the best arrangement to meet the owner's goals for the project. In the private sector, the contracting plan should not have to conform to long-standing corporate policy if a different arrangement offers advantages.

It is wise for an owner to make a complete analysis and evaluation of all contracting alternatives and project factors to arrive at a sound project strategy. Major items to consider are

- project needs
- requirements of the project-execution plan
- key schedule milestone dates
- the scope of the contractor's services
- possible contracting alternatives
- local project conditions
- contracting-market conditions

The owner is looking for the contractor who will perform best by delivering a quality facility, on time, at the lowest overall cost. In striking the ideal match between owner and contractor, the owner should select a contractor from a list of firms that really want the job. Factors affecting a contractor's desires are present workload, a prestige project, repeat business, market position, profitability, and the like. Incentives are necessary on both sides of the contracting equation to obtain superior performance and results. It is the merging of the owner's goals with those of the contractor that leads to a successful contract award.

Contractor selection is much like a miniproject; it needs a project-management approach. This means that a total project management approach to planning, organizing, and controlling the selection effort is necessary. Figure 2.3 shows a typical schedule for major activities in the selection process. You will have to add the applicable time, scale, and other activities that suit your particular project. In many cases, setting the proposal opening date controls the timing of the preceding activities. The owner must be sure to allow enough time for the contractors to prepare a sound bid and proposal.

The contracting basis used causes the time scale for the proposal to vary widely. A proposal based on cost plus a fixed fee requires less

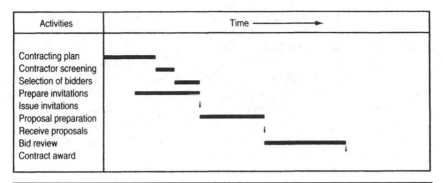

Activities	Time ⟶
Contracting plan	
Contractor screening	
Selection of bidders	
Prepare invitations	
Issue invitations	
Proposal preparation	
Receive proposals	
Bid review	
Contract award	

Figure 2.3   Sequence of contracting activities.

time to prepare than a lump-sum proposal. Fixed-price proposals require preparation of detailed cost estimates by the contractor and bidding documents by the owner. Documentation for a proposal for cost plus a fixed fee is much less complicated and time consuming for both parties.

Contractor Screening

The contractor-screening process again depends upon the type of owner involvement. In government projects, the owner must publicly advertise the project to open the work to all interested firms. The projects are also listed in various builder's exchanges, which post all construction projects within specified geographic areas—a service to which many in the construction industry subscribe. The problem here is keeping the list of bidders to a reasonable length. Government owners normally include the taking of public bids in the A&E's scope of services as part of the preparation of the bidding documents. And most government projects requires bidders to provide a bid bond indicating that they have the wherewithal to perform the work if they are adjudged the low bidder.

In private work, the owner can restrict the list of bidders to only those firms deemed qualified or desirable to perform the work. The responsibility then rests with the contractor's business-development group to get their firm onto the preferred-bidders lists.

The screening process for private work begins with the issuance of a request for statements of interest from a prequalified list of five to ten firms or more, depending upon the size of the project. The goal of the screening is to arrive at a short list of bidders ranging in size from three to six, there again depending upon the size of the project.

This short-list approach has several goals. The first is reducing the selection team's work in reviewing the detailed proposals. Second, the approach helps to control expenses for proposal preparation. Contractor proposals for medium and large projects often require an investment of several hundreds to hundreds of thousands- of dollars. Since there can be only one winner, the unsuccessful bidders consume a great deal of overhead when they prepare a complex bid.

It has become a common practice among public-project bidders to have the public agency offer a stipend to the second and third bidders on large and complex projects that require a large number of work hours in the preparation of the bid. This somewhat offsets the cost for the contractors who submitted noncompetitive bids, and encourages them to continue bidding on these types of projects.

Selecting a short list of eager, qualified, and competitive contractors should assure owners of a good, representative selection of proposals. Some major points to consider in the screening process are as follows:

- Screen as wide an area as needed to satisfy your needs.
- Use a written screening document—letter, fax, or e-mail—that can be printed out as documentation.
- Prequalify the screening list.
- Use a simple screening request document.
- Indicate whether or not a payment and performance bond will be required. Ask bidders to provide evidence that they will be able to provide one, if required.

Although concise, the screening documents should cover all of the critical points needed to select a good bidders list of quality firms. The screening process really requests miniproposals to get some preliminary facts from the contractors. Some of the key points to cover are

- a brief project description
- a statement of the scope of services
- key dates for proposals, contract award, project start, and desired completion
- a tentative project plan and schedule
- a statement of contractor's interest
- location of the work
- a request for the status of the contractor's existing workload
- a list of the contractor's current technical-personnel capability
- pertinent experience in the area of the project site
- any other project-pertinent factors worthy of evaluation

The screening analysis requires a thorough review of the positive replies to the screening document. A major item to check is the contractor's capability to handle the project within the required time frame. Since these firms have been approached and are assumed to be capable of handling the work, the owner's main concern is with their ability to staff the project properly. For this, the selection team needs to study the contractor's loading curves, submitted with the screening reply. Figure 2.4 provides a typical format indicating a contractor's current workload data.

The figure shows construction-personnel loading curves for two prospective contractors. Contractor A's curve shows the current backlog tapering off in time to add the expected project workload with some capacity to spare. The curve for Contractor B shows that

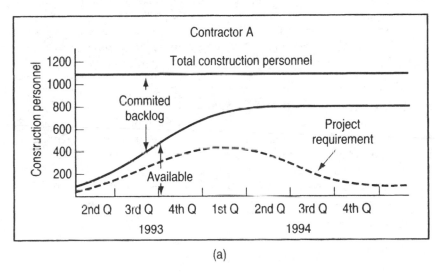

(a)

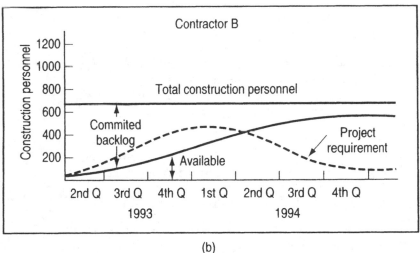

(b)

FIGURE 2.4 Contractor personnel loading curves.

a backlog will not permit the commitment of enough personnel from the contractor's pool to staff the new project to meet the schedule. The only alternative for Contractor B is to increase staff. Such options must be investigated thoroughly before selecting that contractor.

At this stage, the owner is making only a preliminary evaluation of the workload; a more thorough investigation will be done later on. Also, at this stage contractors will try to show themselves in the most favorable light, so the personnel-loading curves will be somewhat optimistic. Furthermore, the contractor may book some additional work during the proposal-selection process. The remaining responses need to be checked for other pertinent factors, such as experience, location, dedication, interest, project-execution plan, and possible value engineering or other cost-saving suggestions. Scoring these individual factors on a weighted scale of 1 to 5 points gives a total score for each proposal. The short list for proposals consists of the three to five highest scores.

Preparing the Request for Proposal (RFP)

As shown on the proposal schedule in Fig. 2.3, the work for preparing the RFP documents would also include preparing invitations to bid to be sent to selected bidders. Thus there should be no delay in issuing the RFP as soon as the proposal slate is ready.

At a minimum, the RFP documents should include the following major sections:

- Proposal instructions
- Form of proposal
- Scope of work and services
- Pro forma contract
- Coordination procedure and job standards

The size of RFP documents can range from a few pages to several books, depending on the type and size of the project. A small HVAC job, for example, might require several pages, while a paper mill, tunnel, bridge, or high-rise commercial building might need several volumes.

The proposal instructions should include certain subjects to ensure that uniform proposals are received, permitting a meaningful comparison. Receiving proposals in a variety of formats makes comparison difficult and can even lead to a poor selection and unmet project goals. Important areas to include in the instructions to contractors are:

- *General project requirements*—project description, index of proposal documents, scope of work, site location and description, etc.

- *Special project conditions*—owner contacts, site-visitation requirements, proposal meetings, etc.

- *Technical-proposal requirements*—as needed for contractor's technical input on design-build projects

- *Project-execution proposal requirements*—could include a table of contents of key information that the owner desires (e.g., corporate organization chart, experience record, project-execution plan, project organization chart, key personnel assignments, current workload charts, preliminary schedule, description of project control functions, and any other pertinent data desired)

The proposal form is at the heart of the contractor's commercial proposal. Depending upon the type of contract being contemplated, the proposal form may request all of the contractor's business terms, such as pay scales, fringe benefits, overhead, fees, travel-expense policy, heavy-equipment charges, and computer or IT expenses, if applicable. This section also asks contractors to state any exceptions to the terms of the proposal agreement, the pro forma contract, and any other terms of the proposal invitation.

The pro forma contract is a draft of the contract that the owner plans to execute with the successful firm. The number of exceptions taken to the pro forma contract will be directly proportional to the volume of the document. Standard short-form contracts issued by most professional A/E/C (A=archtitect, E=Engineering, C=Contractor firms can offer some particular advantages in this regard. At the other end of the scale are the long-form contracts generated by large corporate legal staffs and government agencies. The long-form contracts often generate more problems than they solve; most legal people do not recognize the benefits of the KISS principle.

The coordination procedures include the basic standards and practices that owners expect contractors to follow in executing the project. This section may include the owner's design practices and standards, local government requirements, safety regulations, machinery standards, procurement procedures, mechanical catalogs, operating manuals, accounting standards, and acceptance of the work. This section is probably the largest volume producer in the proposal documents, depending on how much of this material an owner normally uses. Owners with limited standards can select the standards needed from a menu provided by the contractor and still have a good project.

Most of the material contained in the RFP is an outline for the execution of the project. Design it to get the contractors more deeply involved in thinking about your project. One goal of the RFP is to get contractor's juices flowing to offer any unique construction technology or procedures and to get management input to improve the project's cost or schedule. Good contractors are eager to display their competence and improve their standing with the owner, as a way of improving their chances of winning the contract. They may offer some design changes that would produce savings. Any such suggestions or alternates must be reviewed by the owner's design consultants before any consideration of acceptance is made.

Evaluating the Proposals

The purpose of having strict requirements in the RFP documents is to generate consistent proposals, leading to easy and accurate comparison on a common basis. Nonconforming proposals are subject to revision or rejection, which adversely reflects on the errant proposer. The three critical sections of the proposal request requiring special attention are

- the business proposal
- the technical proposal
- the project-execution proposal

These sections are assigned to the specialists engaged by the owner on the proposal team for the purpose of extracting those people's expert input. Reviewing the business and technical portions separately avoids any influence of the commercial terms on the technical portion of the proposal. The final step is to combine the commercial and technical evaluations in order to select the best overall proposal. The selection team can then present their findings to upper management for approval.

The commercial evaluation focuses on the total cost of the various contractor services. If the proposals are on a lump-sum basis, the commercial evaluation is quite simple and involves comparing a few lump-sum figures. The technical and performance evaluations are then more critical to ensuring that the offerers are pricing the full scope of services. That is necessary to ensure that the owner's project expectations will be satisfactorily met.

The evaluation team must reconstruct the total cost of the services from the estimated project labor hours, multiplied by the average hourly rates for each proposal. Since the labor-hour estimate may vary for each proposal, evaluations must compare the proposals against a standard set of labor hours. The selection team then multiplies the estimated standard labor hours by each offerer's total personnel costs to arrive at the expected total cost for services. This puts the expected labor costs for all proposals on a common-cost basis. Remember, this type of cost evaluation is only approximate, so it does not carry the same weight in an evaluation as a lump-sum proposed price.

It is worth noting that the cost of labor includes many elements besides the trade worker's basic pay. The various additional costs to the basic labor rate may vary from contractor to contractor, and it may benefit owners in analyzing bids to include in the RFP a request for labor rates for any self-perform work.

Before making the final selection, the owner's selection team should visit the contractor's offices for an on-site examination of the firm's proposed personnel, systems, equipment, and facilities. If at all possible, the owner's representative should plan to visit at least one construction site for each short-listed contractor, to observe their field organizations in action.

Schedule that office and site visit well in advance so that the contractor's top management and key personnel are present for a personal interview. The appraisal team should study the organization from top to bottom to get a feel for its ability to meet the project's needs. Especially important are the top management's commitment, capable personnel, modern facilities and equipment, workable systems and procedures, job history, and a reference list. Now is the time to recheck the workload curves to see that they are factual. Review each contractor's work in progress as well as any outstanding proposals beside your own. Also, now is a good time to pursue the details of any technical-proposal features that may need clarification or further discussions with the contractor's top technical people.

If the bidder is offering a design-build proposal, visiting both the design consultant and the constructor's offices will be necessary, unless both parties occupy the same office. Give special attention to how they propose to coordinate the design, procurement, and build interfaces.

The contractor will supply much of the information an owner seeks as part of a dog-and-pony show presented for the owner. The owner's representative should display politeness as the presentation proceeds but should not accept it as the final answer to all questions. Remember, the main purpose of the presentation, from the contractor's point of view, is to downplay any weaknesses in the proposal (if there are any), highlight the contractor's strengths, and do a selling job.

With the final office visits completed and evaluated, the owner is ready to make the final selection. The three main areas to rate are the commercial, technical, and project-execution proposals. To get the final rating, combine the scores from the three major areas with the results from the office and site visits. That will finalize the recommendation to top management for approval.

It is a good practice to rank the top three contenders in a 1-2-3 order of preference before starting final negotiations with number one. Then if an agreement cannot be reached with number one, drop back to the number-two contender, and so on. It is usually not a problem to reach an agreement with the prime candidate, if all matters were fairly presented by all parties. Using a well-organized selection process such as the one described previously should ensure a good selection and a happy project.

The final work of the selection team is to tell the unsuccessful candidates of the final decision as soon as possible. Expect that the unsuccessful offerers will request a debriefing as to why their proposal was not accepted. If the request is made in the spirit of wanting to improve their performance on the next proposal, a fair and open discussion of nonconfidential matters is in order. This provides the contractor with at least a partial payback for its efforts and investment.

Developing Construction Proposals

The contractor's project- and construction-management people meanwhile are performing their function on the other side of the proposal-preparation process.

Many project and construction managers regard proposal assignments as fill-in work to be done only when they are not assigned to an active project. This is indeed a shortsighted attitude toward a function that is the lifeblood of their organization. After all, it is successful proposals that make everyone's job in the company possible.

The construction manager who gives business development and top management a strong performance when it comes to selling jobs makes a valuable contribution to the company and to him- or herself. A truly professional construction manager cannot afford to take a proposal assignment lightly.

The Proposal Goals

Managers of construction proposals have the goal of turning out winning endeavors on a tight budget within the time allotted. As with owners' project managers, contractors' construction managers must treat the proposal work as a miniproject using the same planning, organizing, and controlling functions. Miniproject schedules allow no room for false starts, missed targets, or mistakes along the way.

The contents of a construction proposal are highly variable, depending on the type of contracting strategy used. Most will require some sort of bid price or cost estimate as part of the commercial portion.

The detailed cost estimating falls into the estimating department's domain, subject to review and approval by the construction manager and top management. We will go into more detail on the various forms of estimating in later chapters.

In fixed-price bidding for publicly funded projects, having a competitive price is all-important for gaining consideration for your proposal. However, it is also important to have all of the boilerplate items in the RFP covered. Does the proposal address the RFP requirements for affirmative action and does it include specific goals to be achieved? Are there small-business set-asides, subcontractor qualifications, progress evaluation and reporting systems, and the plethora of government regulations involved in public projects? Is compliance with the attached Davis–Bacon pay scale acknowledged? Although the applicable government regulations will be included in the agency's bid documents, it is incumbent upon the contractor to be familiar with these myriad regulations. Failure to address any one of these areas gives the public-agency owner cause to reject a bid. Also, the construction estimate must be based upon the construction-execution plan as developed by the construction manager and upper management. All of this adds up to the conclusion that the right price must also include the costs to comply with the various government programs,when it comes to winning a project.

In private work, where having the lowest price may be somewhat less important, developing a powerful project-execution approach for the proposal is even more compelling. In the private sector, owners look much harder at contractor performance, quality, and schedule than at price alone.

The Construction Manager's Role in Proposal Development

Responsibility for proposal generation varies from contractor to contractor, depending on company policy. In larger construction companies. the business-development group is usually responsible for the actual production and delivery of the proposal documents. That group also coordinates the live presentation to the owner and visitation from the prospective client.

If a turn-key contract approach is proposed, a major technical part of the proposal will be devoted to the design and procurement services involved. In that case, a project manager (director) may be assigned to coordinate the technical input for the proposal.

NOTE: *For those unfamiliar with the term turn-key, it differs slightly from a conventional design-build project. This type of contract, frequently employed in cogeneration projects, refineries, and power-plant construction, is just that: turn-key. The owner turns the key, opens the door, and finds the project ready for operation.*

The construction manager (we use this terminology interchangeably with *contractor*) is responsible for developing a dynamic project-execution proposal. The construction manager must be certain that the construction-technology approach to the work is sound and is presented in such a way as to match the goals of the prospective client and the requirements of the RFP. In developing the construction plan, the construction manager marshals the best company resources available, i.e., uses the expertise of the appropriate company-department heads. The task of editing and polishing the specialist inputs to create a unified proposal format also falls to the construction manager. This means that the construction manager must also be an excellent communicator.

The construction manager, with the approval of top management, also develops the construction organizational chart and key personnel assignments for the proposal. The organization must present a can-do project-executive team when the client's selection committee makes its office visit. The construction manager should also prepare and rehearse the proposed project team for any live presentations to the client. The training program and rehearsal culminate in a dry run before a devil's-advocate tribunal. The tribunal should include top management, so the presentation had better be right! Any final adjustments to it should be made based on the comments received during the dry run.

The Construction-Site Survey

A key proposal function of the construction manager is to lead the construction-site inspection team and to oversee the preparation of the site-survey report. The amount of effort put into the site inspection is a function of the project's size, complexity, and location. The input can vary from one person on a small project to a dozen or more on a very large project. The cost of the latter can be considerable, but the benefits to the proposal and to the project initiation after a successful proposal can also considerable. Our discussion will be based on the requirements for a large project; smaller jobs can be scaled back to suit their size requirements.

If the construction site is located in an area where your firm works regularly, you already know more about the local conditions than do your outside competitors. However, a site visit is still necessary to get the information pertaining to the specific site and to make a quick review of local conditions to confirm that your existing data are still valid. If the site is distant, the site survey should be set to coincide quite closely with the contractor-orientation meetings scheduled by the owner or the architect/engineer.

An effective site survey requires a sound plan, good organization, and constant control by the survey team's leader. Make sure that all the participants have thoroughly reviewed the bid documents, plans, specifications, and project data before they see the site. Team members should develop lists of questions and points that need clarification if they are to properly bid the job.

Another reason for conducting a thorough site inspection is that most construction contracts contain a clause requiring the contractor to verify that he or she has visited the site and is familiar with its characteristics. If a contractor is successful in obtaining a contract for construction and, upon the commencement of construction, observes—as an example—that a well pipe is in the exact location of a proposed footing, the contractor cannot request extra costs for lowering the footing to virgin soil after the well pipe has been removed if this well pipe would reasonably have been noticed during a site inspection.

The construction manager should hold an organization meeting with all team members present, to ensure that everyone understands the goals of the proposal and the site survey. Each team member's specific assignment should be discussed and coordinated to eliminate possible overlaps and confusion during the inspection trip. A professional approach to the survey can make a very favorable impression on the owner and on local agencies, and this will enhance the contractor's image immensely.

A site-inspection checklist for all types of projects is too voluminous to include here, but the following points should be investigated at a minimum:

1. Project location
   a. Remote or developed area
   b. Climatology and history of natural disasters
   c. Site access road, lay-down areas, and existing traffic patterns
   d. Available transportation systems
   e. Soil conditions and site-survey maps
   f. Location of neighbors
   g. Existing facilities and ongoing operations
   h. Off-site warehouse facilities
   i. Local legal and licensing requirements
   j. Applicable local codes and regulations
   k. Availability of construction-staff housing (if required)
   l. Local economic climate and cost of living
   m. Availability of local source for construction supplies, services, and equipment

2. Labor survey
   a. Labor posture (union or open shop)
   b. Availability and quality of local craft labor
   c. Local labor practices and history
   d. Availability of local-hire office and service staff
   e. Quality and availability of subcontractors
   f. Prevailing wages, fringes, and other labor costs
   g. Area labor productivity
   h. Competing projects in the area (ongoing and future)
   i. Local craft-training posture
   j. Date when union contracts are up for renegotiation (if union labor)

3. Site development

   a. Greenfield or existing plant

   b. Demolition work required

   c. Evaluation of site development and drainage design

   d. Construction-waste disposal—soil, liquid, hazardous waste, etc.

   e. Applicable government regulations

4. Temporary site facilities

   a. Availability of temporary construction utilities

   b. Temporary offices, trailers, and portable sanitary facilities

   c. Construction warehouses: existing, temporary, and other

   d. Heavy-equipment rental and services

   e. Communication systems: voice, data, and wireless

   f. Site safety and security considerations

   g. Availability of local services: banking, hospital, health services, and police

5. Public relations

   a. Check on how the project is being received in the community.

   b. Contact involved local government agencies.

   c. Collect local newspaper articles relating to the project.

   d. Check on any activism against the project that is anticipated.

   e. Evaluate any potential environmental problem areas.

   f. Visit ongoing projects in the area to assess potential local problem areas.

6. Client relations

   a. Establish rapport with the client organization.

   b. Obtain any necessary missing project information.

   c. Discuss the client's project goals for the construction program and facility.

   d. Review your planned project-execution approach with the client to eliminate any potential conflicts.

   e. Check out the client's facility labor posture.

The construction manager should ensure that the team members have the necessary tools and data to carry out an effective survey. Portable recorders, cameras, measuring devices, bid documents, and a portable builder's level will make the best use of the time allotted to conduct the site visit.

The final site-survey report should be presented in an organized and timely manner for use by the proposal team. This means that it should be clearly written and published as soon as the trip has been completed. A poorly prepared report will waste the considerable investment made in the survey trip and will not reflect well on the team leader.

Reviewing the Pro Forma Contract

The construction manager should take the lead in reviewing and responding to any administrative problem areas in the pro forma contract presented in the RFP. The legal staff will handle some of this review, as it affects liabilities and other risk-management issues. The construction manager is responsible for reviewing the day-to-day operating requirements called for in the contract. The specific areas of interest in the contract for the project manager or construction manager are

1. Scope of work
2. Payment terms
3. Change-order procedures
4. Quality control
5. Guarantees and warranties
6. Schedule requirement
7. Suspension and termination

Your key role in the proposal team puts you in a position to come up with the hooks that will make the selection committee pick your proposal. This is an opportunity for creative thinking, so do not pass it up. You must reduce the potential impact of any negative proposal factors—human-resources shortages, poor office or site location, adverse costs, etc.—that might hurt your chances of winning the proposal. Let us assume that you have been good enough at your craft to win the proposal. You developed the most creative solutions, organized the strongest project team, orchestrated an outstanding dog-and-pony show, and made the fewest mistakes. On that happy note, you are now ready to play a key role on the contract-negotiation team. This requires a solid background in the area of construction contracting—which happens to be our next subject.

Construction Contracting

The basic format of construction contracting has not changed much in the last several decades. The fixed-price approach has been with us for centuries, but in recent years more collaborative contract formats have come to the fore, with the purpose of diluting adversarial relationships and reducing the potential for litigation. We will discuss this in more detail a little later in this chapter.

The managers on both sides of the construction-contract equation play influential roles in the negotiation and acceptance of the contract document. These are the people responsible for fulfilling the terms of the contract and making it work. If all goes well—and it often does—the lawyers and top management involved may never have to look at the contract again.

The reason for a contract is to clearly and fairly set forth the responsibilities of each party involved in the project before the work begins. It also spells out the remedies one party may seek from the other for failure to perform or for default on any of the terms. When such a situation occurs, the contract becomes the primary document for settling any later arbitration or litigation claims.

Contracts for construction of capital projects have assumed more importance in recent years because we have become a litigious society. Years ago, the executed contract sometimes arrived closer to the end of the project than to the beginning. The trend toward a tightening up of contracting procedure has not been all bad, for it has forced project participants to finish their homework *before* starting the project.

Due to the complexity of some major projects, the contract cannot cover all possible situations that are likely to arise. This means that there must be a good deal of give-and-take as well as reasonableness on both sides. Having a cooperative approach to contract administration wins at least half the battle, with the other half being won by having a sound and equitable contract document to start with.

This discussion of contracts is from the standpoint of the construction manager and the general contractor. An approach that allows all parties to meet their project goals usually works out best. Here are some of the major areas of contractual risk that must be considered when selecting a construction-contract strategy:

- Degree of project definition
- Possibility of ongoing design changes
- Possibility that the facility will not meet its performance goals
- Potential escalation of labor and material costs
- Unknown labor productivity in project locale
- Difficult climate conditions
- Inadequate or unqualified labor supply
- Occurrence of labor strikes
- Possibility that bids will exceed the estimate
- Unfeasible strategic completion date
- Process design that has not been adequately proven
- Inadequate licensing package

- Casualty loss not covered by insurance
- Liability to third parties

The owner and contractor must evaluate the applicable items on this list to determine which party will assume which risks. Some of the owner's risks can be readily covered by insurance and bonds that are available based upon the amount of the contract sum. Clauses to that effect must be included in the contract.

The owner eventually assumes the lion's share of the risks in selecting the basic design of the facility, developing the conceptual design, and selecting the site for the work. The contractor's risk occurs in the construction-management areas of labor supply, productivity, schedule, and local site conditions.

Contract Formats

The variations on standard contract formats are virtually endless. The major issuers of construction contracts are the American Institute of Architects (AIA), the Associated General Contractors of America (AGC), the Construction Management Association of America (CMAA), the Design-Build Institute of America (DBIA), and, for engineering-driven projects, the Engineers Joint Contract Document Committee (EJCDC). We can classify the basic contract formats as those listed in Table 2.1. We will expand on these various types after we describe their basic function. The letter of intent is a different type of agreement, not issued in any special format by any organization, prepared for a limited scope of work.

Letter of Intent: A short-form type of contract; exactly what it says—a letter expressing the owner's *intent* to enter into a contract for a limited scope of work at a predetermined price, to be completed within a specific time frame. These Letters of Intent are often used as a precursor for additional work, as the initial scope is in progress. This could apply when a commercial owner has re-leased some property but must first demolish the existing renter's layout so the new tenant can develop a new floor plan. Or this can be used when the owner wishes a limited scope of work to commence, at a fixed price, while weighing the commitment to the design of a new building. A letter of intent may be issued to clear and grub the potential building site but not proceed beyond that scope of work until the owner has completed the project financing.

Cost of the Work plus a Fee: A rather uncomplicated form of contract in which the owner agrees to pay the contractor for the cost of the work plus a fee, generally represented as a percentage of that cost. The contractor should present with this contract a list of which costs are to be reimbursed and which costs are not. This type of contract is generally employed when emergency work is required—e.g., repair

Primary Advantages	Primary Disadvantages	Typical Applications	Comments
Cost-Plus			
PD: Minimal (Scope of work does not have to be clearly defined.)			
1. Eliminates detailed scope definition and proposal preparation time. 2. Eliminates costly extra negotiations if many changes are contemplated. 3. Allows client complete flexibility to supervise design and/or construction.	1. Client must exercise tight cost control over project expenditures. 2. Project cost is usually not optimized.	1. Major revamping of existing facilities. 2. Development projects where technology is not well defined. 3. Confidential projects where minimum industry exposure is desired. 4. Projects where minimum time schedule is critical.	Cost-plus contracts should be used only when client has sufficient engineering staff to supervise work.
Cost-Plus, with guaranteed maximum			
PD: General specifications and preliminary layout drawings			
1. Maximum price is established without preparation of detailed design drawings. 2. Client retains option to approve all major project decisions. 3. All savings under maximum price remain with client.	1. Contractor has little incentive to reduce cost. 2. Contractor's fee and contingency are relatively higher than for other fixed-price contracts, because price is fixed on preliminary design data. 3. Client must exercise tight cost control over project expenditures.	When client desires fast time schedule, with a guaranteed limit on maximum project cost.	
Cost-Plus, with guaranteed maximum and incentive			
PD: General specifications and preliminary layout drawings			
1. Maximum price is established without preparation of detailed design drawings. 2. Client retains option to approve all major project decisions. 3. Contractor has incentive to improve performance, since he shares in savings.	Contractor's fee and contingency are relatively higher than for other fixed-price contracts, because price is fixed on preliminary design data.	When client desires fast time schedule, with a guaranteed limit on maximum cost and assurance that the contractor will be motivated to try for cost savings.	Incentive may be provided to optimize features other than capital cost, e.g., operating cost.

Table 2.1 Types of Contracts

Cost-Plus, with guaranteed maximum and provision for escalation — PD: general specifications and preliminary layout drawings		
1. Maximum price is established without preparation of detailed design drawings. 2. Client retains option to approve all major project decisions. 3. Protects contractor against inflationary periods.	1. Project involving financing in semi-industrialized countries. 2. Projects requiring long-time schedules.	1. Escalation cost-reimbursement terms should be based on recognized industrial index. 2. Escalation clause should be negotiated prior to contract signing.

Bonus/penalty, time, and completion — PD: Variable, depending on other aspects of contract		
1. Extreme pressure is exerted on contractor to complete project ahead of schedule. 2. Under carefully controlled conditions, will result in minimum design and construction time.	Usually applied to lump-sum contracts when completion of project is absolute necessity to client in order to fulfill customer commitments.	1. Defining the cause for delays during project execution may involve considerable discussion and disagreement between client and contractor. 2. Application of penalty under certain conditions may result in considerable loss to contractor. 3. Pressure for early completion may result in lower quality of work.	1. Project execution should be carefully documented to minimize disagreements on reasons for delay. 2. The power to apply penalties should not be used lightly; maximum penalty should not exceed total expected contractor profit.

Bonus/penalty, operation, and performance — PD: Variable, depending on other aspects of contract		
Directs contractor's peak performance toward area of particular importance to client.	When client desires maximum production of a particular by-product in a new process plant to meet market requirements.	1. Application of penalty under certain conditions may result in considerable loss to contractor. 2. Difficult to obtain exact operating conditions needed to verify performance guarantee.	Power to apply penalties should not be used lightly.

Primary Advantages	Primary Disadvantages	Typical Applications	Comments
Lump sum, based on definitive specifications **PD: General specifications, design, drawings, and layout—all complete**			
1. Usually results in maximum construction efficiency. 2. Detailed project definition assures client of desired quality.	1. Separate design and construction contracts increase overall project schedule. 2. Noncompetitive design may result in use of overly conservative design basis. 3. Responsibility is divided between designer and constructor.	When a client solicits construction bids on a distinctive building designed by an architectural firm or when a federal government bureau solicits construction bids on a project designed by an outside firm.	Clients are cautioned against use of this type of contract if project is not well defined.
Lump sum, based on preliminary specifications **PD: Complete general specifications, preliminary layout, and well-defined design**			
1. Competitive engineering design often results in cost-reducing features 2. Reduces overall project time by overlapping design and construction. 3. Single-party responsibility leads to efficient project execution. 4. Allows contractor to increase profit by superior performance.	1. Contractor's proposal cost is high. 2. Fixed price is based on preliminary drawings. 3. Contract and proposal require careful and lengthy client review.	1. Turnkey contract to design and construct fertilizer plant. 2. Turnkey contract to design and construct foreign power generation plant.	1. Bids should be solicited only from contractors experienced in particular field. 2. Client should review project team proposed by contractor.
Unit price contracts, flat rate **PD: Scope of work well defined qualitatively, with approximate quantity known**			
1. Construction work can commence without knowing exact quantities involved. 2. Reimbursement terms are clearly defined.	1. Large quantity estimate errors may result in client's paying unnecessarily high costs or contract extra. 2. Extensive client field supervision is required to measure installed quantities.	1. Gas transmission piping project. 2. Highway building. 3. Insulation work in process plants.	Contractor should define the methods of field measurement before the contract is awarded

Unit price contracts, sliding rate **PD: Scope of work well defined qualitatively**			
1. Construction work can commence without knowing quantity requirements. 2. Reimbursement terms are clearly defined.	Extensive client field supervision is required to measure installed quantities.	1. Gas transmission piping project. 2. Highway building. 3. Insulation work in process plants.	Contractor should clearly define the methods of field measurement before the contract is awarded.
Convertible contracts **PD: Variable, depends on type of contract conversion**			
1. Design work can commence without delay of soliciting competitive bids. 2. Construction price is fixed at time of contract conversion, when project is reasonably well defined. 3. Overall design and construction schedule is minimum, with reasonable cost.	1. Design may not be optimum. 2. Difficult to obtain competitive bids, since other contractors are reluctant to bid against contractor who performed design work.	1. When client has confidential project requiring a balance of minimum project time with reasonable cost. 2. When client selects particular contractor based on superior past performance.	Contractors selected on this basis should be well known to client.
Time and materials **PD: General scope of project**			
1. Client may exercise close control over contractor's execution methods. 2. Contractor is assured a reasonable profit. 3. Reimbursement terms are clearly defined.	1. Project cost may not be minimized. 2. Extensive client supervision is required.	Management engineering services supplied by consulting engineering firm.	Eliminates lengthy scope definition and proposal preparation time.

PD = Project Definition

Source: Reproduced with permission from "A Fresh Look at Engineering and Construction Contracts," *Chemical Engineering*, Sept. 11, 1967, pp. 220–221.

TABLE 2.1 Types of Contracts (*Continued*)

of fire or theft damage—and it is impractical to establish a fixed price for work of undeterminable scope. Or it can be employed where the exact nature of the work, generally minor in nature, cannot be accurately determined.

Cost of the Work plus a Fee, with a Guaranteed Maximum Price (GMP): Similar to a cost-plus contract, but the contractor guarantees the final cost of the project, except where adjustments are made to either increase or decrease the scope of the work. These GMP contracts are often used when an owner is anxious to start a project but the contract documents are not 100 percent complete. Generally, GMP contracts can be employed when the plans and specifications are about 70 percent complete. It then behooves both owner and contractor to prepare an "Assumption and Qualification Scope Exhibit," which includes items that both parties reasonably assume will appear on the final 30 percent or so of completed drawings. The contractor assumes the cost of work that the remaining 30 percent of the design is anticipated to include. Because the owner, designers, and contractor may each have a different idea of the nature of work in the remaining design, this exclusion/inclusion list, when added to or deleted from as design progresses, will represent the combined concept of that yet-to-be designed work.

Lump Sum or Stipulated Sum: A completed set of plans and specifications, along with specific bidding instructions, presented to a list of bidders who are to prepare their bids based upon the scope of work outlined in these bid documents—no more, no less. Any exceptions to the scope of work so presented may be cause for the contractor's bid to be rejected. This type of contract is most frequently used for public-works projects. When it is used in the private sector, the owner may select one of the bidders and negotiate both scope and price, then issue a contract on that basis.

Joint-Venture Agreement: The mobility of contractors nowadays will find international contractors working in local markets, and it is common to find large U.S. construction companies operating in many states at the same time on various types of projects. In some cases, local contractors can be of great assistance in obtaining preferred subcontractor pricing or knowledge of local labor markets, preferred vendors of materials and equipment, or intimate knowledge of specific sites of a questionable nature. A joint venture will prove advantageous to both parties if both can prosper from joining forces for a specific project. The joint-venture agreement will spell out the obligations, rights, and responsibilities of each participant, specifying the risks each will assume and the fees each will garner from a successful project.

Turn-Key: A type of contract where the private contractor, upon learning of the owner's project requirements, designs and builds to those requirements, turning over the project in a ready-to-use state. Although this type of contract is frequently used in chemical or petroleum-processing plants, it has also been utilized in such diverse projects as housing where the residential units are fully equipped and ready for occupancy and in motels or hotels where all items are furnished ready for operation once the key is turned over to the owner.

Build-Operate-Transfer

This concept gained some popularity in the last three decades of the twentieth century. It is a process of building infrastructure, whereby the developer agrees to design, build, and finance a revenue-producing piece of infrastructure—a toll road, tunnel, or bridge. It operates and maintains the structure for a period of time agreed upon with the government in whose jurisdiction it lies, in return for collecting tolls and ancillary revenue from the project. At the end of the agreed-upon term, the developer transfers authority back to the government for a minimum sum. In this way, a government desirous of building that highway, bridge, or tunnel can get the project built, operated, and maintained at very little cost, since the turnover fee collected at the end of the period is usually in the thousands of dollars. This is how the Channel Tunnel (between Great Britain and France) was built, the Queen Elizabeth Bridge in England, and the tunnels between Hong Kong and Kowloon.

As an added benefit, quality is usually very high, since the developer is responsible for all maintenance during a period generally in the range of 75 to 90 years.

American Institute of Architects Contracts

The following is a partial list of AIA contracts; a more complete listing can be obtained by accessing the institute's Web site (www.aia.org/contractdocs).

- A101 2007—Standard Form of Agreement Between Owner and Contractor Where the Basis of Payment Is a Stipulated Sum
- A102 2007—Standard Form of Agreement Between Owner and Contractor Where the Basis of Payment Is the Cost of the Work plus a Fee with a Guaranteed Maximum Price
- A103 2007—Standard Form of Agreement Between Owner and Contractor Where the Basis of Payment Is the Cost of the Work plus a Fee Without a Guaranteed Maximum Price
- A105 2007—Standard Form of Agreement Between Owner and Contractor for a Residential or Small Commercial Project
- A107 2007—Standard Form of Agreement Between Owner and Contractor for a Project of Limited Scope

AIA Integrated Project Delivery (IPD) Contracts

Between 2007 and 2010, the AIA and the AGC began to promote a new set of construction documents. These contracts stress collaboration between all parties to the construction process that will "harness the talents and insights of all participants to optimize project results, increase the value to the owner, reduce waste and maximize efficiency through all phases of design, fabrication and construction." These contracts are

- A195 2008—Standard Form of Agreement Between Owner and Contractor for Integrated Project Delivery, as developed by the American Institute of Architects (AIA.), Washington, D.C.
- A2952008—General Conditions of the Contract for Integrated Project Delivery

There is a series of six other contracts employing the IPD concept. The C191 2009—Standard Form Multi-Party Agreement for Integrated Project Delivery, contains some basic concepts of this new form of collaborative contract. A complete copy of C191 2009 can be purchased from the AIA via their Web site-www.aia.org.

In this contract, the contract sum is established in a much different way; the parties agree to a *target* cost, and to the extent that the actual costs are less than the target cost, the parties share in the savings.

Article 4, Compensation, is also unique in that it states that the owner will reimburse the parties for their cost of work, which shall continue *regardless of whether actual costs exceed the target cost*; however, when labor costs exceed actual costs, the owner shall (1) reimburse the parties for all labor costs, based upon negotiated hourly labor costs, and (2) not be required to reimburse the parties for any further labor costs incurred (though unstated, this may occur until the actual costs get back on track with the target cost). Article 8 deals with risk sharing and includes a clause stating that all parties waive claims against each other with the exception of seven egregious conditions, including willful misconduct and damages arising from liens, claims, or encumbrances against the project filed by entities not a party to the agreement.

A full understanding of the nature of these IPD contracts can only be obtained by reading one of them in its entirety, but the basis of all such contracts is to develop a team approach to the construction process, where the sharing of profits and risks is enjoyed by owner, design consultants, and contractor, and the willingness to enter into a give-and-take relationship is memorialized.

A Partial Listing of AIA IPD Contracts

- A195 2008—Standard Form of Agreement Between Owner and Contractor for Integrated Project Delivery

- A295 2008—General Conditions of the Contract for Integrated Project Delivery
- B195 2008—Standard Form of Agreement Between Owner and Architect for Integrated Project Delivery
- C191 2009—Standard Form Multi-Party Agreement for Integrated Project Delivery (this "multi-party" concept includes owner, architect, contractor, and key subcontractors all signing this one contract)
- C1952008—Standard Form of Single Purpose Entity Agreement for Integrated Project Delivery (this is similar to C191 2009, where a limited-liability company is contemplated)

Construction-Management Contracts: The CM acts as the owner's agent and actually more like the owner's construction professional assigned to work with the owner's architect and engineer, offering advice on constructability issues, budgets, and labor and materials availability. The scope of work assigned to the CM can begin with working with the design consultants from conceptual design through development to final, completed plans and specifications, and on to general-contractor selection, supervision of construction, project closeout, and turnover. The CM's fee is usually a percentage of the cost of construction. The CMAA publishes the following contracts:

- CMAA Document GMP-1—Standard Form Between Owner and Contractor Where a Guaranteed Maximum Price is Provided
- CMAA Document A-2—Standard Form of CM Contract Between Owner and CM
- CMAA Document A-3—General Conditions to the Standard Owner–CM Contract
- CMAA Document A-4—Contract Between Owner and Designer
- CMAA Document GMP-2—Contract Between Construction Manager and Contractor
- CMAA Document GMP-3—General Conditions for the GMP-2 Contract

Program Manager: This carries the CM's agency concept much further. The program manager assists the owner in obtaining financing if required, obtains all necessary permitting for the project, works with public-relations firms and public agencies to secure legislative initiatives, and may even operate and maintain the facility for the owner. This project-manager concept has gained more popularity, generally only on very large projects where the owners neither has the staff to fulfill all of these functions nor wishes to add staff to do so.

Associated General Contractors of America (AGC) ConsensusDocs Contracts

AGC has developed 90 ConsensusDocs contracts divided into six series, covering standard formats such as general conditions, cost of the work plus a fee, cost-plus with a guaranteed maximum price, construction manager, and design-build.

ConsensusDocs 200 Agreement and General Conditions Between Owner and Contractor (Where the Contract Price Is a Lump Sum) provides some insight into the intent of the contract formats—for parties to try to work together and resolve problems amicably. One statement emphasizes the need to develop a clear communication path between owner and contractor. The design professionals are excluded from the dispute process, leaving problems to be resolved between owner and contractor. If that cannot be attained, a neutral dispute-review board will be mutually appointed by both sides, who will share costs and arrive at a resolution. If this is not possible, the dispute moves on to mediation and finally to arbitration.

The owner is responsible for design and design coordination, and the contractor is responsible for design elements only as specifically called for in the contract. Both parties designate an authorized representative, and the contractor designates a safety director.

The basis for all of these ConsensusDocs is to eliminate the adversarial relations that oftentimes creep into a project when one party to the contract believes it is being treated unfairly. Here again, a more comprehensive understanding of the nature of these AGC contracts can only be appreciated when one is read from start to finish.

A Partial Listing of ConsensusDocs Contracts

- ConsensusDocs 200—Standard Agreement and General Conditions Between Owner and Contractor (Where the Contract Price Is a Lump Sum).

- ConsensusDocs 200.1—Potentially Time and Price Impacted Materials. An interesting contract for contractors engaged in multiyear projects where the potential for inflation-impacted material prices may occur.

- ConsensusDocs 300—Standard Form of Tri-Party Agreement for Collaborative Delivery. The three parties are owner, design consultants, and contractor, meant to operate in an environment of collaboration.

- ConsensusDocs 400—Standard Design-Build Agreement and General Conditions Between Owner and Design-Builder (Where the Basis of Payment Is the Cost of Work with a GMP)

- ConsensusDocs 500—Standard Agreement and General Conditions Between Owner and Construction Manager (Where the Basis of Payment Is a Guaranteed Maximum Price with an Option for Pre-Construction Services)

Design-Build: Instead of hiring an architect to design a project and then engaging a contractor via either the competitive-bid or negotiated-contract process, the owner engages a one-source entity—a design-builder, who will not only translate the owner's defined project but build it as well. Therefore the owner deals with that single entity, and the entire design and construction process is capable of delivering the completed project considerably faster, since the construction can commence ordering long-lead equipment and materials early in the design process. A successful design-build project depends upon the owner's having a very defined scope of work in mind prior to engaging the design-build team. The Design-Build Institute of America publishes a series of contracts, including the following:

- DBIA Document No. 501—Contract for Design-Build Consultant Services

- DBIA Document No. 520—Standard Form of Preliminary Agreement Between Owner and Design-Builder

- DBIA Document No. 525—Standard Form of Agreement Between Owner and Design-Builder—Lump Sum

- DBIA Document No. 530—Standard Form of Agreement Between Owner and Design-Builder—Cost plus Fee with an Option for GMP

- DBIA Document No. 535—Standard Form of General Conditions to Design-Build Contract Between Owner & Design-Builder

In February 2012, the DBIA issued a teaming agreement. This agreement is intended to be used when the design-builder is the prime contracting entity and will subcontract to other parties such as design consultants, trade contractors, and specialty design-build contractors, together forming the teaming party. There is also a subsequent agreement, which sets the baseline for how the parties will work together during the project and allows them to assess risk and costs at an early date. The teaming agreement contains an optional clause adding a provision for liquidated damages if the other terms and conditions cannot be negotiated in the subsequent agreement.

Bridging—Another Approach to Design-Build

Not so much a distinct contract, bridging approaches the formation of a project—primarily design-build—obliquely, although its concept can be applied to other forms of construction-contract development. It is a testing-the-waters kind of approach.

Bridging is a process in which an owner hires a design firm to produce a partial set of schematic drawings for a proposed new project. These drawings can then be used to test the market for pricing when an owner is not 100 percent sure it wants to proceed with the project. The owner invites a select short list of contractors to review the schematics and submit order-of-magnitude or ballpark pricing. The contractors invited to submit pricing are those who have had experience in building similar projects.

By using this approach, the owner can limit its financial exposure if pricing is much higher than anticipated. All the owner has lost is the cost of the schematic design; however, these plans can be put away for a future date in case the project is resurrected. But if the contractors present preliminary pricing based upon the schematic design that falls within the owner's budget, the owner may want to proceed with completion of the design and obtain a firm contract sum.

By submitting these schematic drawings, an owner can also elicit suggestions from the contractors on changes to the design that could affect savings or improve the design. These bridging documents often consist of six to 12 drawings and a 30-page booklet of basic building specifications: the shape of the building, height, structure and building envelope descriptions, and some site-work details, such as parking provisions, roadways, and landscaped areas.

This bridging process can be divided into two phases:

- The owner requests qualifications, which includes the bridging bid instructions, such as exhibiting experience in similar projects and experience in design-build projects.

- The short-listed bidders are requested to submit proposals to expand the design from the schematics and provide approximate costs and pertinent information, showing that they understand the owner's program.

Upon receipt of the pricing and an understanding by the contractors that they conceptually envision the design expanding, the owner has several choices: abandon the project; direct the architect to complete the design, possibly inviting one of the bidders to work along with the architect in a true design-build mode; or engage another architect to develop a modified or entirely new design.

Now there are some legal issues involved in this process. There is a question of who owns the design and who is the architect of record. If the initial design group is retained and completes the design, there are no legal issues of ownership of design. But if another architect is hired and uses a modified version of the original design, who is the architect of record and who assumes liability for the design?

If the initial design is accepted, a contractor engaged to work with that design group can fast-track the project, which is another advantage of the bridging process.

Lean Construction and the Lean Construction Institute

The founder of the Lean Construction Institute—Greg Howell—after years of studies in the field, found that about 54 percent of construction-work assignments were not completed on time, and he thought that establishing some of the rules applicable to manufacturing might benefit the construction industry. Lean construction makes significant use of tools to facilitate planning and control to maximize value and minimize waste.

In collaboration with McDonough Holland and Allen—a law firm based in Sacramento, California—Howell developed a contract form titled "Integrated Agreement for Lean Project Delivery Between Owner, Architect and CM/GC." The various sections (there are 28) of this 77-page document provide the basic concepts of lean construction:

- Section 3, Relationship of Parties. This section requires all parties to work together and achieve a transparent and cooperative environment for collaboration for the benefit of the project.

- Section 5, Collaboration and Integrated Preconstruction Services. None of the parties to the preconstruction process are to proceed in isolation from the others.

- Section 6, Project Planning and Scheduling. This is one of lean construction's keystones, relying on just-in-time deliveries of materials and a make-ready-look-ahead schedule encompassing a minimum six-week window. Weekly look-ahead planning meetings and updating of milestones are an integral part of the planning process.

- Section 11, Value Engineering, Constructability and Work Structuring. The general contractor (GC) or CM and his or her trade subcontractors are to continually pursue opportunities to create additional value by proposing ideas to cut costs.

- Section 15, Construction Phase Operations. This section sets forth the "5S" plan: *sort*—remove clutter and use just-in-time procedures; *set in order*—identify location where all items will be used and installed; *shine/sweep*—remember that a clean and orderly workplace is a safe and efficient one; *standardize*—distribute a "standard practices" memo for all to follow; and *sustain/self-discipline*—create an environment where the first four elements thrive and constantly evolve. It seems as though these five elements contained in the lean-construction process have applicability in all forms of the construction process.

Engineers Joint Contract Documents Committee (EJCDC)

Engineering firms look to the EJCDC for owner–engineer contracts, which include both design and construction. There are contracts for geotechnical services, land surveying, environmental remediation, and design-build services. The contracts are arranged by groupings:

- C Series—Construction Related Documents
- E Series—Owner-Engineer Documents and Engineer Sub-Consultant Documents
- R Series—Environmental Remediation Documents
- P Series—Procurement Documents
- D Series—Design-Build Documents
- E Series—Owner-Engineer contracts

There is also E-562—Agreement Between Engineer and Engineer's Subcontract, to be used when the engineer subcontracts for services such as lab work, data management, cost reviews, site support, drafting, or technical services.

In June of 2012, EJCDC's legal counsel issued a document titled "Recent Court Decisions of Relevance to Contract Documents" (Table 2.2). Although these decisions were extracted from a myriad of construction-related court findings, they seem to have applicability for other contracts relating to general contracting, and are worth reviewing.

Contract Administration

We have spent a lot of time talking about arriving at a mutually satisfactory contract document, but that is only half the battle. CMs and GCs have the prime responsibility for overseeing the contract for the life of the project.

As a foundation to good owner–contractor relations, it is best for both parties to be proactive in conforming to the contract requirements. A good way to remind yourself of key contractual requirements is to set up a contract tickle file to automatically remind you of key notice dates and performance features required by the contract. This can be accomplished using any number of readily available software programs; you may even already have one available in your current project-management program.

Issue	Citation	Summary	Contract Document Implications
1. Statute of Limitations for design error claim	*Newell Recycling of Atlanta, Inc. v. Jordan Jones and Goulding, Inc.*, Supreme Court of Georgia (2010).	Engineering firm designed a shredding facility for a private recycling company. The services were provided pursuant to a collection of letters, agreements, and a "draft Scope of work." After construction was completed, the facility's concrete drainage control structure began to fail. The recycling company waited to bring suit until more than four years after discovering the drainage control structure's failure. The Georgia Court of Appeals examined both a four-year statute of limitations, on "any implied promise or undertaking," and a six-year statute of limitations on actions based on written contracts. It concluded that even if the paperwork taken together could be viewed as a written contract, nonetheless the four-year limitation period should apply, because the core claim really involved the conduct of professionals acting in their area of expertise— professional malpractice. Apparently the Court of Appeals believed that the predominant obligations of the design professional were implied, unwritten duties, despite the existence of the written documentation. The Georgia Supreme Court reversed, holding that a professional malpractice claim premised on a written contract should be governed by the six-year statute of limitations applicable to written contracts. The shorter statute of limitations would apply only if the claim was based solely on an oral or implied obligation. If there are implied duties but also a written agreement, the six-year statute of limitations would apply.	The case shows that using a standard written contracts can eliminate disputes about whether there is or is not a contract—as opposed to the confusion that occurred here. However, the implications of being governed by laws applying to contracts will vary. Here, the contract statute of limitations helped the owner and harmed the design professional. In some cases the contract statute of limitations, regardless of length in years, will start running well before the parallel tort statute of limitations. Contract limitation periods are frequently based on the time of the breach; tort limitations, by contrast, are often based on the date of the discovery of the injury. Note in this regard that EJCDC E-500, Agreement between Owner and Engineer for Professional Services, provides that any applicable limitation period should stand running no later than the date of substantial completion of the project.

TABLE 2.2 Recent Court Decisions of Relevance to Contractors *(Reproduced with Permission of the Engineers Joint Contract Documents Committee [EJCDC]. Visit www.ejcdc.com.)*

EJCDC: RECENT COURT DECISIONS OF RELEVANCE TO CONTRACT DOCUMENTS June 2012			
Issue	Citation	Summary	Contract Document Implications
2. Application of ten-year statute of repose to complex multi-party construction project: when does the ten-year period begin to run?	*State of New Jersey v. Perini Corporation*, Appellate Division, Superior Court of New Jersey (2012).	The centralized underground hot water distribution system for South Woods State Prison deteriorated after project completion, resulting in the need to shut down the system, and therefore shut down the prison. The state sued four companies responsible for design and construction, and the manufacturer of defective piping. The New Jersey statute of repose is ten years. The state's lawsuit was filed more than ten years after the prison, including the hot water system, was put into service; however, the last of thirty separate certificates of substantial completion was dated just under ten years prior. In a convoluted decision the appellate court held that the statute of repose did not begin to run until all phases of the project were substantially complete. The court acknowledged that multiple statute of repose starting points were possible, but that in this case substantial completion of all phases was necessary because the defendants had all maintained some degree of continued presence at the jobsite until well after occupancy. The court held that the statute of repose does not apply to protect manufacturers of materials.	The EJCDC construction documents allow for partial utilization and multiple substantial completion certifications. Thirty separate certificates is an extreme example, and would present various administrative challenges. Two of the defendants were subcontractors. If they had finished their work and not returned to the jobsite, the individual certificates applicable to their work would have been taken as the starting point of the statute of repose for their liability. Their punch list work extended their exposure, for reasons that are not well explained by the court. Wording and interpretation of statutes of repose varies from state to state. It is typical for the statutes to not apply to manufacturers of incorporated materials.

| 3. Contractor's entitlement to compensation for extra work. | G. Voskanian Construction, Inc. v. Alhambra Unified School District, Court of Appeal of the State of California, Second Appellate District (2012). | Contracts for relocation of school buildings and for installation of fire alarm system required that modifications be in writing, consistent with state public contract law.

On the relocation contract, the school district issued oral orders for extra work, but later endorsed the necessity for the extra work in change orders; these did not address additional compensation. The appeals court held that the contractor could recover for the value of the extra work.

On the fire alarm contract, the extra work was necessary because the plans and specifications incorrectly portrayed the number of rooms requiring alarms; bidders had not been allowed to view the interiors of the buildings because classes were in session during the pre-bid walk-through. The appeals court held that if extra work is necessitated by misleading or inaccurate drawings and specifications, the contractor may recover for the work, regardless of receiving prior written authorization. | The school district did not administer these contracts very well, and the result is not unfair. On the relocation contract the procedure was akin to use of a Work Change Directive proceed with the work, pricing to be determined later. The fact that the directive to proceed (authorization to do the extra work) was put in writing well after the fact was an added wrinkle.

The extra compensation on the fire alarm contract is a bit more debatable. No question that deficient design documents can trigger the right to more compensation; but when the contractor realized that there were many more rooms than depicted, should the school district have been given the opportunity to scale down the scope of the installation? The need was probably driven by code, but quantity increases should be approved in advance; analogous to getting a determination on what to do when a differing site condition is encountered. |

TABLE 2.2 Recent Court Decisions of Relevance to Contractors (*Reproduced with Permission of the Engineers Joint Contract Documents Committee [EJCDC]. Visit www.ejcdc.com.) (Continued)*

EJCDC: RECENT COURT DECISIONS OF RELEVANCE TO CONTRACT DOCUMENTS June 2012

Issue	Citation	Summary	Contract Document Implications
4. Common law indemnification duty in worker injury case.	*McCarthy v. Turner Construction, Inc.,* Court of Appeals of New York (2011).	Manhattan property owner leased a retail space to Ann Taylor, Inc. Ann Taylor entered into a contract with a CM at risk for build-out of the space. The CM subbed out the telephone/data cabling; sub-subbing of actual cable installation. Employee of the sub-sub was injured in a fall from a ladder. A lawsuit followed, based in part on a New York scaffolding law that imposes strict liability on general contractors and property owners when a fall occurs on a jobsite. Ultimately the CM and the property owners both contributed $800,000 to a settlement. The property owners then pursued a common law indemnification claim against the CM. The Court of Appeals found that although the CM had taken responsibility for supervising and directing the work in its contract with the retail tenant, it had not actually supervised or directed the injured worker's activities; CM had delegated that to the subs. The CM was non-negligent and faced liability to the worker only through the vicarious terms of the scaffolding law; thus it did not owe a common law indemnity duty to the property owners. The court rejected an argument that a new rule should be imposed to create an indemnification duty merely for failing to exercise the authority to supervise or control.	The EJCDC indemnification clause in the General Conditions (C-700) applies only to the extent of Contractor's or Subcontractors' negligent acts or omissions; and duty is owed to Owner and Engineer, and not to remote parties such as the Site's owner. This is a good decision for general contractors and CMs who delegate safety and supervision duties to subs. However, note that typically express indemnification duties would appear in contracts, leases, etc., up and down the chain, thus allowing a property owner to obtain compensation from the tenant, which would then push the claim down to the GC. In the New York case, it may be that the lease to Ann Taylor was favorable to tenant and excluded any duty to indemnify the property owners for construction injuries.

5. Delegation of safety obligations to independent contractor.	*Seabright Insurance Co. v. U.S. Airways, Inc.*, Supreme Court of California (2011).	U.S. Airways hired an independent contractor, Aubry Co., to maintain and repair a baggage conveyor. One of Aubry's employees was injured while working on the conveyor, allegedly as the result of lack of safety guards at "nip points." The employee collected from worker's compensation insurance; the worker's comp insurer and the employee then brought an action against U.S. Airways. The California Supreme Court held that when U.S. Airways hired Aubry Co., U.S. Airways implicitly delegated any tort law duty of care it might have had to ensure Aubry's employees' safety. Thus U.S. Airways delegated to Aubry the duty to identify the absence of needed safety guards, and to take steps to address the hazard. The court held that delegation is favored as a matter of policy. Aubry's costs in obtaining workers' compensation insurance were presumably factored into the contract price. It was significant that the independent contractor had sole control over the means and methods of performing the maintenance and repair work. Also noted was the inequity in allowing employees who happen to have a hiring party (e.g. U.S. Airways) to pursue receive a greater recovery than employees who do not work for hired contractors.	The decision is consistent with EJCDC principles regarding allocation of safety duties. The case should stand as a valuable precedent because of the prestige of the court and the clarity of the analysis and reasoning.

TABLE 2.2 Recent Court Decisions of Relevance to Contractors (*Reproduced with Permission of the Engineers Joint Contract Documents Committee [EJCDC]. Visit www.ejcdc.com.*) (*Continued*)

EJCDC: RECENT COURT DECISIONS OF RELEVANCE TO CONTRACT DOCUMENTS June 2012

Issue	Citation	Summary	Contract Document Implications
6. Absolute liability for excavation work.	*Yenem Corp. v. 281 Broadway Holdings*, Court of Appeals of New York (2012).	Plaintiffs were owners and tenants of landmark cast iron and masonry building in Manhattan. Defendants were the owner and contractor of adjacent lot. Excavation was conducted on the adjacent lot to a depth of 18 feet below curb level. The landmark building shifted nine inches out of plumb and was declared unsafe for occupancy. Plaintiffs sought summary judgment based on a city ordinance requiring support during excavation of neighboring buildings. The defendants argued in part that the poor condition of the landmark building excused them from liability. The court noted that violation of certain state laws can create absolute liability, making it easier for claimants to make their case, and to obtain summary disposition. This status is not usually extended to city ordinance violations, but in this case the court determined that the ordinance had its origin in state law, and therefore conferred absolute liability status on the violation. The court held that the pre-excavation condition of the landmark building would be relevant to the measure of damages, but not to the question of liability.	1n C-700, EJCDC addresses the responsibility of the contractor with respect to damage to adjacent properties. Liability could also be determined to rest with the engineer, if the excavation would inevitably destabilize adjacent buildings, regardless of standard precautions taken by contractor.

7. Application of economic loss doctrine to contractor's claims against design professional on design-build project.	*Maeda Pacific Corporation v. GMP Hawaii, Inc.*, Supreme Court of Guam, (2011).	Design-build project to establish water supply system for Air Force base. Design-builder Maeda retained an engineering firm, GMP Associates, Inc., as lead designer and for engineering quality control, and also subbed out design and build of a water reservoir tank to Smithbridge, which in turn subbed out the tank's structural design. The tank roof collapsed during testing of the system, probably as the result of lack of proper venting. Maeda pursued claims against the two design professionals, GMP and Jorgenson. The Supreme Court of Guam examined the issue of whether the economic loss doctrine should apply in Guam. That doctrine precludes parties in a contractual context from pursuing tort claims for purely economic or commercial losses. Rather, such parties must pursue economic loss claims under applicable contract provisions. According to the economic loss doctrine, where what is at stake is in essence "the benefit of the bargain," contract law, rather than negligence/tort law, should control. In the contractual context, the economic loss doctrine "encourages the party best situated to assess the risk [of] economic loss ... to assume, allocate, or insure against that risk." The court adopted the economic loss rule for Guam non-residential construction cases. It further held that the rule should apply to claims against design professionals, and should apply regardless of whether there is privity (a contractual relationship) between the two parties, under the understanding that the other party is doubtless a party to other contracts involving the same overall construction project.	The economic loss doctrine is entirely consistent with EJCDC principles favoring it through written contract that addresses the risks and issues that typically arise on a construction project. Guam joins the majority of jurisdictions, though the scope of the application of the doctrine varies widely. No discussion in the case as to the significance to Maeda of pursuing rights in tort against GMP, with which Maeda had a contract. It is possible that contract rights had lapsed or been waived in some manner. This is a laudably clear and cogent discussion of the economic loss doctrine.

TABLE 2.2 Recent Court Decisions of Relevance to Contractors (*Reproduced with Permission of the Engineers Joint Contract Documents Committee [EJCDC]. Visit www.ejcdc.ccm.*) (*Continued*)

CMs and GCs must document the project according to the terms and conditions of the contract, either manually or on a computer. Keep in mind that the contract sets up only the minimum documentation required; a contract administrator must thoroughly prepare to defend any contract claims that may be likely to arise.

Treat the contract documents as confidential material. Limit the distribution to those members of the project team who have a need to know. Keep a copy available in your file for ready access if your supervisor needs to discuss these matters with you.

Notice to Proceed

Government contracts often include a clause that work is not to commence without an official notification from the owner. This means that signing a contract does not authorize the contractor to commit funds on behalf of the owner. Any money spent before the notice to proceed is issued is at risk to the contractor. The notice to proceed may come in two forms, a first authorization to mobilize and a second notice to proceed to commence construction. The actual start date of the project counting down the number of days to completion,when two Notices to Proceed are issued is that second notice to proceed; if only one notice is issued, that is the official start date of the project.

Summary

In this chapter we have touched on some of the major points in the proposal and contracting environment, as seen from the GC or CM's viewpoint. This is an important area that often gets neglected as people develop a career in construction management. Construction firms live or die on successful projects and contract execution. Your management will be looking for strength in these areas when it is selecting personnel for assignment to key projects and promotion.

Work with experienced people and learn from them. Do not be afraid to ask questions—how else can you learn? And as the new wave of contracts emanates from the AIA, AGC, observe how new trends emphasize collaboration and working together. These are important traits to work toward.

CHAPTER 3

Project Planning and Initiation

We are now ready to explore a total construction project management approach to executing capital projects. Planning is the first step of our total project management philosophy for planning, organizing, and controlling the execution of capital projects.

First we will discuss the methods used to prepare a project-execution plan (or master plan); then we will move on in Ch. 4 to scheduling detailed execution of the work. In later chapters we will cover the financial and project resources for the project.

We have deliberately separated project planning from project scheduling to stress the point that these are two separate and distinct functions. The project and construction manager and their key staff members prepare the master plan; the scheduling people put the plan on a timeline.

Another administrative consideration that has to be settled is that of selecting the sort of construction projects we are going to discuss. Up to this point we have been talking about three general types of construction projects, namely design-build, construction design-bid-build, and construction-management modes.

The design-build mode involves a major personnel change in the collaborative working arrangements between architect and contractor. When an owner decides to pursue this method of construction, he or she will seek out a firm, either an architect-led team or a contractor-led team that has combined skills and experience working together to not only design the owner's project but build it as well—as one entity, instead of the conventional two-party mode. In this arrangement, the owner works with one team, and it is up to the team to determine which member will be the owner's prime contact from conceptual design to project turnover.

With the owner's representative working closely with the design-build representative, planning of design and construction fall under one aegis and can be more specific, because both design and construction-planning endeavors are melded into one process. This varies considerably from the more conventional design-bid-build process and the construction-management approach, so we will first deal with the planning activities for those two methods of project delivery.

Planning Philosophy

Planning must be done logically, thoroughly, and honestly, if you are to have a chance to succeed. Previous experience of many years has honed our basic planning logic for capital projects to a fine art. There is no point in trying to develop a new planning logic for each new project. Also, the owner has already set up a basic project-execution format in the contracting plan.

After selecting the normal planning logic for your project, you should examine the work for exceptional features that would affect the normal logic of your plan. Look for any special problem areas that may be different this time around, such as unusual client requirements, an out-of-the way location, or potential internal or external delaying factors.

Work these potential problem areas over in detail to reduce their negative effects on the master plan and later the schedule. You may leave the details of the schedule to your planning specialists, but you must set the basic scheduling logic for them.

A project manager should opt for simplicity in project management now we want to stress being thorough in the area of planning. Each aspect of your plan calls for individual scrutiny. Small details passed over in the early stages can rise up later in the job to smite you mightily when you least expect it. Enough things will go awry by accident, so there is no need to increase the burden by overlooking problems in the planning stage.

Honesty in planning is also very important. Remember, you are the one responsible for carrying out the project. You will be the one finally held accountable for its success and failure. You may blame a late project on your planning people, but your management will not.

By honesty we mean facing reality. Presenting a plan that has been assembled with overoptimistic assumptions, delivery dates, or productivity factors, merely for the sake of presenting an aggressive plan, is not being honest. If problems truly exist, they must be faced, not folded into an overoptimistic plan.

It is best to maintain a balance between optimism and pessimism during this early planning process. A project plan without some float or contingency is not realistic. Things will go wrong. The plan will spring shut on you eventually unless you have an inordinate amount of luck on your side. Many times it is hard to find any float in the construction schedule, because it has all been used up in the prior conceptual and design phases. Often it falls to the construction manager (CM) or project manager (PM) to deliver the bad news of a slippage of the strategic end date to the owner and designer who were responsible for burning up the float.

It is up to you to convey the need for float in your end of the project—the construction phase. You must explain to an owner how extreme weather, delays in delivery of materials or equipment due to labor disputes, factory shutdowns during vacation, and disruptions of efficient work crews due to injury or sickness can adversely impact a schedule, and that these are events that cannot be predicted; they must be taken into account via a contingency or added float in the schedule.

On the other hand, a "fat" plan is wasteful of time and money. Parkinson's law tells us that work always expands to fill the time allowed. We have always felt that the highest project efficiency often results from being slightly understaffed. Working with a schedule that is a little tighter than average, coupled with suitable incentive programs, also helps.

Good construction managers must learn to work under pressure; if they do not, they will not survive. The workday is limited to eight hours, but there will be many times when much longer hours are required, and possibly Saturday or Sunday work. This is the nature of the construction business.

Types of Planning

Several types of planning are involved in any capital project. Let us define the types and see how the construction manager fits into this overall planning process. The three major types of planning are

1. Strategic planning, which involves the high-level selection of the project objectives

2. Operational planning, which involves the detailed planning required to meet the strategic objectives

3. Scheduling, which puts the detailed operational plan on a time scale set by the strategic objectives

Does the construction team do the strategic planning? No, that is done by the owner's corporate planners. They decide what project to build and what the completion date has to be to meet their project goals. The project-development phase involves a great deal of strategic planning. It requires the input of market analysis, financial planning, project feasibility, and so on. Those areas need thorough study before the project can get the green light. Usually the strategic completion date is fixed before the decision is made to proceed without allowing more time on the end date. That uses up the schedule float on the front end and results in a tight schedule for the remaining work.

The construction team formulates the master construction-execution plan within the guidelines called for in the strategic and contracting plans. This chapter will explain how to make an operational plan for a typical project. The job of putting the plan onto a time schedule falls to the project schedulers. That points up the difference between planning and scheduling, which so often are mistakenly thought to be interchangeable.

Operational-Planning Questions

Operational planning usually raises some interesting questions for resolution in the construction master-planning phase:

- Will the operational plan meet the strategic-planning target date?
- Are sufficient construction resources and services available within the company to meet the project objectives?
- What is the impact of the new project on the existing workloads?
- Where will we get the resources to handle any overload?
- What company policies may prevent the plan from meeting the target date?
- Are equipment or materials involved with unusually long delivery times?
- Are the project concepts and designs firmly established and ready for the start of construction?
- Is the original contracting plan still valid?
- Will it be more economical to use a fast-track scheduling approach?

We must answer these and any other pertinent questions in preparing the construction master plan before the detailed scheduling can start. Preparing a detailed construction schedule before a logical construction-execution plan has been formulated is a waste of time and money.

The Construction Master Plan

The master plan must address how we will plan, organize, and control the major work activities to meet our goals of finishing the work safely, on time, within budget, and as specified in the contract documents.

A major consideration in formulating the master plan is the contracting plan, which helps us to answer a lot of questions: Are we going to operate as a union or open-shop contractor? Are we going to subcontract or self-perform some or all of the work?

Many other questions are *not* answered by the contracting plan: Do we need more resources for design review or only for construction? What are the owner's policies or legal requirements in this regard? What government and social constraints come into play in the execution of this project? How do the contractual requirements affect the master plan?

We must unravel these questions and a host of others during the development of the project-execution plan. Some of these are answerable immediately; others must wait for information that develops as the project progresses.

Project-execution plans are subject to review and evaluation as the work progresses. Minor variations are common, but you should consider major changes with extreme caution. Changes to the plan often bring on a great deal of trauma and should not be made lightly. Because all parts of the plan are so closely interdependent, a change in one major part can affect the interaction of all of the other parts.

When we have completed the master plan and gotten it approved, we can start thinking about more detailed operations plans:

- the time plan (schedule)
- the money plan (budget)
- the project-resources plan (people, materials, and services)

The time plan, which results in the project schedule, will be the subject of Ch. 4. There we will address the activities that we will schedule, the scheduling methods available to us, and the ways to select the best systems to suit conditions.

The Construction Manager's Role

The construction manager—whether a CM or a project manager (GC)—is the prime mover in preparing the construction master plan. This is accomplished through the judicious use of the complete construction team's many talents. The construction manager gets input from the key people in the owner and designer organizations as well as company management and service groups. The CM may even find need for planning input from major subcontractors who have not yet signed on but may be competitive and on the short list. The project manager obtains their input from company staff and through the use of requests for information from the owner via the design consultants.

Lead people from subcontractors, when brought on board, and from service departments such as estimating and cost control, scheduling, personnel, accounting, safety, and engineering (if required) are assigned to the project. They have the responsibility for planning the portions of the work normally performed by them. It is the manager's duty to coordinate the contributing groups and to see that they have a sound basis for their part of the master plan.

The CM or GC organizes and chairs a series of planning meetings early in the project. The various contributors to the plan can interact with one another to resolve any operational differences that may arise. The differences not readily resolved by the meeting participants must be resolved by the CM or GC.

The master-planning effort should include early establishment and dissemination of the specific project goals. That allows the master-plan contributors to formulate their individual plans to meet the overall project goals. The CM or GC is also responsible for handling all of the missing information and must fill in the blanks as well as finalize the decisions required to complete the master-planning effort on time.

After receiving all parts of the plan, the CM or GC integrates them into a written master plan for presentation to the company's management team for approval prior to submitting it to the owner. The CM or GC effects any changes resulting from the final approval stage and issues the plan as part of the field-procedure manual.

Project-Execution Formats for Capital Projects

Unfortunately, there is no single project-execution format that fits all of the various types of projects we have previously discussed. The construction activities that were shown in Table 1.3 are quite different from one type of project to another. When reading Fig. 3.1, readers may have to adapt the blocks to suit non-process-type projects, commercial project, or high-rise residential projects. We hope readers will have some interest in how the different groups of the capital-project industry execute their projects. One never knows when crossing the line into another type of expertise may become necessary. Also, there is a good possibility that a construction manager will find new and valuable techniques to use in his or her operating environment.

The involvement of manufacturing processes creates the differences between process and nonprocess projects. A chemical or mechanical process weaves its way through all the design, procurement, construction, and project activities necessary to complete a process facility. Nonprocess projects tend to be more people oriented, so aesthetics and creature comforts are stressed in the design.

Another way to state the difference is that manufacturing facilities are basically driven by process design, whereas nonmanufacturing projects are driven by architectural or civil design. Architectural and civil-works projects involve such human and infrastructural needs as schools, office buildings, hospitals, highways, dams, and public

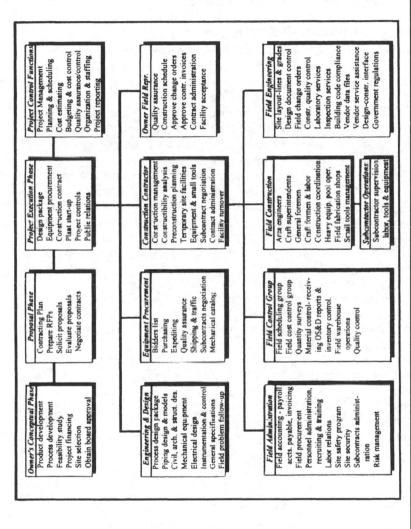

Figure 3.1 Major work activities in a construction project.

The figure shows a hierarchical chart of major work activities in a construction project, organized in three rows of boxes:

Top row:

Owner's Conceptual Phase
- Product development
- Process development
- Feasibility study
- Project financing
- Site selection
- Obtain board approval

Proposal Phase
- Contracting Plan
- Prepare RFPs
- Solicit proposals
- Evaluate proposals
- Negotiate contracts

Project Execution Phase
- Design package
- Equipment procurement
- Construction contract
- Plant start-up
- Project controls
- Public relations

Project Control Functions
- Project Management
- Planning & scheduling
- Cost estimating
- Budgeting & cost control
- Quality assurance/control
- Organization & staffing
- Project reporting

Middle row:

Engineering & Design
- Process design package
- Piping design & models
- Civil, arch. & struct. des.
- Mechanical equipment
- Electrical design
- Instrumentation & control
- General specifications
- Field problem follow-up

Equipment Procurement
- Bidders list
- Purchasing
- Expediting
- Quality assurance
- Shipping & traffic
- Subcontracts negotiation
- Mechanical catalog

Construction Contractor
- Construction management
- Constructibility analysis
- Preconstruction planning
- Temporary site facilities
- Equipment & small tools
- Subcontract negotiation
- Contract administration
- Facility turnover

Owner Field Repr.
- Quality assurance
- Construction schedule
- Approve change orders
- Approve contr. invoices
- Contract administration
- Facility acceptance

Bottom row:

Field Administration
- Field accounting - payroll acts, payable, invoicing
- Field procurement
- Personnel administration, recruiting & training
- Labor relations
- Site safety program
- Site security
- Subcontracts administration
- Risk management

Field Control Group
- Field scheduling group
- Field cost control group
- Quantity surveys
- Material control- receiving OS&D reports & inventory control
- Field warehouse operations
- Quality control

Field Construction
- Area engineers
- Craft superintendents
- General foremen
- Craft formen & labor
- Construction coordination
- Heavy equip. pool oper.
- Field fabrication shops
- Small tools management

Subcontractor Operations
- Subcontractor supervision labor, tools & equipment

Field Engineering
- Site layout-lines & grades
- Design document control
- Field change orders
- Constr. quality control
- Laboratory services
- Inspection services
- Building code compliance
- Vendor data files
- Vendor service assistance
- Design-constr. interface
- Government regulations

transportation systems. In process projects, the civil and architectural input serves to house and support the needs of the processing operations.

The nuances of the processes involved do make for subtle differences in the construction-management techniques used. Although CMs and GCs use the same management techniques in different technical environments, they rarely seem to cross over the line to handle projects outside their fields of specialization.

A Typical Construction-Project Format

The development of any project follows the major phases: conceptual, definition, execution, start-up, and divestment. Our discussion model envisions the owner using it's central engineering group (for an engineering-based project) or an outside design firm to provide civil, architectural, structural, and mechanical/electrical/plumbing design, as well as a builder to build it if it is a commercial or institutional project. The scope of this book is to examine the management of the construction effort for the project.

The major work activities required to execute a typical construction project are shown graphically in Fig. 3.1. Each of the major activities shown breaks down into several individual work activities, so that a large project can have as many as 2000 to 3000 activities to schedule. If the construction team is to meet its goals, it must successfully plan, organize, and control work activities so that they are performed in proper sequence and on time.

A brief description of the major project operations should give readers a common understanding of the actual work performed by each group in a large capital-project team. Many of the activities must happen before the construction phase starts. Starting with the owner's conceptual phase and working through the project-execution phase to the final project turnover, we will encounter the following major areas of project execution:

1. Owner's conceptual phase

2. Proposal phase

3. Project-design phase
   a. Engineering/architectural
   b. Procurement (major equipment)
   c. Project-control functions
   d. Construction input

4. Procurement (subcontractors and vendors)

5. Construction

6. Facility start-up, commissioning, and turnover

The top row of blocks in Fig. 3.1 primarily represents owner functions. The second row introduces design, engineering, equipment procurement, and construction work, all of which are normally contractor functions. The third row shows the detailed field activities performed by the constructor.

Owner's Conceptual Phase

The owner usually assigns a representative to direct and coordinate the conceptual phase of a proposed new facility. The major activities involve in this exercise might be

- product development
- process development or process licensing, if it is an engineering-driven project
- marketing surveys
- determination of project scope and design basis
- estimation of capital costs
- project financing
- economic feasibility studies
- board approval of the project

The owner's marketing or research-and-development group develops a new product or an improved process for making an existing product. If that work offers a chance for increasing the owner's profit, there is an incentive to pursue the venture. An organized and well-thought-out plan to build a new (or revamped) facility to manufacture the product or provide the service should result. But first, a proven process for making the product must be developed and tested by using experimental equipment in the laboratory or a computer model. Alternatively, the owner may choose to buy a process license from an outside firm specializing in that process or product. In that case, the research-and-development steps are not required.

While the research and development is progressing, the sales department will be conducting the marketing survey to determine the project's market volume, probable selling price, market-entry costs, and other expenses involved in bringing the product or service to market. The plans to finance the project will proceed concurrently with the technical work to make available the data necessary to perform a feasibility study. The feasibility study is the culmination of all of the work done during the initial part of the conceptual phase. The conceptual group presents the study to the board of directors for preliminary approval to proceed with further development of the project.

The initial strategic planning results from this first phase of the work. The strategic project-completion date is established, along with the basic project financial plan, in which the project budget plays a major role. When you are presented with this plan, good planning on your part dictates that you take into account some of the contingencies in the financial and time plans to cover the usual errors and omissions that may show up as the plan progresses. An owner would be wise to address construction input to the feasibility study during this preliminary phase. An astute owner may contact a construction group he or she has worked well with in the past, to discuss scheduling, cost estimating, and other constructability matters.

Since we are still very low on the overall project life-cycle curve, the financial risks at this point are still controllable. The owner's project team makes further evaluation of the project as the detailed development work evolves along the life-cycle curve. The final go/no go decision occurs when the detailed design is about 30 percent complete and commitments for major equipment purchases are ready to proceed.

The elapsed time for this part of the project can run from several months for a simple project to several years for a complex one. Since the amount of work in this early stage may not warrant the appointment of a full-time project manager if the owner has sought assistance from a construction company, the CM or GC will probably handle several such conceptual, or feeler, projects simultaneously. The conceptual-project manager rarely carries the project through the execution stage, since his or her main function may be sales development, not project execution.

When the project passes between the two phases conceptual and execution, it is extremely important that the definition of the product, and the feasibility of making and marketing it at a profit, are well proven and documented. Many a project has started down the road to disaster because of incomplete or sloppy work during that critical phase. Some of the worst disasters that we have experienced have occurred because the owner tried to skip the conceptual phase altogether.

Proposal Phase

Once the board has given its approval to proceed with the project, the owner is ready to enter the proposal phase and select a contractor. That phase involves the following activities:

- preparing a contracting plan
- prequalifying the contractor list
- preparing a request for proposal
- receiving and analyzing the proposals
- selecting the best proposal
- negotiating a contract

These activities were described in Ch. 2, so it is not necessary to repeat them here. Suffice it to say that the owner still has not committed a great deal of money to reach this point.

Project-Execution Phase

At this point, the owner's representative is ready to execute the project in accordance with the owner's contracting plan. The main decision affecting our discussion is what type of contract format will be chosen. Will it be a design-bid-build, design-build, or turn-key project? Will the owner go the route of construction management or employ a third-party contractor?

The project life-cycle curve starts to slope upward with the initiation of the project-execution phase. That is the result of significant commitments of the project's financial resources to proceed with the following work activities:

- design phase
- equipment-procurement activities
- early construction activities
- project-control functions

The design phase covers the activities required to generate the plans and specifications for the procurement of any additional equipment or materials and the construction of the facility. The various technical disciplines involved are

- civil, architectural, and structural design
- mechanical design
- any process-equipment design
- electrical design
- plumbing design
- general specifications
- interior design

Preconstruction Activities

The scope and timing of the preconstruction activities depends upon the type of construction contract being considered. One of the virtues of hiring a CM is the ability to use the construction manager's knowledge of costs, constructability issues, and scheduling. Essential construction input to the overall schedule early in the planning and scheduling phase is also possible at this time.

Getting early construction input during the planning and design phases can save the owner time and money, which more than makes up for any extra costs involved. When contractors bid on the completed plans and specifications, much valuable time and construction input is lost if all bids are over budget. There is an old saying that design-bid-build generally becomes design-bid-redesign-bid-build.

The value of a construction manager is evident if the owner is willing to accept the other responsibilities that go along with the hiring of a CM: having subcontract agreements and material and equipment purchases made in the owner's name instead of the general contractor's and—if the CM is a CM for fee instead of a CM at risk—going without an assurance that the stated budget of the project will be met.

Construction Input

When possible, construction people should be brought on board during the detail-design phase to provide construction input to the detail-design team. Bringing this expertise to the design area does not have to involve a large number of personnel hours: It can be one general construction expert assisted by specialists on an as-needed basis. This type of involvement of construction expertise during either conceptual or detail design, as stated previously, is one of the rationales behind the use of a CM.

Constructability Analysis

The constructability analysis consists of reviewing the design documents during the preliminary design phase. Suggestions are made to simplify the design so as to reduce construction costs where possible. The installation of heavy equipment is studied to reduce installation costs and to plan the heavy-lifting equipment required. The design is reviewed with an eye to selecting the best application of construction technology to minimize field costs such as earthmoving, concrete forming and placement, site accessibility, and the like. Input to preconstruction planning must also include the production of the preliminary construction schedule for the overall project-planning effort.

Construction input for the schedule logic and elapsed time for key construction activities in the preliminary Critical Path Method (CPM) schedule is vital for the development of a rational project schedule. That input includes scheduling the ordering and arrival of long-lead equipment and material to suit the field schedule.

The preconstruction input also must consider the contracting plan and how the strategic plan for the project will affect the available human, physical, and financial resources. How well this preliminary work is done has a profound impact on the contractor's performance when the baton is passed to him or her to start field activities.

Procurement Activities

The activities of the owner's procurement group are closely interwoven with those of the design and construction groups. Procurement of owner-related equipment is responsible for getting those materials and equipment to the construction site as specified and on time to meet the construction schedule. A key early output from the procurement group is the project's procurement plan. The procurement plan is also a major part of the overall project materials-management plan.

The owner is responsible for assigning the project's procurement functions as part of the contracting plan purchasing that equipment required for the operation, if the project is a manufacturing facility. In a split design-and-construct approach, the design firm specifies the equipment and the construction contractor purchases it in accordance with the design specifications.

Procurement of both equipment and specialty contractors must be made according to the prescribed schedule; delays in procurement may severely impact the construction schedule.

Project-Control Functions

A team of control specialists performs the necessary project-control functions to ensure that the project goals relating to budget, schedule, and quality are effectively met. Cost managers from both the owner's side and the construction side will be monitoring all project commitments and expenditures to see that they conform to the budget and cash-flow projections. A monthly project-cost report presents the data to the project team.

The contractor's cost report should contain:

- A copy of the detailed budget, showing the amount budgeted for each element of the construction project and including a breakdown of the general conditions costs

- A corresponding cost to date for each item listed in that report, as of the date of the report

- A projected cost to complete for each item in the report

- An analysis of whether an item is over budget or under budget, and a grand total reflecting whether the contract sum is on target or is over or under the budget

Items that exceed the budget need to be examined to determine the reasons why they are and what can be done to bring them back on budget. The scheduling group monitors the project schedule on a regular basis and reports any drift off target at least monthly. The project team should reevaluate the schedule periodically, analyze any tasks that are exceeding their target date, and assign personnel to investigate and report back as to what caused the deviation, whether it is critical, and, if so, what is required to get back on track.

The various groups responsible for each type of material closely monitor and report the status for their particular area on a regular basis. Material and project quality are checked regularly by the responsible managers relating to structural, mechanical, and architectural components.

The contractor's project or construction manager coordinates the control functions and issues monthly progress reports to the owner's management. Any off-target items are highlighted and discussed, along with recommended solutions for any problems.

The onset of the project-execution phase triggers a significant commitment of major human, physical, and financial resources to the project by both owner and contractor. Design costs, for example, can run from 8 to 12 percent of the total project cost. The placement of purchase orders for process equipment and materials involves an even greater commitment of funds by the owner, if the project is a chemical or petrochemical one. A commercial office-building project will be concerned about furniture, office equipment, furnishings (window treatment, artwork, etc.). This is a good time for the prudent owner to look at the project financial plan to see if the return on investment is still valid. Using the expanded design database to make a capital-cost estimate with an expected accuracy of plus or minus 15 percent would be a good investment. That would be a major refinement over the estimate of plus or minus 40 percent associated with the original feasibility study.

The Owner's Field Representative (OFR)

The function of the OFR can fall into several categories, depending on the size and complexity of the project. On large, complex projects, the owner usually assigns a resident engineer or staff with construction experience to monitor the performance of the construction work. On some large projects, the OFR staff includes a quality-control specialist to assist in monitoring the contractor's field activities. In that situation, the owner's field team and the contractor's construction-management team work very closely with each other on a day-by-day basis.

Where the owner has engaged a design team to provide not only the plans and specifications but also field inspections, that A&E representative would perform the functions of the OFR as shown in Fig. 3.1. Although the figure is representative of a process-type project, less emphasis will be placed upon process design, process equipment, and other procurement of engineering-type materials. A project manager can substitute other activities for commercial or institutional projects.

The Role of the Contractor

Construction activities involve the largest single commitment of resources on the project, about 50 to 60 percent of the owner's total project budget. Before starting any field work, the construction team should perform preconstruction activities and review the construction input supplied to the master-planning effort. This preliminary work usually conforms to the contracting philosophy and to the field-labor survey made during the proposal phase.

The construction group's primary mission is to construct the facility in accordance with the owner's contract requirements and the field schedules. Cost-control procedures are established to ensure that the project budget and schedule are met in accordance with that contract.

Project completion requires the construction team to be responsible for final system start up; testing, adjusting, and balancing; and commissioning to ensure that all contract requirements have been met. Tracking quality during construction should reduce the punch list at the end of the job. Once construction is completed, the builder remains responsible for the start-up team with maintenance services until the owner accepts the building and the facility begins operation.

Major Field Activities

The major areas designated by the CM or GC to the key players on the field team are shown in the boxes located in the bottom row of Fig. 3.1 The activities carried out in those areas are an important part of the construction project's management and execution. Adjustments need to be made to portions of Fig. 3.1 to convert from process-type projects to vertical construction projects.

Field construction is where the physical construction work actually gets done. This group will be led by an experienced field superintendent who reports directly to the project manager. The superintendent is assisted in the execution of the construction work by an organization of field engineers, trade superintendents, general forepersons, and subcontractor supervisors. All other members of the field group perform their duties to support the field construction operations.

The field superintendent, in conjunction with the CM or GC, coordinates the field construction activities to staff the project, mobilize the tools and equipment, order materials, direct the subcontractors, monitor field productivity, and perform any other functions required to carry out the construction work. The superintendent works closely with the various staff support groups to maximize their input and promote the success of the construction operations.

The field engineering group is a technical arm of the civil-engineering field operations. Their work starts with setting the lines, grades, and survey monuments for laying out the facility and the ancillary structures on the site. They also are the first unit on the site to supervise the installation of temporary facilities and supervise the site-development subcontractor. They are responsible for determining whether adequate soil bearing capacities exist to meet the design criteria.

The field engineer receives and distributes the technical documents for the field organization. He or she manages the design–construction interface to ensure that the project is built according to the design documents. All design clarifications, design and field changes, change orders, as-built drawings, vendor contracts, and the like pass through either the field engineer or the project superintendent before going on to the project managers.

The field engineer is also responsible for quality control as an assistant to the project superintendent. This includes all quality-control testing services, laboratory reports, radiographic services, etc. The field engineer also maintains copies of all applicable codes and government regulations and interprets their application to the project.

In a process facility, the field engineer's office maintains the documentation for final testing and acceptance of the process systems as they are completed by the construction group. The complete field technical-documentation file is turned over to the owner as a record of the quality control for the facility. The field engineer also supports the field start-up team, participates in the facility-acceptance procedures, and closes out the field files at the end of the job.

The field control group is a staff group that reports directly to the CM or GC. Its main function is to monitor the field schedule, oversee the field cost-control systems, and control the material purchased and received at the site. The scheduling group is responsible for keeping the field schedule current as to task planning and reporting progress to date. They run weekly field schedule meetings for this purpose.

The cost-control group monitors the project's cost commitments, expenditures, indirect costs, and labor productivity to ensure adherence to the field budget. These factors are summarized and reported monthly in the field cost report. Off-target items are discussed with the CM or GC for analysis and to determine methods by which they can be put back on track.

The materials-control group is responsible for implementing and operating the construction-materials management plan. Its function includes field procurement, receiving, and warehousing of all construction supplies and materials.

The field warehousing facilities are set up to receive, store, and control the construction materials and equipment delivered to the site. That includes all construction tools and equipment as well as materials bought in the field. The warehousing function is a major part of the project's materials-management program. An important warehousing-planning decision is whether new project buildings can be made ready for use for warehousing or whether temporary structures will be needed.

The field administration function at the site is an important staff function required to support the construction activity. The accounting function, generally operating out of the CM or GC's central office, handles the field payrolls, accounts payable, project invoicing, and local banking to maintain cash flow to the project.

An effective field personnel group is critical to field operations to ensure a reliable source of local personnel and craft labor in accordance with the project's labor posture. This group handles the site labor relations, which can make or break the performance of the entire field operation. Whether the trade workers are unionized or not, the field personnel must become familiar with the provisions of their labor agreements and operate within the confines of the terms of any such agreements. Planning, organizing, and controlling the site's safety program is an important function of the administration group. The CM or GC plays a key role in setting up the policies and practices to be followed by the site's safety group to plan and enforce an effective safety and health program. Safety has many implications, which we discuss in Ch. 12. Along with the impact on the cost of workers' compensation and with the disruption of crews that have worked efficiently together that comes when one member is injured, owners do not like to have a poor safety record associated with their projects.

The CM or GC is responsible for setting policies and for exercising prudent risk management. The site-security plan should address personnel access, control of physical loss of materials and equipment, contacts with the local fire and police departments, and the location of the nearest hospital.

The administrative group handles the business portion of the subcontractors function. It processes the paperwork for payment, change orders, and subcontract administration.

The risk-management function of the administrative group oversees those functions relating to the site's insurance requirements. These people ensure that insurance policies not only meet the contract requirements but are maintained in force. They process insurance claims and assess risks at the site.

Facility Start-Up Activities

Although the facility start-up is the last activity on the project, it has to be considered in the project master plan. The start-up, or commissioning, plan establishes the order for putting the operating units into service. That, in turn, sets up the strategic date for mechanical completion of the operating units. This agenda for completion of the operating units must be considered in the master plan and the detailed schedule for the project.

The amount of construction participation in the start-up must also be considered in the scope of services, along with the money and resource plan for the overall project. Even though the services are not to be performed until very late in the project, the CM or GC must not overlook them during the planning phase.

Construction-Project Initiation

Getting a construction project off on the right foot is vital to the project's success. If the project initiation is slow or flawed, the adverse effects will be felt throughout the project and may result in unmet project goals. Formulation of the construction master plan and project initiation usually occur simultaneously. We must not allow them to get in each other's way.

Some project-initiation activities actually occur during the proposal phase. To make possible the preparation of the original cost estimate, we have to make some assumptions based on the project scope, our original site survey, and our estimate of indirect costs in the field. If we are awarded the contract and are ready to start the project, these data must be reviewed and confirmed with the current project plan before they are used. The project-initiation activities occur in the home office between the dates of the contract award and the opening of the field office. This is an extremely busy time, when most of the project planning and organizing is accomplished.

The best way to describe the project-initiation procedure is through a project-initiation checklist. The following items must be considered as a minimum approach to kicking off a construction project.

Project-Initiation Checklist

1. Become completely conversant with the contract terms and conditions, especially performance, scope, cost, and schedule.

2. Review the contracting plan, and update it if necessary. Establish a project priority list.

3. Study available design documents to become knowledgeable on the technical aspects of the project.

4. Prepare a master plan for approval by management.

5. Finalize the labor posture for the project and develop a local site agreement if required.

6. Become knowledgeable as to the project's cost estimate, budget, and schedule, if you did not participate in their development.

7. Develop and issue a field-procedure manual (FPM), including a statement of the project goals.

8. Prepare a project organization and staffing plan, in accordance with the project master plan, and include it in the FPM.

9. Bring key staff members on board as required for project initiation.

10. Initiate the project's materials-control plan, including procurement activities, especially if your scope includes all procurement. Concentrate on long-delivery items.

11. Review subcontract proposals and award subcontracts, especially those connected with site development or demolition.

12. Prepare a site layout for installing the temporary site facilities, lay-down areas, and utilities, and start arrangements for their installation.

13. Establish rapport with the client's representatives and their organization.

14. Organize and chair internal and client project kick-off meetings, per your duties and responsibilities.

15. Establish policies and procedures for expense accounts, as well as field-assignment allowances for supervisory staff.

16. Keep your client's management informed of your project-initiation plans.

17. Start schedule and cost-control activities. You are spending money!

18. Set up contract tickler files for early warning on contractual obligations.

19. Ensure that the project's risk analysis and bonding (if required) have been completed and that required insurance coverage and bonds are in place.

20. Establish job files and transfer them to the field office when ready.

21. Develop a safety program applicable to the specific project.

22. Perform (or review) the heavy-lifting requirements for the project, if the project requires them.

23. Ensure that the constructability reviews have been done and that the recommendations have been incorporated into the design after consultation with the owner's design consultants.

24. Initiate arrangements for small or specialty tools required for the project.

25. Establish the interface between the applicable design consultants and contracting organizations.

26. Establish the administrative procedures required by the contract and company policies.

If you were fortunate enough to be assigned as a CM or the contractor's project manager during the proposal phase, your project-initiation duties will be much easier. However, if you are assigned to the project only after the contract award, you will have a lot of ground to make up in a hurry. An even worse situation occurs when you are assigned to take over a project that has already been started by someone else. In that case, you will have to learn the job and handle ongoing problems at the same time. In all those cases, remember the 11th commandment: *Know thy contract as thyself.*

Executing the project-initiation procedures depends on the size and complexity of the project. On small projects you will be doing most of the listed activities yourself. On larger or more complex projects, you will have to delegate some of the duties to available staff specialists. They may be home-office staff people or certain field staff already assigned to the project.

Item 9 on the checklist calls for bringing key field personnel onto the project as required. Care must be exercised here to prevent bringing people on board too early and adversely affecting the indirect budget for the field. Make sure that the people brought on early have continuing work to keep them productively busy until the field opens.

The FPM is an important document to have prepared early in the project, to assist in indoctrinating your new people as they come on board. The best approach to getting at least a preliminary version of the FPM is to model it from a previous job that is similar to yours. You don't have time to reinvent the wheel during this critical phase of the job.

Starting the procurement program is key to getting started early. This is especially true if you have a contract based upon completed plans and specifications on a fixed or lump-sum price basis. You need to get the long-lead delivery items to the field as early as possible to avoid falling behind the schedule and incurring penalties. Early procurement is also necessary to preclude the vendors from raising prices before you can get your orders placed.

Item 23 on the checklist indicates that constructability analysis was made during design review. Preconstruction services can take place only when the constructor is brought on board early in the design phase. That procedure usually is not possible on a fixed or lump-sum price, but only on design-build projects or negotiated projects where a contractor is engaged during the conceptual and design stage.

Because there are a lot of things happening at the same time during this hectic period, it is absolutely essential that you make a priority list for handling the key issues first. Once you get a priority list started, it is a good practice to keep it going throughout the job!

Summary

In this chapter we have discussed the construction manager's and the project manager's role in the overall project planning and initiation effort. Sound planning is the cornerstone of effective construction management. Planning the broad range of activities involved with any type of capital project is essential if you are to meet your project goals. Nobody is ever lucky enough to have a successful project without a well-conceived plan!

Concurrent with the master planning, the CM or GC must get the project started on a solid footing. Project initiation is an art that must be learned early if you want to become a successful construction manager.

CHAPTER 4

Construction Scheduling

I n the preceding chapter we discussed the essential project master-planning effort, which lasts a relatively short time at the beginning of the project. In this chapter, we will discuss the schedule preparation as it specifically relates to project construction activities. Scheduling activities are continuously updated throughout the project life cycle.

Today, even the smallest construction organizations have access to scheduling software, but it is the project manager who has the responsibility to provide the input necessary for the preparation of the initial schedule and the constant updating that occurs as the project proceeds along the path to completion. He or she will be seeking information from some of his or her staff and also from subcontractors and vendors once they have been brought onboard.

The construction manager (CM) or general contractor (GC) makes a major contribution to the construction schedule by way of the master plan discussed in Ch. 3. He or she must also review the project's schedule in detail to check the scheduler's logic approach, the elapsed times for the various events and components expressed in the schedule, and the float time to be apportioned to key elements of the schedule.

Three Basic Types of Schedules

The three basic types of schedules are the bar chart, the critical-path method (CPM), and the program evaluation and review technique (PERT).

The bar chart's proper name is the Gantt chart, after its originator, Henry Gantt. Gantt or bar charts are relatively easy to prepare and are quite sufficient for simple, uncomplicated projects with few elements to be tracked. Each element of the schedule is given a space in the vertical composition of the chart. Along the top horizontal span of the chart are dates from the start of the project to the finish. Each activity is represented by a horizontal line indicating its start and finish according to the timeline across the top of the chart. Gantt charts can become quite unwieldy when more than 30 activities are to be included.

CPM was originated by M. R. Walker at DuPont in conjunction with J. E. Kelley at Remington Rand in 1957; with the construction of the UNIVAC 1 computer, CPM was first put to test in the design of a new chemical plant in 1958. It is now used on most construction projects of all sizes, since the costs of both scheduling hardware and computer hardware are now affordable to most contractors.

PERT was devised in 1958 for the U.S. Navy's Polaris missile program. It provided a focus around which managers could brainstorm ideas, solidify them, and put them into the program. The managers posed such questions as "Which critical activities or tasks in the project could delay the entire project?" This is also the same type of question raised while a CPM construction schedule is being prepared.

There are six steps common to both CPM and PERT, as well as any other process of creating schedules:

1. Definition of the project and all of its major or significant work tasks.

2. Development of the relationships among the activities, determining which activities are to proceed before others and which activities are to follow as the project moves toward completion.

3. Preparation of the network schematic. Initially, when CPM schedules were drawn by hand, a method of using arrows to show the start and finish of activities was common, with boxes or rectangular figures containing the activities. Nowadays, CPM software indicates these relationships plus many others in a more sophisticated fashion.

4. Assignment of a time sequence for each activity, and in some cases also a cost, if the schedule is to be cost loaded.

5. Computation of the longest time path through the schedule. This is referred to as the critical path.

6. Use of this network of activities to plan, schedule, monitor, and control the project.

Five Key Questions When Preparing a Schedule

The scheduler must keep in mind five basic questions as he or she begins to construct a schedule:

- Is this a start activity?
- Is this a finish activity?
- What activity precedes this one?
- What activity follows or succeeds this one?
- What activity runs concurrently with this one?

Other questions need to be addressed in addition to those basic ones:

- What is the original duration of this activity?
- As the CPM chart becomes revised during construction, what is the remaining duration of this activity?
- Can we provide an early start for this activity, and if so, what would be the early finish for the activity and how will it impact the activity that follows?
- How is float being addressed when either early start or early finish is achieved?

A simple CPM schedule involving 13 finishing operations is shown in Fig. 4.1; one page of a complex CPM schedule containing thousands of activities is reflected in Fig. 4.2. The components of both are much the same and include:

- Activity identification
- Original duration of that activity
- Remaining duration as construction progresses through that activity
- Early start
- Early finish
- Float
- The actual calendar for the period of construction showing the start and finish as initially planned and the identification of the activity

Scheduling Approach

Although in private-sector work, as we discussed in Ch. 2, the contract with the owner will establish the end date and the execution of the contract will establish the start date, in many public projects the notice to proceed will establish the start date and the completion date may be spelled out in the form of X number of days from the start date determined by the notice to proceed.

When a design-build contract is awarded, the design-build team—in concert with the project owner—will establish a project delivery date which will account for both the design and the construct cycles.

Owner's Schedule for Design-Bid-Build Contracts

Owner project-management organizations come in all shapes and sizes. Some obvious purposes of the schedule of such an organization would be to determine the length of construction financing, order long-lead pieces of equipment, or plan the moves of personnel from

JHU STUDENT HSG - REV 6

ST. PAUL

3rd FL

Activity ID	Orig Dur	Rem Dur	AP21 Early Start	AP21 Early Finish	AP23 Early Start	AP23 Early Finish	Early Variance	Total Float
SPIN3430	10	1	01MAY06A	19JUL06	01MAY06A	27JUL06	-7	-51
SPIN3520	5	5	19JUL06	25JUL06	28JUL06	03AUG06	-7	-51
SPIN3530	8	3	26JUL06	04AUG06	24JUL06A	08AUG06	-2	-51
SPIN3540	3	3	07AUG06	09AUG06	09AUG06	11AUG06	-2	-51
SPIN3550	3	3	10AUG06	14AUG06	14AUG06	16AUG06	-2	-51
SPIN3560	7	7	15AUG06	23AUG06	17AUG06	25AUG06	-2	-51
SPIN3490	4	4	24AUG06	28AUG06	28AUG06	31AUG06	-2	-51
SPIN3480	3	3	30AUG06	06SEP06	01SEP06	05SEP06	-2	-51
SPIN3500	3	3	04SEP06	06SEP06	06SEP06	08SEP06	-2	-51
SPIN3050	4	4	07SEP06	12SEP06	11SEP06	14SEP06	-2	-51
SPIN3510	4	4	13SEP06	18SEP06	15SEP06	20SEP06	-2	-51
SPIN3660	7	7	19SEP06	27SEP06	21SEP06	29SEP06	-2	-51
SPIN3690	4	4	28SEP06	03OCT06	02OCT06	05OCT06	-2	-51
SPIN50010	0	0		04OCT06		05OCT06	-1	-51

JHU STUDENT HSG - REV 6

ST. PAUL

3rd FL

SPIN3430 COFFERED CEILING FRAMING
SPIN3520 LIGHT FIXTURE R/I
SPIN3530 GWB COFFERED CEILINGS
SPIN3540 PRIME POINT COFFERED CEILINGS
SPIN3550 ACT GRID COFFERED CEILINGS
SPIN3560 LIGHT FIXTURE R/I COFFERED CEILINGS
SPIN3490 GWB CEILINGS
SPIN3480 PRIME PAINT AND POINT
SPIN3500 1ST COAT FINISH PAINT
SPIN3050 FLOORING
SPIN3510 ACT CEILING TILE
SPIN3660 LIGHT FIXTURES AND TRIM
SPIN3690 2ND COAT FINISH PAINT
SPIN50010 COMPLETE - ST PAUL BLDG

Start Date	14MAY04
Finish Date	05OCT06
Data Date	27JUL06
Run Date	18NOV06 09:19

© Primavera Systems, Inc.

| Early Bar | TARGET |
| Progress Bar | Critical Activity |

AP23

STRUEVER BROS. ECCLES & ROUSE, INC.   Sheet 1 of 1

JHU STUDENT HSG - REV 6

LONGEST PATH W/TARGET

Date	Revision	Checked	Approved

Figure 4.1 A simple CPM schedule containing 13 activities.

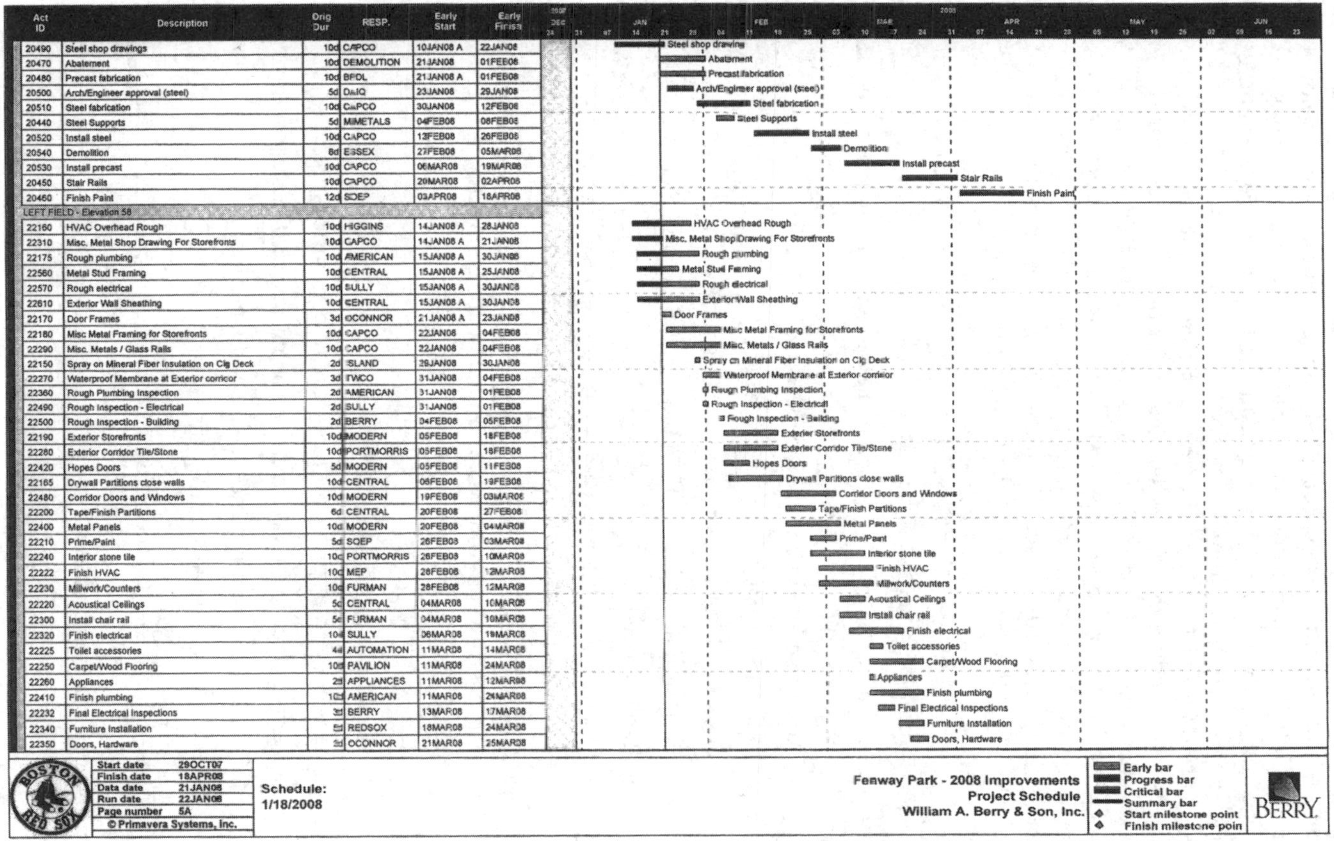

Act ID	Description	Orig Dur	RESP.	Early Start	Early Finish
20490	Steel shop drawings	10d	CAPCO	10JAN08 A	22JAN08
20470	Abatement	10d	DEMOLITION	21JAN08	01FEB08
20480	Precast fabrication	10d	BFDL	21JAN08 A	01FEB08
20500	Arch/Engineer approval (steel)	5d	DAIQ	23JAN08	29JAN08
20510	Steel fabrication	10d	CAPCO	30JAN08	12FEB08
20440	Steel Supports	5d	MIMETALS	04FEB08	08FEB08
20520	Install steel	10d	CAPCO	13FEB08	26FEB08
20540	Demolition	8d	ESSEX	27FEB08	05MAR08
20530	Install precast	10d	CAPCO	06MAR08	19MAR08
20450	Stair Rails	10d	CAPCO	20MAR08	02APR08
20460	Finish Paint	12d	SOEP	03APR08	18APR08
LEFT FIELD - Elevation 58					
22160	HVAC Overhead Rough	10d	HIGGINS	14JAN08 A	28JAN08
22310	Misc. Metal Shop Drawing For Storefronts	10d	CAPCO	14JAN08 A	21JAN08
22175	Rough plumbing	10d	AMERICAN	15JAN08 A	30JAN08
22560	Metal Stud Framing	10d	CENTRAL	15JAN08 A	25JAN08
22570	Rough electrical	10d	SULLY	15JAN08 A	30JAN08
22610	Exterior Wall Sheathing	10d	CENTRAL	15JAN08 A	30JAN08
22170	Door Frames	3d	OCONNOR	21JAN08 A	23JAN08
22180	Misc Metal Framing for Storefronts	10d	CAPCO	22JAN08	04FEB08
22290	Misc. Metals / Glass Rails	10d	CAPCO	22JAN08	04FEB08
22150	Spray on Mineral Fiber Insulation on Clg Deck	2d	ISLAND	29JAN08	30JAN08
22270	Waterproof Membrane at Exterior corridor	3d	TWCO	31JAN08	04FEB08
22360	Rough Plumbing Inspection	2d	AMERICAN	31JAN08	01FEB08
22490	Rough Inspection - Electrical	2d	SULLY	31JAN08	01FEB08
22500	Rough Inspection - Building	2d	BERRY	04FEB08	05FEB08
22190	Exterior Storefronts	10d	MODERN	05FEB08	18FEB08
22280	Exterior Corridor Tile/Stone	10d	PORTMORRIS	05FEB08	18FEB08
22420	Hopes Doors	5d	MODERN	05FEB08	11FEB08
22165	Drywall Partitions close walls	10d	CENTRAL	06FEB08	19FEB08
22480	Corridor Doors and Windows	10d	MODERN	19FEB08	03MAR08
22200	Tape/Finish Partitions	6d	CENTRAL	20FEB08	27FEB08
22400	Metal Panels	10d	MODERN	20FEB08	04MAR08
22210	Prime/Paint	5d	SOEP	26FEB08	03MAR08
22240	Interior stone tile	10d	PORTMORRIS	26FEB08	10MAR08
22222	Finish HVAC	10d	MEP	28FEB08	12MAR08
22230	Millwork/Counters	10d	FURMAN	28FEB08	12MAR08
22220	Acoustical Ceilings	5d	CENTRAL	04MAR08	10MAR08
22300	Install chair rail	5d	FURMAN	04MAR08	10MAR08
22320	Finish electrical	10d	SULLY	06MAR08	19MAR08
22225	Toilet accessories	4d	AUTOMATION	11MAR08	14MAR08
22250	Carpet/Wood Flooring	10d	PAVILION	11MAR08	24MAR08
22260	Appliances	2d	APPLIANCES	11MAR08	12MAR08
22410	Finish plumbing	10d	AMERICAN	11MAR08	24MAR08
22232	Final Electrical Inspections	3d	BERRY	13MAR08	17MAR08
22340	Furniture Installation	5d	REDSOX	18MAR08	24MAR08
22350	Doors, Hardware	3d	OCONNOR	21MAR08	25MAR08

FIGURE 4.2 A complex CPM schedule with many activities.

one facility to another. Large corporations handling multiple larger projects as well as many renovation and remodeling projects often maintain a scheduling department. Smaller companies may employ personnel with construction experience who can prepare a simple schedule, or they can engage a scheduling firm to prepare a more elaborate one if it is required for several purposes.

A public-project owner must create a schedule, whether it be a milestone schedule or a more detailed one, prior to advertising the project as required by law. This macro type of schedule is normally included in the bid documents to alert the bidders to the amount of time the public agency has allotted for construction. The owner's major project-execution activity milestones, to be scheduled after project conception and approval, are as follows:

1. Award the design contract.
2. Complete the design contract.
3. Approve the design documentation—the final plans and specifications.
4. Advertise for construction bids.
5. Open and evaluate construction bids.
6. Compare bids with the architect or engineer's estimate.
7. Redesign (by A&E) if initial bids exceed the budget.
8. Rebid the new design.
9. Open and evaluate revised construction bids.
10. Select and approve a contractor.
11. Negotiate the contract, possibly incorporating some value-engineering suggestions submitted by the contractor.
12. Submit the notice to proceed in either one or two phases, as discussed in Ch. 2.
13. Monitor field schedule performance.
14. Monitor and approve or deny schedule modifications.
15. Perform final inspection, commissioning, and facility acceptance.
16. Make final payment and release of the retainage.
17. Monitor the project during the warranty period.

These 17 activities can be easily scheduled on a manually created bar chart as shown in Fig. 4.3. They can also be scheduled with any number of different scheduling-software programs available on the market today.

There is no float in this schedule, and it is highly unlikely that each activity will flow to the next as represented in the figure. If the owner, in either the milestone schedule or the bid documents, does

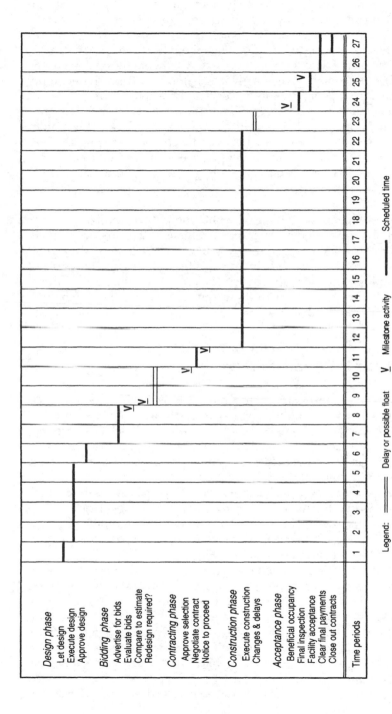

FIGURE 4.3 Owner's bar-chart schedule—design-bid-build project.

Legend: ===== Delay or possible float   ∨ Milestone activity   ▬▬▬ Scheduled time

not include float, the successful bidder will need to broach that subject with the owner. Even owners who consider all of the details required for their project may fail to consider all of the activities involved or may be overly optimistic about the strategic end date.

The planned float for each activity, whether in the owner's or the contractor's schedule, is not always maintained for each operation. When early operations run out of float, it is conveniently *borrowed* from ensuing activities, which is the only way to keep to the strategic end date. So float is a key item in any schedule, and the builder may need to educate an owner who has prepared his or her own schedule without any float. Unless this is explained to the owner, it could lead to the builder's accepting an overly tight construction schedule in an attempt to maintain that owner's unrealistic strategic end date.

In competitive bidding on public projects, if the bidder takes exception to any of the terms and conditions of the bid documents in the bid submission—including an unrealistic schedule—the bidder stands a good chance of being disqualified. Through a consensus of all bidders who agree that the public-project schedule is unrealistic, and who join together, perhaps the public-project owner can be convinced to modify that schedule.

Owner's Schedule for Third-Party Constructor Contracts in the Private Sector

Now let us consider a private owner's schedule for a design-bid-build project, i.e., a third-party arrangement. Let us assume the design has been finalized and the owner is proceeding on a "cost plus a fee, not to exceed" basis—a guaranteed-maximum-price contract. Once the contractor has been selected via a competitive-bid process, the owner's list of major execution activities can be presented to the contractor for review and comment.

The owner's list of activities will be not as expansive as the contractor's schedule, but it gives a starting point to determine if the schedule is reasonable and will be accepted by the builder as the builder's scheduling team fills in the blanks. The owner's preliminary schedule may look something like the bar chart in Fig. 4.4.

Many contracts require the contractor to provide a detailed schedule to the architect within a specified period of time after the contract is awarded. Under the previous edition of the American Institute of Architects' document A201, General Conditions of the Contract for Construction, the contractor was to submit this schedule for the architect's *approval*; but the 2007 edition of A201 merely states that the schedule is to be submitted for the architect's *information*. This schedule, when submitted, will be referred to as the *baseline schedule*, and all future changes will be referred to as revisions to this baseline, i.e., Revision 1, Revision 2, etc.

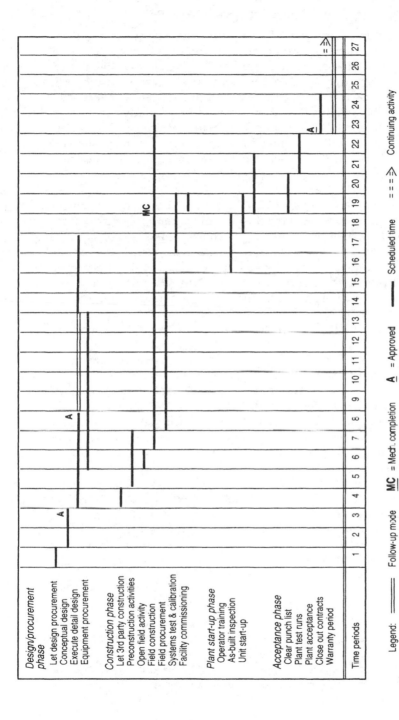

Figure 4.4 Owner's bar-chart schedule—third-party contract.

Project-Execution and Scheduling Philosophy

Construction-execution and scheduling philosophy are two items that greatly affect the preparation of the construction schedule. The construction-execution philosophy is actually the master plan that was discussed in Ch. 3. The plan lays down the basic ground rules for construction execution. It also answers such questions as: What is the construction scope? Will design be completed beforehand or run concurrent with construction? This happens frequently with guaranteed-maximum-price contracts, since the contract between the owner and contractor has most likely been executed prior to the 100 percent completion of the plans and specifications.

Who does procurement? Will the contractor purchase all materials and equipment or will the owner provide certain items? Will the contractor self-perform work or subcontract all or major portions of the work? This will determine whether the contractor has full control over the procurement schedule or whether the subcontractor will pursue the standard method of supplying materials and equipment—submit shop drawings for the architect's approval and receive approval prior to placing the order. Obviously, time in the schedule must be provided for the shop-drawing submission-and-review time.

Each answer affects the selection of a suitable scheduling approach and format. *Scheduling philosophy* refers to the selection of the scheduling system. For example: Will we use bar charts or CPM? (The only time we have seen PERT used was on a military project.) How often will the schedule be cycled and updated? How many activities will be included in the schedule? Do we have trained people in-house to prepare the schedule or do we need to bring in an outside contractor? What are our contractual obligations with the owner to review and update the schedule? (Generally on large projects this becomes part of the weekly or biweekly project-meeting agenda.)

The reliability of the construction schedule is a function of the degree of design completion available when the schedule is made or the ability, when a contract for construction is signed, to get input from subcontractors and vendors after they have been selected.

On small projects, a milestone schedule may be the only one used or needed. On larger projects, the construction schedule serves as the basis for making more detailed weekly work plans in the field for each major activity. In any case, the approved construction schedule is the fundamental working document used by the owner and the contractor to set major milestones and monitor the actual construction progress.

Proposal Schedules

Often, contractors submitting bids in a competitive environment will prepare milestone schedules; in fact, this is frequently required as part of the bid submission. However, the bidder should look very closely at the time frame set aside for construction in the bid documents. Is it really a practical amount of time allotted for this complex project? If so, should the builder devote a little more time in developing a somewhat more detailed analysis, in case he or she is the successful bidder and becomes saddled with a very tight schedule that is not feasible based upon the initial analysis?

Detailed Field Scheduling

Even a detailed CPM schedule with an extended work-breakdown structure is not always suitable for scheduling the day-to-day activities in the field. Detailed field planning and labor availability relating to critical operations ought to be monitored by the field managers.

Figure 4.5 is a weekly summary of key subcontractor work crews, prepared daily for consolidation and review at the weekly or biweekly project meetings. If there is a deficiency in the number of workers, causing a work task to fall behind schedule, this type of

Project: Arlington University
Location: Arlington,Virginia

Report Number: 139
Date:

Project No.: 11937

Prepared By: D. Hedrick

SUMMARY OF WORKFORCE FOR WEEK 7/1/2006 THRU 7/7/2006

SUMMARY OF WEEK'S	SAT	SUN	MON	TUE	WED	THU	FRI
WEATHER CONDITION							
COMPANY/TRADE	SAT	SUN	MON	TUES	WED	THURS	FRI
Centerline	3		96		96	98	99
Henry J. Knott Masonry	17		19		17	16	21
Fidelity	2				89	90	93
MBR	31	28	25		50	48	52
Scriba					3	3	3
Simplex Grinnell	5	6	9		10	8	9
Otis Elevator							
CHS	6				5		
TSI							
ISEC			6		6	6	6
Fewster	7		5		7	6	7
Business Flooring							
Manolis	12		7		11	10	9
Normac							
EDI							
Heidler							
Champion Elevator							
Benfield Electric	3		5		5	5	3
Davenport Fireproofing							
Ryan Restoration			5		4	4	
Cherry Hill						5	
Machado	11		5		3	4	4
TOTAL	97	34	182	0	303	306	306

FIGURE 4.5 Weekly summary of key subcontractors' work.

summary is a very effective tool in pointing the finger at the sub-contractor who has failed to supply sufficient labor to maintain his or her portion of the scheduled work—which may also impact the schedule of those work activities to come after completion of that predecessor activity.

Scheduling Systems

Bar Charts

Bar charts are the simplest form of scheduling, and have been in use longer than any other such system. They offer the advantage of being inexpensive to prepare, easy to read, and easy to update. They are, however, difficult to follow when there are multiple work tasks requiring several pages of bar charts. They are also unable to show the interrelation of one task to another.

CPM Schedules

The use of CPM scheduling has become ubiquitous in the construction industry today, taught in nearly all college and community-college project-management courses. A CPM schedule has two keystones: the proper and orderly list of activities that precede and succeed each other—predecessors and successors—and the duration of each activity.

The CPM schedule, when professionally prepared with significant input from the CM or GC, offers the following advantages:

- Inclusion in the schedule of a list of each component of construction
- Concise information regarding which activities must occur before the next one can begin—predecessor and successors
- A carefully thought-out prediction of the time required for overall project completion and the time required to reach milestone dates to achieve that end goal
- A calendar date for the start and finish of each of the detailed work tasks and activities
- A matrix that can be manipulated to change the project's completion date, if required, by the addition or deletion of items of work
- A basis for scheduling subcontractors, materials, and equipment
- A method for balancing schedules, personnel, equipment, and costs (if the schedule is resource loaded)
- A means to evaluate time requirements when changes to the project are being considered

- A method of recording project progress or lack thereof as documentation, to either pursue or defend against a claim

CPM Pitfalls to Avoid

The prime advantage of the CPM schedule is its ability to include the duration of each event or activity and the series of activities that creates the *critical path*. A common practice in preparing a CPM schedule is to obtain input from subcontractors and suppliers of materials and equipment. An accurate assessment of activity durations and predecessor events can be gleaned from those subcontractors who have years of experience in performing their tasks over and over and have a treasure trove of productivity figures.

There is the question of lag time, the amount of time that must be allowed from the completion of a previous task to the commencement of the successor. Think about concrete: Cast-in-place concrete foundations will have had test cylinders taken during the pour; the first test at seven days may not meet the strength requirements of the structural engineer, who would prefer to wait until 10 days after the pour for another test that, in most cases, will evidence sufficient strength for subsequent stresses to be placed on the foundations. This is another reason for float.

This concept of float applies to other operations as well, and has nothing whatsoever to do with subcontractors meeting their task durations. Think about drywall: There must be a lag time between the installation or hanging of the Sheetrock and the time allotted for taping and sanding before painting can begin.

Activity Durations

The success of the CPM process rests on having not only the proper predecessors and successors but also durations. Subcontractors and vendors—being fully aware of the commitment they make when they provide durations to the CM or GC with duration times and being risk averse—may supply durations that are much longer than actually required. According to one expert in the field, some of these durations are one and one half times longer than they should be. Durations submitted by subcontractors and vendors need to be carefully reviewed, and picked apart to uncover any unreasonably long activities. This is the beginning of relationship building with the subcontractors and vendors.

The CPM schedule is almost a living, breathing thing, because it changes so frequently; some operations that prove to have shorter durations than stated may be countered by another operation that exceeds the stated duration. So this juggling act will probably continue throughout the life of the project.

It is important that predecessors and successors be kept current and any changes made, when required, to reflect any overall milestone dates. The weekly or biweekly project meeting is the forum for review and modification, if required, of the CPM schedule. Prior to that meeting, either the project manager or the field superintendent must carefully walk through the project, schedule in hand, noting those operations that appear to be on schedule, those that appear to be behind schedule, and those that appear to be ahead of schedule. This information is essential for a realistic review of the CPM schedule at the meeting.

Float: What It Is and the Importance of Who Owns It

It is a given that not all tasks will begin and end according to the baseline schedule, and because Murphy's law is alive and well in the construction industry, changes will occur. Some may be due to severe weather or longer periods of inclement weather in the early stages of site work; there may be changes to the scope of the work, with some items deleted or added by the owner; or there may be a temporary shortage of critical materials, as there was with gypsum drywall in the 1980s.

To compensate for those unknown but sure-to-exist delays, a contingency should be added to the schedule. This contingency is called float.

An important provision that goes back to the negotiation of contract terms with the owner is who owns the float. When the contract includes a liquidated-damages clause assessing the contractor in some cases thousands of dollars per day for late delivery of the project, the question of float takes on an added importance.

If the owner owns the float, it can be used to delay making decisions requested by the contractor. It can also be used when the architect or engineer fails to respond to the contractor's request for information, to answer a question that will affect job progress, or to promptly review a critical shop drawing.

If the contractor owns the float, it is included in the CPM schedule as such, and it is not questioned by the owner or architect, it is safe to assume that the contractor's float as published is accepted.

If the question of float is not addressed in the contract, it should be brought up at the first project meeting attended by the owner's representatives and the design consultants. It is better to hash it out than wait for the shoe to drop later on.

Shop-Drawing Submission Schedules

Often overlooked, a schedule for shop-drawing submissions from each applicable subcontractor and vendor is another key issue that could be incorporated into the CPM schedule or could be created as a separate document. It is a rather simple document to create, listing the provider of the drawing, a description of the drawing, the date the drawing is to be submitted, the review time allotted to the architect or engineer (either by contract or by negotiation), the due date for the return from the A/E, the actual date the drawing is received, and the action taken (approved, approved as noted, or disapproved—resubmit).

This document, created with information supplied by the subcontractor or vendor, submitted to the architect for review, should include a time allowance for this review process and ought to be reviewed as frequently as the CPM schedule. If there is a lag in either submission by the subcontractor or vendor or review by the A/E, this needs to be documented in the project-meeting minutes.

Precedence Diagramming Method (PDM)

Although it is rarely used nowadays, it might be worth mentioning PDM scheduling, in case a CM or GC comes across that terminology while discussing scheduling procedures. PDM grew out of arrow diagrams and actually grew out of the Gantt or bar-chart design. This scheduling was referred to as "ladder feed," a process of creating a schedule with a linear time frame, unaffected by potential delays for equipment or vendor delivery schedules. This type of schedule is applicable to linear projects such as pipeline installation, highways, and canals. It was used in the 1970s when the use of computers had not yet pervaded the industry. Figure 4.6 is an example of PDM scheduling. As computer capabilities and availability expanded rapidly in the 1980s, additional functions were added to the basic PDM schedule such as multiple types of relationships, lead and lag times, dependencies, and multiple resources. As shown in Fig. 4.7, the basic terminology and presentation of a PDM schedule reflects this lateral or linear approach. Figure 4.8 contains a glossary of scheduling terms.

Summary

From simple to complex projects, construction schedules are necessary for a number of reasons. Either mandated by contract or required to properly monitor the multitude of activities occurring daily at the construction site, the schedule provides the road map to get from one place to another. For rather uncomplicated projects of limited scope, a Gantt or bar chart may be perfectly suitable, but for more complex projects, a CPM schedule is a necessity. Many contractors have in-house CPM schedule-preparation capability, but for some projects a professional scheduling outfit is required. The contractor's professional staff is required to provide information to the scheduler as to predecessors, successors, durations, and float, aided by their experience on previous projects and data obtained from their subcontractors and vendors. Float—not only who owns it, but how it is properly applied to key components—will assist in maintaining the critical path. Combined with frequent and careful monitoring and adjusting of the CPM schedule, it means a successful project lies ahead.

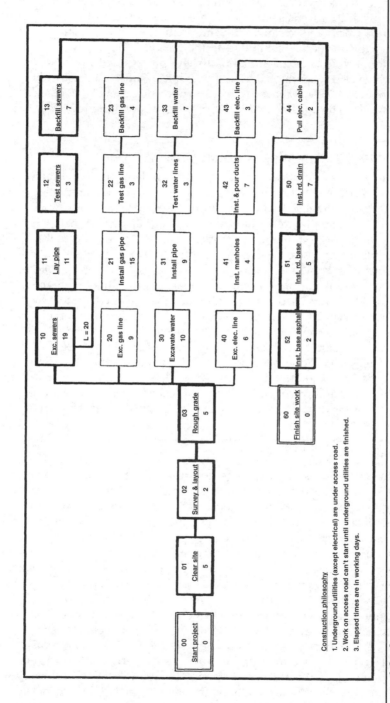

Figure 4.6 Contractor's logic diagram for site development.

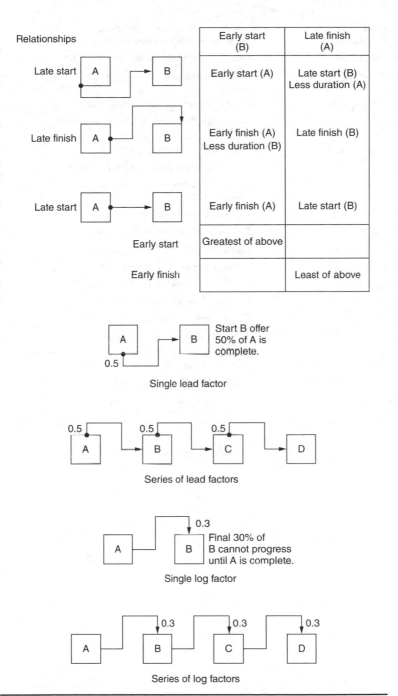

Figure 4.7 Basic PDM terminology.

PERT method	Program evaluation research technique
CPM schedule	Schedule using the critical path method
PDM schedule	Schedule using the precedence diagraming method
Arrow diagram	CPM diagraming method using arrows
Logic diagram	Arrow diagram of complete project or section of a project
Time-scaled chart	Logic diagram with a time scale
Activity	Any significant item of work on a project
Activity list	List of work items for a project; also work breakdown structure
Activity duration	Elapsed time to perform an activity
Optimistic time	Earliest completion: shortest time
Pessimistic time	Latest completion: longest time
Realistic time	Normal completion: average time
Activity number	Number assigned to each activity
Early start date	Earliest date activity could start
Late start date	Latest date activity could start
Float	Measure of spare time on activity
Free float	Time by which activity can be changed without affecting next activity
Total float	Total free time on any activity or project
Negative float	Time a critical activity is late

FIGURE 4.8   Glossary of scheduling terms.

CHAPTER 5

Estimating, Budgeting, and Cost Control

Thehe project money plan is the financial forecast for the project; it sets the basis for the control of project costs and cash flow. Developing the money plan involves the functions of cost estimating, budgeting, cash flow, cost control, and project profitability. Taken together, these functions make up the field of cost engineering.

This chapter considers cost engineering and control from the viewpoint of the construction manager (CM) or general contractor (GC). Although he or she may or may not have participated in the preparation of the budget, once the contract for construction has been executed by the owner, the CM or GC will then be dealing with those costs throughout the project. In the early stages, if the CM or GC participated in the estimating process, he or she will be interfacing with project estimators. In the execution stage, the CM or GC will be dealing with controlling costs and issuing periodic cost reports.

Those of you who prepare your own estimates on small projects will find your intimacy with the costs to be an advantage as the project progresses and you assume the task of cost control.

Estimating Sources

There are many ways in which to obtain an estimate for your project:

- An in-house estimator utilizing standard or customized estimating software as its database

- Cost data purchased from such sources as RSMeans or McGraw-Hill's Sweets cost guides, among others

- An on-line estimating service or an independent local estimating service

- A combination of any of these, e.g., an in-house estimating service preparing an estimate for a project in which certain components are unique and require the purchase of costs not included in the company's database nor familiar to the company's estimator

The need for estimating will occur throughout your involvement in the construction industry. Not only will an estimate be required when you submit a hard bid for new work, but from time to time, previous clients considering a new project may call and request ballpark figures for a new project being proposed.

When negotiating a project with an owner where a contract for a cost plus a fee with a guaranteed maximum price is being considered, you will prepare many estimates as the design travels from conceptual or developing to final and the owner weighs design considerations against the design components prior to finalizing both design and budget.

The design-build process may find one of your estimators spending a great deal of time with the design consultants as the design-versus-cost process proceeds and the owner presents its requirements. And even when the project is underway, the introduction of changes to the original contract scope and price will require estimates to document the addition or deletion of scope of work.

The Need for Identifying and Isolating Construction Components in the Estimate

The Construction Specifications Institute, based in Alexandria, Virginia (csi@csinet.org)—referred to simply as CSI—was founded in 1948 by the specification writers of government agencies to improve the quality of construction specifications. Shortly thereafter, the private sector recognized the value of this endeavor, and today CSI is the organization responsible for the creation of specifications best practices, developing the standards and formats that we find in any professionally prepared specifications manual.

Along the way, CSI created the numbering system for the various components of construction familiar to all as the divisions in the specification manuals (Fig. 5.1). This numbering system, referred to as the CSI MasterFormat Groups, Subgroups, and Divisions is almost universally used by estimators as they prepare their cost breakdowns; Fig. 5.2 shows the latest iteration (2010). The 2010–2011 update, shown in Fig. 5.3, includes more precise categories, such as process-equipment systems and components which were not included in the initial 16 categories.

This division numbering system is also used in the preparation of budget categories in the request for payment or requisition form

MASTERFORMAT 1995 EDITION

Before November 2004, MasterFormat was composed of 16 primary divisions; there are now 50 divisions.

- Division 01 — General Requirements
- Division 02 — Site Construction
- Division 03 — Concrete
- Division 04 — Masonry
- Division 05 — Metals
- Division 06 — Wood and Plastics
- Division 07 — Thermal and Moisture Protection
- Division 08 — Doors and Windows
- Division 09 — Finishes
- Division 10 — Specialties
- Division 11 — Equipment
- Division 12 — Furnishings
- Division 13 — Special Construction
- Division 14 — Conveying Systems
- Division 15 — Mechanical
- Division 16 — Electrical

FIGURE 5.1   Construction Specifications Institute (CSI) manual specification divisions.

presented to the owner for the contractor's monthly payment requests. It is common practice to combine several categories within a specific division for billing purposes. Division 3, Concrete, comes to mind, where footings, foundations, and other ancillary concrete work are being billed. General concrete floor slabs are given a specific category, since billing is most likely based on a square foot of slab in place.

UniFormat®

CSI also offers another method of arranging construction-cost information via a system called UniFormat, which separates costs into elements or parts of a building based upon their function (Fig. 5.4). For example, an estimate using UniFormat breakdowns would be used in preliminary project-component descriptions when discussing schematic design costs. An estimator can display a cost for components of the substructure, superstructure, and mechanical/electrical/plumbing.

The original UniFormat was developed in 1972 by a task force composed of the U.S. government's General Services Administration, CSI, the Association for Advancement of Computing in Education, the Tri-Service Committee, RSMeans, and the Canadian Institute of Quantity Surveyors. UniFormat was issued by the American Society for Testing and Materials in 1993 as ASTM E1557. It is a method of arranging construction information based upon functional elements of a structure or parts of a project, characterized by their functions

MasterFormat GROUPS, SUBGROUPS, AND DIVISIONS

PROCUREMENT AND CONTRACTING
REQUIREMENTS GROUP
Division 00 – Procurement and Contracting
                Requirements
    Introductory Information
    Procurement Requirements
    Contracting Requirements

SPECIFICATIONS GROUP

GENERAL REQUIREMENTS SUBGROUP
Division 01 – General Requirements

FACILITY CONSTRUCTION SUBGROUP
Division 02 – Existing Conditions
Division 03 – Concrete
Division 04 – Masonry
Division 05 – Metals
Division 06 – Wood, Plastics, and Composites
Division 07 – Thermal and Moisture Protection
Division 08 – Openings
Division 09 – Finishes
Division 10 – Specialties
Division 11 – Equipment
Division 12 – Furnishings
Division 13 – Special Construction
Division 14 – Conveying Equipment
Division 15 – Reserved for Future Expansion
Division 16 – Reserved for Future Expansion
Division 17 – Reserved for Future Expansion
Division 18 – Reserved for Future Expansion
Division 19 – Reserved for Future Expansion

FACILITY SERVICES SUBGROUP
Division 20 – Reserved for Future Expansion
Division 21 – Fire Suppression

Division 22 – Plumbing
Division 23 – Heating, Ventilating, and Air-
                Conditioning (HVAC)
Division 24 – Reserved for Future Expansion
Division 25 – Integrated Automation
Division 26 – Electrical
Division 27 – Communications
Division 28 – Electronic Safety and Security
Division 29 – Reserved for Future Expansion

SITE AND INFRASTRUCTURE SUBGROUP
Division 30 – Reserved for Future Expansion
Division 31 – Earthwork
Division 32 – Exterior Improvements
Division 33 – Utilities
Division 34 – Transportation
Division 35 – Waterway and Marine Construction
Division 36 – Reserved for Future Expansion
Division 37 – Reserved for Future Expansion
Division 38 – Reserved for Future Expansion
Division 39 – Reserved for Future Expansion

PROCESS EQUIPMENT SUBGROUP
Division 40 – Process Integration
Division 41 – Material Processing and Handling
                Equipment
Division 42 – Process Heating, Cooling, and Drying
                Equipment
Division 43 – Process Gas and Liquid Handling,
                Purification, and Storage Equipment
Division 44 – Pollution and Waste Control Equipment
Division 45 – Industry-Specific Manufacturing
                Equipment
Division 46 – Water and Wastewater Equipment
Division 47 – Reserved for Future Expansion
Division 48 – Electrical Power Generation
Division 49 – Reserved for Future Expansion

FIGURE 5.2   CSI's 2010 *MasterFormat* Groups, Subgroups, and Divisions.
(*By permission of the Construction Specifications Institute, Alexandria, Virginia.*)

without regard to the materials but as systems or assemblies. This system is frequently used by estimators called upon to present cost estimates during the schematic-design stage, pricing building components or assemblies such as substructure, building envelope, interiors, and services. Uniformat is also used in software for building information modeling, since it allows basic costs to be applied as the structure's properties are being defined. It is also seen in performance specifications and preliminary project descriptions.

MasterFormat Transition Matrix - 2010 Update to 2011 Update

For more information on MasterFormat visit www.masterformat.com

Contents reflect current MasterFormat titles and numbers as of April 2011.

CSC

* NOTE: This matrix compares the 2010 Edition of *MasterFormat* to the expanded 2011 Edition of *MasterFormat*. All revised 2010 numbers and titles are listed along with their corresponding 2011 edition location.

2010 Edition *MasterFormat*			2011 Edition *MasterFormat* *			Change Notes
10 SECTION	10 TITLE	2010 LEVEL	2011 SECTION	2011 TITLE	2011 LEVEL	
			00 54 33	Digital/Electronic Data Protocol Exhibit	3	New Title
			00 54 36	Building Information Modeling Exhibit	3	New Title
			02 41 19.19	Selective Facility Services Demolition	4	New Title
			03 15 13.13	Non-Expanding Waterstops	4	New Title
			03 15 13.16	Expanding Waterstops	4	New Title
			03 15 13.19	Combination Expanding and Injection Hose Waterstops	4	New Title
			03 15 13.21	Injection Hose Waterstops	4	New Title
			03 15 16	Concrete Construction Joints	3	New Title
03 21 00	Reinforcing Steel	2	03 21 00	Reinforcement Bars	2	Title Change for clarification
03 21 11	Black Steel Reinforcing	3	03 21 11	Plain Steel Reinforcement Bars	3	Title Change for clarification
03 21 13	Galvanized Reinforcing Steel	3	03 21 13	Galvanized Reinforcement Bars	3	Title Change for clarification
03 21 16	Epoxy-Coated Reinforcing Steel	3	03 21 16	Epoxy-Coated Reinforcement Steel Bars	3	Title Change for clarification
03 21 19	Stainless Steel Reinforcing	3	03 21 19	Stainless Steel Reinforcement Bars	3	Title Change for clarification
			03 21 21	Composite Reinforcement Bars	3	New Title
			03 21 21.11	Glass Fiber-Reinforced Polymer Reinforcement Bars	4	New Title
			03 21 21.13	Organic Fiber-Reinforced Polymer Reinforcement Bars	4	New Title
			03 21 21.16	Carbon Fiber-Reinforced Polymer Reinforcement Bars	4	New Title
03 22 00	Welded Wire Fabric Reinforcing	2	03 22 00	Fabric and Grid Reinforcing	2	Title Change for clarification
			03 22 19	Composite Grid Reinforcing	3	New Title
03 23 00	Stressing Tendons	2	03 23 00	Stressed Tendon Reinforcing	2	Title Change for clarification
			04 05 21	Masonry Strengthening	3	New Title
			04 43 13.13	Anchored Stone Masonry Veneer	4	New Title
			04 43 13.16	Adhered Stone Masonry Veneer	4	New Title
			05 16 13	Cable Bow Truss Assemblies	3	New Title
			05 17 00	Structural Rod Assemblies	2	New Title
06 42 13	Solid Lumber Paneling	3	06 42 13	Wood Board Paneling	3	Title Change for clarification
			06 42 14	Stile and Rail Wood Paneling	3	New Title
06 42 16	Wood-Veneer Paneling	3	06 42 16	Flush Wood Paneling	3	Title Change for clarification
			07 27 36	Sprayed Foam Air Barrier	3	New Title
07 61 91	Tinplate and Template Roofing	3	......	......	......	Title removed. Considered to be archaic and no longer used.
			07 64 00	Sheet Metal Wall Cladding	2	New Title
			07 64 13	Standing Seam Sheet Metal Wall Cladding	3	New Title
			07 64 16	Batten Seam Sheet Metal Wall Cladding	3	New Title
			07 64 19	Flat Seam Sheet Metal Wall Cladding	3	New Title
			08 44 18	Glazed Steel Curtain Walls	3	New Title
09 54 26	Linear Wood Ceilings	3	09 54 26	Suspended Wood Ceilings	3	Title Change for clarification
			09 75 13	Stone Wall Facing	3	New Title
			09 75 19	Stone Trim	3	New Title
			09 77 53	Vegetated Wall Systems	3	New Title
			10 22 14	Expanded Metal Partitions	3	New Title
			10 26 41	Bullet Resistant Panels	3	New Title
			10 51 29	Phenolic Lockers	3	New Title
			10 71 19	Flood Barriers	3	New Title
			11 52 23	Audio-Visual Equipment Supports	3	New Title
			11 53 26	Laboratory Freezers	3	New Title
			11 53 29	Laboratory Controlled-Environment Cabinets	3	New Title
			12 25 09	Window Treatment Control System	3	New Title
12 36 61.19	Quartz Surfacing Countertops	4	12 36 61.19	Quartz Agglomerate Countertops	4	Title Change for clarification
			12 55 86	Detention Control Room Furniture	3	New Title
			12 56 33.13	Fixed Classroom Tables	4	New Title
			12 56 86	Institutional Control Room Furniture	3	New Title
			12 57 86	Industrial Control Room Furniture	3	New Title

FIGURE 5.3   CSI's MasterFormat 2010–2011 update. (*By permission of the Construction Specifications Institute, Alexandria, Virginia.*)

Old Number	Old Title		New Number	New Title		Change
			13 21 26.13	Walk-in Coolers	4	New Title
			13 21 26.16	Walk-in Freezers	4	New Title
			13 42 13	Bathroom Unit Modules	3	New Title
			23 82 41	Water-to-Water Heat Pumps	3	New Title
26 43 00	Transient Voltage Suppression	2	26 43 00	Surge Protective Devices	2	Title Change for clarification
			26 55 83	Broadcast Lighting	3	New Title
28 33 00	Fuel-Gas Detection and Alarm	2	28 33 00	Gas Detection and Alarm	2	Title Change for clarification
28 33 13	Fuel-Gas Detection and Alarm Control GUI, and Logic Systems	3	28 33 13	Gas Detection and Alarm Control, GUI, and Logic Systems	3	Title Change for clarification
28 33 23	Fuel-Gas Detection and Alarm Integrated Audio Evacuation Systems	3	28 33 23	Gas Detection and Alarm Integrated Audio Evacuation Systems	3	Title Change for clarification
28 33 33	Fuel-Gas Detection Sensors	3	28 33 33	Gas Detection Sensors	3	Title Change for clarification
			31 09 16.26	Bored and Augered Pile Load Tests	4	New Title
			31 32 17	Water Injection Soil Stabilization	3	New Title
			31 66 15	Helical Foundation Piles	3	New Title
			31 68 16	Helical Foundation Anchors	3	New Title
			32 16 13.16	Steel Faced Curbs	4	New Title
			32 31 17	Expanded Metal Fences and Gates	3	New Title
			32 31 32	Composite Fences and Gates	3	New Title
			32 94 19	Landscape Surfacing	3	New Title
			34 81 29	Pontoon Bridge Machinery	3	New Title
			34 81 32	Bascule Bridge Machinery	3	New Title
			35 01 40.51	Waterway Dredging	4	New Title
			35 01 50.51	Marine Dredging	4	New Title
			35 49 26	Floodgate Machinery	3	New Title
			41 53 23.13	Mobile Storage Racks	4	New Title
			43 01 10.13	Gas Blowers Maintenance and Rehabilitation	4	New Title
			43 01 10.16	Gas Compressors Maintenance and Rehabilitation	4	New Title
			43 01 20.13	Liquid Pumps Maintenance and Rehabilitation	4	New Title
			43 01 20.16	Liquid Process Equipment Maintenance and Rehabilitation	4	New Title
			43 01 40.13	Non-pressurized Tanks Cleaning, Maintenance, and Rehabilitation	4	New Title
			43 01 40.16	Pressurized Tanks Cleaning, Maintenance, and Rehabilitation	4	New Title
			43 05 00	Common Work Results for Process Gas and Liquid Handling, Purification, and Storage Equipment	2	New Title
			43 05 10	Common Work Results for Gas Handling Equipment	3	New Title
			43 05 20	Common Work Results for Liquid Handling Equipment	3	New Title
			43 05 30	Common Work Results for Gas and Liquid Hi-Purification Equipment	3	New Title
			43 05 40	Common Work Results for Gas and Liquid Storage	3	New Title
43 11 00	Gas Fans, Blowers, and Pumps	2	43 11 00	Gas Fans, Blowers, Pumps, and Boosters	2	Title Change to include Boosters
43 11 13	Gas Handling Fans	3	43 11 13	Separately Geared Single Stage Centrifugal Blowers	3	Title Change for clarification
43 11 13.13	Axial Gas Handling Fans	4	43 11 26	Axial Fans	3	Title Change for clarification
43 11 13.16	Centrifugal Gas Handling Fans	4	43 11 19	Centrifugal Fans	3	Title Change for clarification
			43 11 14	Direct Drive Integral Shaft Single Stage Centrifugal Blowers	3	New Title
			43 11 17	Horizontally Split Multistage Centrifugal Blowers	3	New Title
			43 11 18	Vertically Split Multistage Centrifugal Blowers	3	New Title
43 11 23	Gas Handling Blowers	3	43 11 23	Axial Blowers	3	Title Change for clarification
43 11 26	Gas Handling Jet Pumps	3	43 11 26	Axial Fans	3	Title Change for clarification
			43 11 31	Rotary Helical Screw Blowers	3	New Title
			43 11 33	Rotary Lobe Blowers	3	New Title
			43 11 34	Regenerative Rotary Blowers	3	New Title
			43 11 36	Rotary Vane Blowers	3	New Title
43 11 26	Gas Handling Jet Pumps	3	43 11 41	Gas-handling Venturi Jet Pumps	3	Number and Title change for clarification
43 11 29	Gas Handling Vacuum Pumps	3	43 11 43	Gas-handling Vacuum Pumps	3	Number and Title change for clarification
			43 11 46	Gas Boosters	3	New Title
43 12 13	Centrifugal Gas Compressors	3	43 12 13	Diagonal or Mixed-Flow Compressors	3	Title Change for clarification
43 12 13	Centrifugal Gas Compressors	3	43 12 11	Centrifugal Compressors	3	Number and Title change for clarification

FIGURE 5.3   (Continued)

Old Code	Old Title	Lvl	New Code	New Title	Lvl	Change
43 12 16	Piston Gas Compressors	3	43 12 16	Axial-flow Compressors	3	Title Change for clarification
			43 12 33	Single-acting Reciprocating Compressors	3	New Title
			43 12 34	Double-acting Reciprocating Compressors	3	New Title
			43 12 37	Diaphragm Reciprocating Compressors	3	New Title
43 12 19	Positive Displacement Gas Compressors	3	43 12 51	Rotary Screw Compressors	3	Number and Title change for clarification
43 12 23	Rotary-Screw Gas Compressors	3	43 12 51	Rotary Screw Compressors	3	Number and Title change for clarification
43 12 26	Vane Gas Compressors	3	43 12 53	Rotary Vane Compressors	3	Number and Title change for clarification
			43 12 56	Rotary Liquid-ring Compressors	3	New Title
			43 12 57	Rotary Scroll Compressors	3	New Title
43 13 13	Process Gas Blenders	3	43 13 13	Gas Blenders	3	Title Change for clarification
43 13 16	Process Gas Meters	3	40 90 00	Instrumentation and Control for Process Systems	2	Title was removed. Process Gas Meters now specified in Division 40.
43 13 19	Process Gas Mixers	3	43 13 19	Gas Mixers	3	Number and Title change for clarification
43 13 23	Process Gas Pressure Regulators	3	43 13 23	Gas Pressure Regulators	3	Title Change for clarification
			43 13 31	Gas Separation Equipment	3	New Title
			43 13 33	Gas Dehydration Equipment	3	New Title
			43 13 36	Combined Gas Separation and Dehydration Equipment	3	New Title
			43 13 39	Gas Recovery and Condensing Equipment	3	New Title
			43 13 43	Waste Gas Burner System	3	New Title
			43 13 46	Gas Control and Safety Equipment	3	New Title
			43 15 00	Process Air and Gas Filters	2	New Title
			43 15 13	Blower Intake and Turbine Air Filters	3	New Title
			43 15 13.13	Static Prefilters	4	New Title
			43 15 13.16	Static Final Filters	4	New Title
			43 15 13.19	Static HEPA Filters	4	New Title
			43 15 13.23	Pulse Filters	4	New Title
			43 15 33	Grease Filters	3	New Title
			43 15 43	Mist Eliminators	3	New Title
			43 15 63	High-temperature Air Filters	3	New Title
			43 15 73	Multiple-application Air Filters	3	New Title
			43 15 73.11	Air Filter Media	4	New Title
			43 15 73.12	Permanent Washable Filters	4	New Title
			43 15 73.13	Poly-ring Air Filters	4	New Title
			43 15 73.15	Rigid Cell Filters	4	New Title
			43 15 73.17	Automatic Roll-type Air Filters	4	New Title
			43 15 76	Chemical Media for Air and Gas Filters	3	New Title
			43 22 69	Liquid Grease Receiving and Dewatering Systems	3	New Title
			43 22 73	Liquid Fillers	3	New Title
			43 22 76	Liquid Screeners	3	New Title
			43 22 79	Liquid Clarifiers	3	New Title
			43 22 83	Liquid Classifiers	3	New Title
			43 22 86	Liquid Homogenizers	3	New Title
			43 22 89	Liquid Presses	3	New Title
			43 22 96	Liquid Versators	3	New Title
			43 22 99	Liquid Votators	3	New Title
43 22 43	Liquid Filters	3	43 27 00	Process Liquid Filters	2	Number and Title change for clarification and expansion
43 22 43.13	Cyclonic Liquid Filters	4	43 27 13	Cyclonic Liquid Filters	3	Number and Title change for clarification and expansion
			43 27 16	Centrifugal Horizontal Pressure Leaf Liquid Filters	3	New Title
			43 27 23	Liquid Bag Filters	3	New Title
			43 27 33	Mechanically Cleaned Liquid Filters	3	New Title
			43 27 43	Tubular Backwashing Liquid Filters	3	New Title
			43 27 53	Vacuum Belt Liquid Filters	3	New Title
43 22 43.16	Media Liquid Filters	4	43 27 63	Woven Media Liquid Filters	3	Number and Title change for clarification and expansion
			43 27 63.13	Disc Liquid Filters	4	New Title
			43 27 63.23	Rotating Drum Liquid Filters	4	New Title
			43 27 63.33	Layer Liquid Filters	4	New Title
			43 27 63.43	Candle Liquid Filters	4	New Title
			43 27 63.53	Vertical Leaf Liquid Filters	4	New Title

Figure 5.3 (Continued)

43 41 13.13	Ferrous Gas and Liquid Pressure Vessels	4		43 42 33	Stainless Steel Pressure Tanks	3	Title removed. Ferrous Gas and Liquid Pressure Vessels now specified by varying container material
43 41 13.13	Ferrous Gas and Liquid Pressure Vessels	4		43 42 36	Aluminum Pressure Tanks	3	Title removed. Ferrous Gas and Liquid Pressure Vessels now specified by varying container material
43 41 13.13	Ferrous Gas and Liquid Pressure Vessels	4		43 42 41	Metallic Specialty Pressure Tanks	3	Title removed. Ferrous Gas and Liquid Pressure Vessels now specified by varying container material
43 41 13.16	Nonferrous Gas and Liquid Pressure Vessels	4		43 41 41	Polyvinyl Chloride Tanks	3	Title removed. Nonferrous Gas and Liquid Pessure Vessels now specified by varying container material
43 41 13.16	Nonferrous Gas and Liquid Pressure Vessels	4		43 41 43	Polythylene Tanks	3	Title removed. Nonferrous Gas and Liquid Pessure Vessels now specified by varying container material
43 41 13.16	Nonferrous Gas and Liquid Pressure Vessels	4		43 41 53	Wood Stave Tanks	3	Title removed. Nonferrous Gas and Liquid Pessure Vessels now specified by varying container material
43 41 13.16	Nonferrous Gas and Liquid Pressure Vessels	4		43 41 63	Precast Concrete Tanks	3	Title removed. Nonferrous Gas and Liquid Pessure Vessels now specified by varying container material
43 41 13.16	Nonferrous Gas and Liquid Pressure Vessels	4		43 41 73	Ceramic Tanks	3	Title removed. Nonferrous Gas and Liquid Pessure Vessels now specified by varying container material
43 41 13.16	Nonferrous Gas and Liquid Pressure Vessels	4		43 41 83	Non-metallic Specialty Tanks	3	Title removed. Nonferrous Gas and Liquid Pessure Vessels now specified by varying container material
43 41 13.16	Nonferrous Gas and Liquid Pressure Vessels	4		43 42 83	Non-metallic Specialty Pressure Tanks	3	Title removed. Nonferrous Gas and Liquid Pessure Vessels now specified by varying container material
43 41 13.19	Fiberglass Gas and Liquid Pressure Vessels	4		43 42 53	Fiberglass Reinforced Plastic Pressure Tanks	3	Number and Title change for clarification and expansion

43 41 16	Atmospheric Tanks and Vessels	3		43 41 00	Non-pressurized Tanks and Vessels	2	Number and Title change for clarification and expansion
43 41 16	Atmospheric Tanks and Vessels	3		43 41 45	Fiberglass Reinforced Plastic Tanks	3	Number and Title change for clarification and expansion
43 41 16.13	Horizontal Atmospheric Tanks and Vessels	4		43 40 00	Gas and Liquid Storage	2	Title removed. Atmospheric Tanks and Vessels now specified by container material rather than position.
43 41 16.16	Vertical Atmospheric Tanks and Vessels	4		43 40 00	Gas and Liquid Storage	2	Title removed. Atmospheric Tanks and Vessels now specified by container material rather than position.

FIGURE 5.3   (Continued)

Construction-Cost Estimating

The foundation of the project's money plan is a sound project-cost estimate: the *predicted* cost of executing the work. The construction estimate should be neither *optimistic* nor *pessimistic*, and should be produced at a reasonable cost.

Accurate forecasting of costs is vital to the survival of a construction business. An estimator, to be successful, should embrace the following tasks:

- Review all sections of the plans and specifications, along with any bidder's instructions and proposed contract format, to gain an accurate perspective of the total scope of the project.

- Review the general and special conditions of the contract to determine the cost impact they may have on the project.

Project: Uniformat Template	Qty	Element (Level 3)			Group (Level 2)	Major (Level 1)
"File > Make Copy" to start your own Uniformat calculation. Edit, add and delete items to adapt your project. Use "sub" sheets like the A20 example to specify your calculation.		Unit	Rate	Amount	Amount	Amount
A. SUBSTRUCTURE				0		11,120
A10 Foundations				0	1,120	
A1010 Standard Foundations	100.30	CY	10.00	1,000		
A1020 Special Foundations	2.20	CY	10.00	20		
A1030 Slab on Grade	2.20	CY	50.00	100		
A20 Basement Construction				0	10,000	
B2010 Basement Excavation	1.00	from A20	1,000.00	1,000		
B2020 Basement Walls	1.00	from A20	9,000.00	9,000		
B. SHELL				0		500
B10 Superstructure				0	500	
B1010 Floor Constructions	10.00		50.00	500		
B1020 Roof Construction				0		
B20 Exterior Closure				0	0	
B2010 Exterior Walls				0		
B2020 Exterior Windows				0		
B2030 Exterior Doors				0		
B30 Roofing				0	0	
B3010 Roof Covering				0		
B3020 Roof Openings				0		
C. Interiors				0		0
C10 Interior Construction				0	0	
C1010 Partitions				0		
C1020 Interior Doors				0		
C1030 Specialties				0		
C20 Staircases				0	0	
C2010 Stair Construction				0		
C2020 Stair Finishes				0		
C30 Interior Finishes				0	0	
C3010 Wall Finishes				0		
C3020 Floor Finishes				0		
C3030 Ceiling Finishes				0		
D. Services				0		0
D10 Conveying Systems				0	0	
D1010 Elevators				0		
D1020 Escalators & Moving Walks				0		
D1030 Material Handling Systems				0		
D20 Plumbing				0	0	
D2010 Plumbing Fixtures				0		

The Engineering ToolBox

Uniformat

The Engineering ToolBox
home   www.EngineeringToolBox.com

Figure 5.4 Uniformat® Code Assignments. (By permission of engineeringtoolbox@gmail.com)

115

Uniformat Template

Project:

"File > Make Copy" to start your own Uniformat calculation. Edit, add and delete items to adapt your project. Use "sub" sheets like the A20 example to specify your calculation.

Uniformat Template	Element (Level 3)				Group (Level2)	Major (Level 1)
	Qty	Unit	Rate	Amount	Amount	Amount
D2020 Domestic Water Distribution				0		
D2030 Sanitary Waste				0		
D2040 Rain Water Drainage				0		
D2050 Special Plumbing Systems				0		
				0	0	
D30 HVAC						
D3010 Energy Supply				0		
D3020 Heat Generating Systems				0		
D3030 Cooling Generating Systems				0		
D3040 Distribution Systems				0		
D3050 Terminal & Package Units				0		
D3060 Controls & Instrumentation				0		
D3070 Special HVAC Systems & Equipment				0		
D3080 Systems Testing & Balancing				0		
				0	0	
D40 Fire Protection						
D4010 Fire Protection & Sprinkler Systems				0		
D4020 Stand-Pipe & Hose Systems				0		
D4030 Fire Protection Specialities				0		
D4040 Special Electrical Systems				0		
				0	0	
D50 Electrical						
D5010 Electrical Service & Distribution				0		
D5020 Lighting & Branch Wiring				0		
D5030 Communication & Security Systems				0		
D5040 Special Electrical Systems				0		
D5090 Other Electrical Systems				0		
				0	0	
E. Equipment & Furnishings						0
E10 Equipment						
E1010 Commercial Equipment				0		
E1020 Institutional Equipment				0		
E1030 Vehicular Equipment				0		
E1040 Other Equipment				0		
				0	0	
E20 Furnishings						
E2010 Fixed Furnishings				0		
E2020 Movable Furnishings				0		
				0	0	
F. Special Construction & Demolition						0
F10 Special Construction						
F1010 Special Structures				0		
F1020 Integrated Constructions				0		
F1030 Special Construction Systems				0		
F1040 Special Facilities				0		
F1050 Special Controls & Instrumentation				0		
				0	0	
F20 Selective Building Demolition						

116

Project:	Uniformat Template		Element (Level 3)			Group (Level2)	Major (Level 1)
		Qty	Unit	Rate	Amount	Amount	Amount
	"File > Make Copy" to start your own Uniformat calculation. Edit, add and delete items to adapt your project. Use "sub" sheets like the A20 example to specify your calculation.						
	F2010 Building Elements Demolition				0		
	F2020 Hazardous Components Abatement				0		
G. Building Sitework					0		0
	G10 Site Preparation				0	0	
	G1010 Site Clearing				0		
	G1020 Site Demolition & Relocations				0		
	G1030 Site Earthwork				0		
	G1040 Hazardous Waste Remedation				0		
	G20 Site Improvements				0	0	
	G2010 Roadways				0		
	G2020 Parking Lots				0		
	G2030 Pedestrian Paving				0		
	G2040 Site Development				0		
	G2050 Landscaping				0		
	G30 Site Civil/Mechanical Utilities				0	0	
	G3010 Water Supply & Distribution Systems				0		
	G3020 Sanitary Sewer Systems				0		
	G3030 Storm Sewer Systems				0		
	G3040 Heating Distribution				0		
	G3050 Cooling Distribution				0		
	G3060 Fuel Distribution				0		
	G3070 Other Civil/Mechanical Activities				0		
	G40 Site Electrical Utilities				0	0	
	G4010 Electrical Distribution				0		
	G4020 Site Lighting				0		
	G4030 Site Communications & Security				0		
	G4040 Other Site Electrical Utilities				0		
	G50 Other Site Construction				0	0	
	G5010 Service Tunnels				0		
	G5020 Other Site Systems & Equipment				0		
	Building Trade Cost - without Design Allowance				11,620	11,620	11,620
					0		Note! Some values in all three columns verifies the accumulations above
	Z10 Design Allowance	11,620.00		5.00%	581	581	581
	Z10 Design Allowance				0		
	Building Trade Cost	12,201.00			12,201	12,201	12,201
					0		
	Z20 Overhead & Profit	12,201.00		6.00%	732	1,952	1,952
	Z2010 Overhead	12,201.00		10.00%	1,220		
	Z2020 Profit						

FIGURE 5.4 (Continued)

117

Project: Uniformat Template	Element (Level 3)				Group (Level 2)	Major (Level 1)
	Qty	Unit	Rate	Amount	Amount	Amount
"File > Make Copy" to start your own Uniformat calculation. Edit, add and delete items to adapt your project. Use "sub" sheets like the A20 example to specify your calculation.						
				0		
Building Construction Cost without Inflation				14,153	14,153	14,153
				0		
Z30 Inflation Allowance	14,153.16			0	425	425
Z30 Inflation Allowance			3.00%	425		
Building Construction Cost (BCC)				14,578	14,578	14,578

FIGURE 5.4 (Continued)

118

- Develop a system of forms to define material, labor-hour, and equipment quantities and related costs for review by management.

- Ascertain the level of contingency necessary to deal with any future inconsistencies or differences of opinion as to what the plans and specifications contain and how the estimator interpreted them.

- Maintain consistency in putting together the quantity surveys according to industry standards, by using classification systems such as MasterFormat.

- Document all parts of the estimate in a logical, consistent, and legible manner for review by management in case changes to some portion of the estimate need to be made.

- When portions of the estimate are obtained from subcontractor bids, check the bid for compliance with the project requirements for those portions of the work to determine if the scope of work is all-inclusive.

- When self-performed work is being estimated, rely on the accuracy of historic related field costs assembled by the field supervisors and reviewed by the project manager.

- During the course of the estimate, review the plans and specifications with an eye kept open to spot any potential value-engineering suggestions or other cost savings in case they are needed to successfully negotiate the project with the owner.

In any event, a contractor should conduct a pre-estimate meeting before agreeing to submit a competitive fixed-price bid and committing estimating time and money to this project. He or she needs to discuss the pros and cons of submitting a bid. Management will take into account constructability problems, applicable construction technology, labor climate, experience on prior similar projects, and the like. After weighing all of the technical, marketing, and financial input, the firm's top management will make the decision whether to prepare a bid or not. How badly do we want this job? Who are our competitors? Do we have any past relations with this owner or their design consultants that might weigh in our favor?

Types of Estimates

Most cost estimates fall into one of five levels:

1. Order of magnitude
2. Conceptual or schematic design
3. Design development

4. Construction documents (at various stages of completion from 50 to 100 percent)

5. Hard bid submission

Level 1: Order of Magnitude

This is one of those back-of-the-envelope types of estimates. A typical situation is when you are having lunch with a previous client and he or she says, "You know, we are thinking about adding about 20,000 square feet onto that building you built for us two years ago. Can you give me a ballpark figure for that work?" Since you were the project manager and you recall the overall cost and the square footage, you can come up with a square-foot cost and increase it by what you perceive to be the rise in costs over that two-year period.

This level 1 estimate requires you to recall historical data for a similar project and update to current prices, generally on a square-foot basis. Other-type projects will use something other than the square-foot basis. Hotels and motels are often ballparked based upon a per-room cost, which would include all ancillary parts, elevators, lobby, dining-room areas, etc.

Level 2: Conceptual Estimating

Upon receipt of some preliminary design information, such as size and quality levels, along with some basic construction components— including structural steel framework or a cast-in-place structure, exterior masonry-wall design possibly accompanied by an elevation drawing, sections through an exterior wall, and a set of performance or preliminary specifications—a conceptual estimate can be assembled. The estimator will be able to refine costs beyond the order-of-magnitude level and even provide the design consultants with costs of alternative systems or components so that they can weigh the costs of one scheme against those of another.

Level 3: Design Development

Design development is accompanied by drawings that are at least 25 percent complete and include a typical floor plan; a typical wall section; a finish schedule that contains quality levels of doors, frames, and hardware; some one-line diagrams showing mechanical/electrical/plumbing risers; and some distribution networks. This allows the estimator to further tighten the tolerances of the design and cost. This more detailed project definition and more refined costs will allow the design consultants, along with the estimator, to look at value-engineering alternatives so that the owner can determine that the design is tracking in the right direction and will fit the budget or, conversely, that the project cannot meet the owner's budget and that further design must cease, at least for the present. But if the design and accompanying estimate are in line with the proposed budget, design will probably proceed to the next level, which is the preparation of construction documents.

Level 4: Construction Documents

At this stage, the project owner feels comfortable enough that the project can be designed within the budget, so the design consultants are given the authority to complete the construction documents—the plans and specifications. The estimator's involvement may still be required if this is not a design-bid-build project. Many project managers have experienced what can be referred to as design creep: As the design is being refined, the architect may decide that a coffered ceiling in the lobby will enhance the prestige of the building, but in the process adds another $15,000 to the job. The mechanical/electrical/plumbing engineers may decide that redundancy in a power source is necessary, and why did they not think about that before? There goes another $50,000 add. There are also constructability issues that may creep into the design that were not anticipated previously in the estimate.

Level 5: Hard Bid Submission

Before the completed drawings are subjected to competitive bidding, if the owner has not had some professional estimator tracking the change from level 4 to level 5, it would be wise to have an estimate prepared to determine if the bids will be in the ballpark or if some serious scope or money must be taken out of the project. Quite often the architect's estimate, obtained from his or her own database or consultation with a friendly builder, is not on the mark. The architect may not have had access to an estimating service or the detailed and current cost information available to a general contractor who deals in these costs on a daily basis.

Many owners of private projects see the value of having an estimator on board during the process from level 1 to level 4. That was one of the reasons the CM concept gained such popularity, especially on very large projects. By hiring the CM during the design stage, the owner has the advantage of having an estimator tracking costs as design travels that path, thereby avoiding any embarrassing cost overruns that lead to more costs to redesign and rebid.

And of course, an owner who has decided to negotiate a project with a contractor will have access to the contractor's estimating department as the plans and specifications develop. There are also professional estimating services that are available to project owners for confirming costs as their design consultants work through design development. All of these costs are well spent.

Percentage of Accuracy

The use of percentage of accuracy is perhaps the most misunderstood concept in the field of cost estimating. The figure does not involve any contingencies included in or added to the estimate; rather, it is an indication of the probability of overrunning or underrunning the estimated cost. The estimate should already have included a contingency allowance to cover unexpected errors or omissions.

Because of the various additions that may have been made as the degree of completion of the design work increases, the percentage of accuracy becomes progressively more refined. The increase in accuracy, obviously, is due to the improvement in project definition and to more refined pricing data that can now be applied by the estimator. The lower percentage-of-accuracy numbers show that the probability of overrunning the estimate is becoming higher.

Figure 5.5 is a graphic representation of the classes of estimates and where they occur in the life cycle of the project. Although this particular diagram pertains to a process-engineering project, the information in it would also be true of vertical-construction projects. Please remember that the nomenclature, the timing of the estimates, and the duration of the project are highly variable in different project

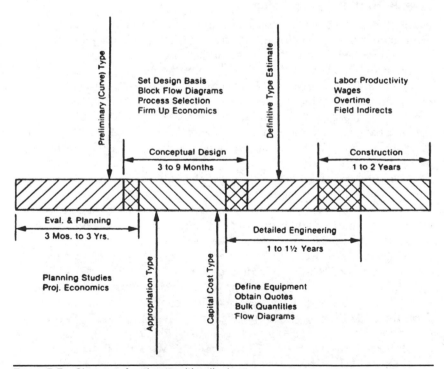

Figure 5.5   Classes of estimates (timeline).

environments. You will have to tailor the diagram to your particular construction-project environment.

In general terms, the percentages of accuracy for the various levels of an estimate are:

- Level 1: plus or minus 25 to 30 percent.

- Level 2: plus or minus 15 to 25 percent.

- Level 3: plus or minus 10 to 15 percent.

- Level 4: plus or minus 5 percent.

- Level 5: The estimate required for submission as a competitive bid should be right on the money.

Appropriation Estimates

As we move to the right along the timeline in Fig. 5.5, more detailed project definition becomes available. This permits some refinement of our estimating procedures. Because we are going for board approval of a capital-project request, our estimate had better be accurate. A specific site may have already been selected, or at least be under consideration, and preliminary plot plans are being developed. If the project is a process-type construction project, flow diagrams, major equipment specifications, building sizes, and overall layouts will be available. Process-flow diagrams with preliminary heat and material balances will have been developed, thereby permitting the preliminary sizing and specifications of major process-equipment items.

If the project is a commercial or institutional building, heat and cooling loads are being established and equipment requirements are being developed in order to determine the electrical requirements and switchgear sizing.

The appropriation estimate is considered a budget type of estimate because it contains a breakdown of several major project-cost elements. For example, it may have been made from separate cost estimates for design, site development, foundation, structural, electrical, and mechanical systems to arrive at the total figure. It is usually the first estimate with enough detail to use the applicable CSI code of accounts to break down and organize the various cost categories. A well-organized estimate can be converted into a preliminary project budget. The design team also uses that budget to control cost factors for subsequent phases of the detailed design.

Cost Trending for Large Projects

Cost trending is a procedure that has been used over the years to predict how project costs will move during the time period between estimates. On larger projects, the project team is likely to find itself with blind spots when predicting project costs during the design-development stage. The project scope may undergo a series of growth changes during the period between the appropriations estimate and the capital-cost estimate. That can lead to financial shock when a later estimate turns out to have grown beyond the feasibility cost for the project.

That rude financial shock can be prevented by using a cost-trending procedure. This involves a group of project and cost engineers who set up and monitor the cost of any design changes that may occur during the design-development stage. The trending group issues a biweekly or monthly cost-trend report which in effect updates the latest estimated project cost to the current design basis. This relatively inexpensive system gives the estimator early warning of any adverse cost trends while there is still time to act on them. It also permits the owner to be more confident about the project-cost forecast when responding to management's inquiries between estimates.

The series of estimates and the cost-trending procedures give owners a set of control points at which they can recheck the financial feasibility of the project while it is still low on the project-resources commitment curve. With design costs running generally from 6 to 12 percent of the total project cost, the owner can cancel the project or scale down its scope before passing the financial point of no return. That can be done before equipment orders are placed or the start of construction is considered. When a project is canceled after the commitment of equipment and construction funding, however, substantial cancellation costs are likely to be incurred.

Phased release of funding based on these key estimating milestones is fairly common in large capital projects. The need for this type of controlled project release is often set out in the contract documents and should be covered in the field-procedure manual.

Estimating the Cost Estimate-Effectiveness Concerns

The cost of performing a project estimate is generally directly proportional to the degree of accuracy produced by the designer. As pointed out in our previous discussions, a project may require several estimates if the desired control of the project costs is to be maintained. Therefore, making those detailed estimates can be a significant cost factor in the home office's construction-services budget.

In the case of a construction firm that normally bids its work lump-sum on fully designed projects, the estimating department is the lifeblood of the organization. Much of that work may be on public projects where competitive bids are required by law. A strong base of historical costs and a cadre of subcontractors with whom the firm has dealt fairly in the past are the two critical components of the estimating department's success when competing in the public sector.

The Prime Elements of an Estimate

Based upon an estimate's being prepared for design documents that are 100 percent complete—in other words, bid documents—the important elements of an estimate can be broken down as follows:

1. Quantity takeoff. Particularly if the contractor is to self-perform some work, a takeoff of the quantities of material will also provide the basis for the labor hours per unit of installation. For example, if the contractor generally self-performs partition framing, drywall installation, taping, and sanding, the quantities as shown on the drawings will be reviewed with any information in the specifications pertaining to gauge of metal studs, drywall specifications, and number of coats of taping depending upon whether the area is open to public view or is concealed or inside utility closets. This contractor will have had a history of daily labor reports reflecting hours and quantities for each one of these operations that can now be applied to quantity takeoff.

   Even if portions of the work are to be subcontracted, quantity takeoffs can be very helpful in comparing the estimator's concept of cost per unit with the subcontractor's or even in checking the accuracy of the subcontractor's takeoff, which may be overly optimistic.

2. Labor hours and rates. These will vary considerably depending on whether the contractor is a union shop or an open shop using both union and nonunion labor on the project. Not only will the basic wage rate differ, but labor burdens will vary considerably between union and nonunion labor. Beside the normal burden of FICA (Social Security) and both state and federal unemployment insurance and workers' compensation costs, the basic hourly wage for union workers can be driven up as much as 100 percent when travel allowance, education funds, holiday and vacation costs, and other fringe benefits are added.

3. Material prices. Depending upon market conditions (supply and demand), some material costs can fluctuate weekly or monthly. During the period of the housing boom in the early part of this century, drywall was scarce and prices fluctuated weekly. Material prices can be also be affected by:

   • Peak or slack time of the year for the manufacturer

   • The size of the order and the size of each delivery, if a blanket order has been placed

   • Physical requirements for the delivery, including site access

- Past payment history with the supplier (slow payers may find the cost of materials slightly higher)
- Single-source suppliers whose pricing evidences the lack of competition
- Variations in the exchange rate, for imported materials

There is another factor to consider when dealing with subcontractors employing union labor: Be aware of the expiration date of the labor agreement, since either wages or fringe benefits may increase afterward and should be considered in the cost of labor if the project extends into this new labor period.

4. Equipment costs. Depending upon the size, the number of units purchased, the delivery schedule (all units delivered in one shipment or in several deliveries), ease of access to the site, equipment costs will vary. Distance from the manufacturer or distributor and a requirement to unload at the direction of the project superintendent may also impact the purchase price.

5. Subcontractor quotes. The day of the master builder who self-performed many if not all trades, from site excavation to mechanical and electrical work, is a thing of the past. Subcontractors have better pricing, due to lower repetitive production costs, and are available to be on-site as demanded. Developing good relations with subcontractors will, in many instances, provide you a more competitive price than other general contractors who may have poor relations with subcontractors—not paying on time, not running projects efficiently and thus not allowing the subcontractors to operate productively, and in general the subcontractors a rough time.

6. Indirect costs. Costs to properly and efficiently supervise the project are generally included in the general-conditions portion of the estimate—field-office rental and supplies, cost of temporary utilities, communications and computer operations, and so forth.

The Role of the Project Manager (PM) or CM in Making a Detailed Estimate

Depending upon the size of the company and the project, the role played by the PM or CM can vary from being an estimate coordinator to leading the overall estimating effort. How the assignment is handled depends largely on the stature and leadership exhibited by the firm's estimating department and the degree of authority with which the PM or CM has been endowed.

We have already cited some of the roles the PM or CM plays in preparing the estimate. The following checklist of estimating activities is a minimum:

- Make an in-depth review of the bidding documents and contract requirements.
- Participate in the pre-estimate strategy meeting.
- Lead the construction-site survey team and the preparation of a report of the site visit.
- Attend site visitations offered by the owner's design team, management meetings, and client meetings.
- Review the estimating plan with the lead estimator.
- Establish the site's labor posture with top management (union, merit shop, or nonunion).
- Develop the scope of indirect services and costs for the field.
- Develop the most efficient construction-technology plan.
- Evaluate heavy-equipment and small-tools requirements.
- Develop a contracting plan with upper management.
- Suggest construction-cost reductions where possible.
- Evaluate the proposed construction schedule.
- Evaluate the project's regulatory requirements and associated cost impacts.
- Develop the field organization and chart.
- Prepare a project initiation and execution plan.
- Assist in developing the supplier or subcontractor sourcing plan and negotiating the purchase orders or subcontract agreements.
- Respond to information and decision requests from the estimator.
- Periodically review estimate progress and methods.
- Make an in-depth review of the final estimate.
- Participate in the final pricing meeting.
- Attend the bid opening, if required, and take notes in case there is justification to disqualify a bidder.

This is an imposing list of activities for the PM or CM on a project of any size, either as the estimate leader or only as a contributor. On larger projects, the PM or CM will need to delegate some of the duties to others if they are to be covered in sufficient depth. The list focuses on the fact that the PM or CM's management is the chief representative in the preparation of a sound, business-oriented construction estimate. If input from management is missing, the estimate may miss its intended goal—that of procuring a new project.

Some of the items on the list are self-explanatory, while others warrant a few lines of explanation. A prime example of the latter is the PM or CM becoming thoroughly expert in the ramifications and requirements of the bidding documents and the pro forma contract so as to be able to advise upper management in the pre-estimate meetings.

Bidding-document packages on large projects can be in the range of 100 or more drawings and up to 12 volumes of specifications. All of this information needs to be digested in order to determine which provisions may add costs to the project that might not be evident from a quick once-over of these documents.

The author experienced a good example of the result of not thoroughly reviewing the specification requirements that will illustrate the point. He was involved in a commercial office project with 290,000 square feet of concrete floor slabs. A top-notch concrete subcontractor was engaged and the floor slabs looked great. The owner, facing a $35,000 extra for an upgrade in the finish hardware, came to the site and agreed that the slab work was very good, but he said, "You know, the specs call for an *F* [flatness factor] equivalent to plus or minus one eighth of an inch flatness in 10 feet." Neither the writer nor the subcontractor had really read the concrete specs thoroughly. When we placed a level, the flatness was one quarter inch in 10 feet—pretty darn good, but it did not meet the spec. The owner said, "I guess you'll have to rip out all the slabs, but that seems to be rather harsh. You know about that extra for hardware?" We, of course, withdrew the $35,000 extra and all learned a valuable lesson: Read the specifications carefully!

Attendance at any prebid meetings or site meetings will provide the PM or CM with not only technical knowledge but also familiarity with the owner and designer's team.

Construction-Site Survey and Report

If the new project is located on a greenfield site or the site of a previous building, a geotechnical report will most likely be included in the bid documents. If the civil engineer had no previous knowledge of subsurface conditions or prior buildings on the site, the geotechnical report can be viewed as his or her best judgment as to what site conditions may likely be encountered by the bidders.

At least initially, the greatest potential for risk will be associated with the site work and site preparation, unless sufficient caveats are included in the bidding documents and the bidder does sufficient analysis of the geotechnical report and his or her own investigation of the site, within limits of time, costs, and access provided by the owner. At times, the owner will allow a bidder to dig a few test holes to gain more knowledge of subsurface conditions. If this is allowed, such holes should be backfilled promptly in case a competitor decides to visit the site shortly thereafter.

What Is a Geotechnical Report?

The geotechnical report, prepared for the owner by the owner's civil engineer and provided to the bidding contractors, is a method of communicating the design and construction recommendations to those bidders.

Test borings are taken in limited areas around the site to provide the bidders with an indication of conditions they may find while excavating the site for substructures and underground utilities. But contractor beware. Many claims involving differing site conditions arise when the contractor, say, finds an absence of rock at test boring number 5 but, when excavating close to that area, discovers a great deal of rock and thus requests an extra to remove the rock. Many court decisions have taken the position that the soil conditions at a given test boring represent the soil conditions at that exact location, not ones that might exist 10 feet away.

A typical log of a test boring is shown in Fig. 5.6, revealing the soil conditions at the various depths of this boring B-1 and a brief description of the soil conditions and the type of soil per the Unified Soil Classification System. In this particular boring, no groundwater was encountered in the entire 35-foot depth of the boring.

While the geotechnical report may vary as to its size and the depth of the information provided, all such reports should contain the same basic information:

- A summary of all subsurface exploration data, including a subsurface soil profile, exploration logs, or in situ test results
- Groundwater information—if present, at what depth
- Interpretation and analysis of the subsurface data
- Specific engineering recommendations for design (most likely more detailed in the report to the design's structural engineer)
- Discussion of conditions for solution of anticipated problems
- Recommended geotechnical special provisions
- A disclaimer that no investigations claim to be representative of the entire site

Table 5.1 sets forth guidelines prepared by the U.S. Department of Transportation for the number and depth of test borings appropriate

LOG OF BORING No. B-1

CLIENT:		DATE: 6-22-99	#02995604	RIG: CME 75
SITE:		PROJECT:		

GRAPHIC LOG	DESCRIPTION	DEPTH ft.	USCS SYMBOL	NUMBER	TYPE	RECOVERY in.	SPT-N BLOWS/ft	WATER CONTENT %	DRY UNIT WT pcf	UNCONFINED STRENGTH qu psf	ATTERBERG LIMITS LL, PL, PI
	6" GRAVEL LEAN CLAY, silty trace organics, gray brown, trace dark brown and red brown, medium (Possible Fill)				PA						
			CL	1	SS	14	7	34.1		2000*	45,21,34
		5			HS						
	LEAN CLAY, calcareous, trace sand and limestone gravel dark brown, brown, very stiff (Possible Fill)										
		10	CL	2	SS	6	5	18.6		7000*	45,23,22
					HS						
		15	CL	3	SS	24	9	24.1		5500*	
					HS						
	LEAN CLAY, trace silt, gray brown, trace dark gray, red brown and dark brown, stiff to very stiff										
		20	CL	4	SS	24	10	22.3		3500*	44,20,24
					HS						
		25	CL	5	SS	24	5	27.6		2500*	
					HS						
	LEAN CLAY, silty, gray brown, trace dark brown, stiff to very stiff										
		30	CL	6	SS	24	19	26.5		5000*	42,18,24
					HS						
	Trace limonites at 34.0' LEAN TO FAT CLAY, gray brown, trace dark brown, very stiff	35	CL-CH	7	SS	24	14	23.5		5000*	
					HS						

FIGURE 5.6   A typical test-boring log.

for various types of subsurface structures and the measures and tests that are to be taken to accompany these test borings and subsurface explorations.

Table 5.2 reveals the classifications of various types of soil under the Unified Soil Classification System (Unified) and the American Association of State Highway and Transportation Officials (AASHTO) system.

The most important step in geotechnical design is to conduct an adequate subsurface investigation. The number, depth, spacing, and character of borings, sampling, and testing to be made in an individual exploration program are so dependent upon site conditions and the type of project and its requirements, that no "rigid" rules may be established. Usually the extent of work is established as the site investigation progresses in the field. However, the following are considered reasonable "guidelines" to follow to produce the minimum subsurface data needed to allow cost-effective geotechnical design and construction and to minimize claim problems. (Reference: "Subsurface Investigations" FHWA HI-97-021)

Geotechnical Feature	Minimum Number of Borings	Minimum Depth of Borings
Structure Foundation	1 per substructure unit under 30 m (100 ft) in width 2 per substructure unit over 30 m (100 ft) in width Additional borings in areas of erratic subsurface conditions	Spread footings: 2B where L< 2B, 4B where L > 2B and interpolate for L between 2B and 4B Deep foundations: 6 m (20 ft) below tip elevation or two times maximum pile group dimension, whichever is greater. If bedrock is encountered: for piles core 3 m (10 ft) below tip elevation; for shafts core 3D or 2 times maximum shaft group dimension below tip elevation, whichever is greater.
Retaining Structures	Borings spaced every 30 to 60 m (100 to 200 ft). Some borings should be at the front of and some in back of the wall face.	Extend borings to depth of 0.75 to 1.5 times wall height When stratum indicates potential deep stability or settlement problem, extend borings to hard stratum
Bridge Approach Embankments over Soft Ground	When approach embankments are to be placed over soft ground, at least one boring should be made at each embankment to determine the problems associated with stability and settlement of the embankment. Typically, test borings taken for the approach embankments are located at the proposed abutment locations to serve a dual function.	Extend borings into competent material and to a depth where added stresses due to embankment load is less than 10% of existing effective overburden stress or 3 m (10 ft) into bedrock if encountered at a shallower depth Additional shallow explorations (hand auger holes) taken at approach embankment locations to determine depth and extent of unsuitable surface soils or topsoil.

TABLE 5.1 Guidelines for Minimum Boring Sampling and Test Criteria

131

Geotechnical Feature	Minimum Number of Borings	Minimum Depth of Borings
Centerline Cuts and Embankments	Borings typically spaced every 60 m (200 ft) (erratic conditions) to 120 m (400 ft) (uniform conditions) with at least one boring taken in each separate landform. For high cuts and fills, should have a minimum of 3 borings along a line perpendicular to centerline or planned slope face to establish geologic cross-section for analysis.	Cuts: (1) in stable materials extend borings minimum 5 m (15 ft) below depth of cut at the ditch line and, (2) in weak soils extend borings below grade to firm materials or to twice the depth of cut whichever occurs first. Embankments: Extend borings to a hard stratum or to a depth of twice the embankment height.
Landslides	Minimum 3 borings along a line perpendicular to centerline or planned slope face to establish geologic cross-section for analysis. Number of sections depends on extent of stability problem. For active slide, place at least on boring each above and below sliding area	Extend borings to an elevation below active or potential failure surface and into hard stratum, or to a depth for which failure is unlikely because of geometry of cross-section. Slope inclinometers used to locate the depth of an active slide must extend below base of slide.
Ground Improvement Techniques	Varies widely depending in the ground improvement technique(s) being employed. For more information see "Ground Improvement Technical Summaries" FHWA SA-98-086R.	
Material Sites (Borrow sources, Quarries)	Borings spaced every 30 to 60 m (100 to 200 ft).	Extend exploration to base of deposit or to depth required to provide needed quantity.
Sand or Gravel Soils		
SPT (split-spoon) samples should be taken at 1.5 m (5 ft) intervals or at significant changes in soil strata. Continuous SPT samples are recommended in the top 4.5 m (15 ft) of borings made at locations where spread footings may be placed in natural soils. SPT jar or bag samples should be sent to lab for classification testing and verification of field visual soil identification.		
Silt or Clay Soils		
SPT and "undisturbed" thin wall tube samples should be taken at 1.5 m (5 ft) intervals or at significant changes in strata. Take alternate SPT and tube samples in same boring or take tube samples in separate undisturbed boring. Tube samples should be sent to lab to allow consolidation testing (for settlement analysis) and strength testing (for slope stability and foundation bearing capacity analysis). Field vane shear testing is also recommended to obtain in-place shear strength of soft clays, silts and well-rotted peat.		

Rock

Continuous cores should be obtained in rock or shales using double or triple tube core barrels. In structural foundation investigations, core a minimum of 3 m (10 ft) into rock to insure it is bedrock and not a boulder. Core samples should be sent to the lab for possible strength testing (unconfined compression) if for foundation investigation. Percent core recovery and RQD value should be determined in field or lab for each core run and recorded on boring log.

Groundwater

Water level encountered during drilling, at completion of boring, and at 24 hours after completion of boring should be recorded on boring log. In low permeability soils such as silts and clays, a false indication of the water level may be obtained when water is used for drilling fluid and adequate time is not permitted after boring completion for the water level to stabilize (more than one week may be required). In such soils a plastic pipe water observation well should be installed to allow monitoring of the water level over a period of time. Seasonal fluctuations of water table should be determined where fluctuation will have significant impact on design or construction (e.g., borrow source, footing excavation, excavation at toe of landslide, etc.). Artesian pressure and seepage zones, if encountered, should also be noted on the boring log. In landslide investigations, slope inclinometer casings can also serve as water observations wells by using "leaky" couplings (either normal aluminum couplings or PVC couplings with small holes drilled through them) and pea gravel backfill. The top 0.3 m (1 ft) or so of the annular space between water observation well pipes and borehole wall should be backfilled with grout, bentonite, or sand-cement mixture to prevent surface water inflow which can cause erroneous groundwater level readings.

Soil Borrow Sources

Exploration equipment that will allow direct observation and sampling of the subsurface soil layers is most desirable for material site investigations. Such equipment that can consist of backhoes, dozers, or large diameter augers, is preferred for exploration above the water table. Below the water table, SPT borings can be used. SPT samples should be taken at 1.5 m (5 ft) intervals or at significant changes in strata. Samples should be sent to lab for classification testing to verify field visual identification. Groundwater level should be recorded. Observations wells should be installed to monitor water levels where significant seasonal fluctuation is anticipated.

Quarry Sites

Rock coring should be used to explore new quarry sites. Use of double or triple tube core barrels is recommended to maximize core recovery. For riprap source, spacing of fractures should be carefully measured to allow assessment of rock sizes that can be produced by blasting. For aggregate source, the amount and type of joint infilling should be carefully noted. If assessment is made on the basis of an existing quarry site face, it may be necessary to core or use geophysical techniques to verify that nature of rock does not change behind the face or at depth. Core samples should be sent to lab for quality tests to determine suitability for riprap or aggregate.

TABLE 5.1 Guidelines for Minimum Boring Sampling and Test Criteria (*Continued*)

Soil Classification			Embankment and Cut Slopes		Structure Foundations (Bridges and Retaining Structures)		Retaining Structures (Conventional, Crib and MSE)	
Unified	AASHTO[1]	Soil Type	Slope Stability[2] Analysis	Settlement Analysis	Bearing Capacity Analysis	Settlement Analysis	Lateral Earth Pressure	Stability Analysis
GW	A-1-a	GRAVEL Well-graded	Generally not required if cut or fill slope is 1.5H to 1V or flatter, and underdrains are used to draw down the water table in a cut slope.	Generally not required except possibly for SC soils.	Required for spread footings, pile or drilled shaft foundations.	Generally not needed except for SC soils or for large, heavy structures.	GW, SP, SW & SP soils generally suitable for backfill behind or in retaining or reinforced soil walls.	All walls should be designed to provide minimum F.S. = 2 against overturning & F.S. = 1.5 against sliding along base. External slope stability considerations same as previously given for cut slopes & embankments.
GP	A-1-a	GRAVEL Poorly-graded						
GM	A-1-b	GRAVEL Silty						
GC	A-2-6 A-2-7	GRAVEL Clayey			Spread footings generally adequate except possibly for SC soils	Empirical correlations with SPT values usually used to estimate settlement		
SW	A-1-b	SAND Well-graded	Erosion of slopes may be a problem for SW or SM soils.				GM,GC, SM & SC soils generally suitable if have less than 15% fines. Lateral earth pressure analysis required using soil angle of internal friction.	
SP	A-3	SAND Poorly-graded						
SM	A-2-4 A-2-5	SAND Silty						
SC	A-2-6 A-2-7	SAND Clayey						

ML	A-4	SILT Inorganic silt Sandy	Required unless non-plastic. Erosion of slopes may be a problem.	Required unless non-plastic.	Required. Spread footing generally adequate.	Required. Can use SPT values if non-plastic.	These soils are not recommended for use directly behind or in retaining or reinforced soil walls.
CL	A-6	CLAY Inorganic Lean Clay	Required	Required			
OL	A-4	SILT Organic	Required	Required			

[1]This is an approximate correlation to Unified (Unified Soil Classification system is preferred for geotechnical engineering usage, AASHTO system was developed for rating pavement subgrades).

[2]These are general guidelines, detailed slope stability analysis may not be required where past experience in area is similar or rock gives required slope angles.

TABLE 5.2 Unified Soil Classification System

Unclassified Site Work

When the bid documents indicate that the site work is designated as "unclassified," this basically means that the contractor assumes all risks for whatever appears in the subsurface conditions. Because it is impossible to conduct an investigation of underground conditions over the entire site, this places a great deal of risk on the contractor; how much or how little the builder includes in its estimate to compensate for the unknown will depend upon how badly the builder would like to obtain the award. Whatever the contractor finds in the way of unsuitable soils, rock, debris, or trash, he or she is obliged to remove the unsuitable material and replace it with material approved by the architect or engineer. This would apply to the following site-work operations:

- Bulk excavation for the building foundations and pad, parking lots, access roads, sidewalks, and any site-improvement structures

- Trench excavation and backfilling with acceptable material for the installation of all underground site utilities

- Off-site disposal of unsuitable materials uncovered during these operations

- Purchase of off-site material acceptable to the architect or engineer to replace the unsuitable materials hauled off-site

When you are dealing with unclassified sites, the relationship you have established with the owner and the owner's design consultants can often produce a reasonable solution to problems created by this "you own it all" situation. We will describe a situation where this did occur.

We had a contract with a owner that had developed large projects not only in the United States but in Europe and Asia as well. The owner's representative was an experienced, hard-nosed individual who played strictly by the book. During contract negotiations we got to know each other a little better, and I thought I saw a bit of reasonableness when we dealt with some contentious provisions of the contract that we did our best to settle as amicably as possible. We could not change the unclassified-site-work provision.

We commenced site work on this building with its 250,000-square-foot footprint and everything was going well until we hit one area while excavating for footings that exposed soils with poor bearing capacity. We continued to excavate, and when we were 15 feet below the design depth and still had only marginally acceptable soils, we requested a special meeting to discuss this problem.

We said, "Well, we guess you are telling us that we have to dig all the way down to China to find soils with proper bearing capacity, is that right?" The owner's representative gave us a cold stare and silence for what seemed to be an hour and finally said, "No, I guess that's not reasonable." So we came to a mutual agreement as to the additional depth when we could stop excavating and then come to an understanding of reasonable extra costs to provide proper bearing capacity for the footing.

Developing a collaborative environment, no matter what type of contract you are administering, can only pay off, in many ways; attacking problems with a willingness to give-and-take will also pay off.

Labor Posture and Contracting Plan

These two interrelated items are usually developed with the contractor's upper management. The labor posture refers to the decision as to whether the project will be built with union or open-shop labor. Usually the contractor works one way or the other, but some firms offer both union and nonunion operations, albeit under different company names. This process is known as having double-breasted operations.

Geographic location plays a big part in the decision to operate a specific project as a union or nonunion (open-shop) operation. Major cities usually have strong union bases, and contractors will conduct their projects by engaging union subcontractors. In smaller cities, the open-shop or merit-shop operation may be a viable option.

The contracting plan is part of the overall project initiation and execution plan developed by the contractor. This "contracting plan" will have a strong bearing on how the estimate is prepared, so it must be formulated and approved before estimating starts. It will be a key part of the pre-estimate meeting and play an important role in the bid/no-bid decisions. The contractor formulates preliminary execution plans right after the bidding documents have been assimilated. The project-execution plan may even be finalized and approved as part of the pre-estimate meeting.

When reviewing the bidding package, the contractor is automatically evaluating the best available construction technologies for building the project at minimum cost. If it is necessary to sell the proposed methods to management, the ideas must be well researched and presented by the PM or CM to get them incorporated into the project-execution plans. Looking at areas of improving labor productivity is fertile ground for that type of idea.

Project regulatory requirements are a vital area of interest for PMs, CMs, and estimators. These regulatory requirements have grown to be a major area of cost over recent years, as well as contributing to lower overall efficiency of the construction process. A good example is erosion control and requirements to keep any runoff from leaving the site. Their effects must be investigated early in the estimating process and their cost effect reduced as much as possible.

Areas in the temporary roadway leading into the jobsite will need to have antitracking materials placed at the entrance to the site so that dirt from the wheels of the trucks entering the project is not deposited on the public roads outside the building site.

What Works to Improve Efficiency?

We discuss efficiency and productivity in more detail in Ch. 8, referring to the construction-management consulting firm FMI. In one of their surveys, respondents were asked "What have you done to improve your company's efficiency?" They gave the following answers:

- Issue weekly status reports with budget incentives to field personnel to provide motivation to beat the labor budgets.
- Improving communication (presumably between field and office).
- Explaining project expectations prior to the start of the job coupled with regular inspections.
- Cost coding and productivity monitoring.
- Establishing a baseline productivity standard and then measuring changes against it.
- Establishing key performance indicators and conducting weekly meetings with field personnel to review job costs and productivity data.
- Forming building teams that work well together.
- Planning those activities that are to be measured, establishing goals, communicating those goals, and measuring feedback in a timely fashion.
- Increase training of project managers and field managers.

Estimating Indirect Costs in the Field

A significant factor in the overall construction estimate are the indirect costs in the field, which are known as field overhead costs for running the project and are generally included in the general-conditions division of the estimate. Some major items of this account are:

- Temporary site improvement
- Field supervisory-staff salaries, burdens, and fringes
- Field-office equipment and supplies
- Site utility systems and costs

- Site safety and security costs
- Materials management and storage requirements
- Vehicles and construction equipment (forklifts, etc.)
- Small tools, supplies, and consumables
- Staff relocation and living allowances if the project is far enough away from the home office
- Temporary communications and cost of service
- Temporary buildings, roads, and their costs to maintain
- Laboratory, field-testing, and inspection costs
- Temporary heating, cooling, dewatering, and weather-related costs—labor and materials
- Government agency compliance-, OSHA, EPA, to name a few

Even this list would not be all-inclusive for a very large construction project, so it is easy to see that the indirect costs in the field are a substantial contributor to the overall construction costs. Effectively predicting these costs has a substantial effect on the estimate and on the total cost for constructing the project. This is especially true in the area of competitive fixed price bidding.

In a small construction firm, organizing and estimating the cost of these items could fall to the construction manager. In a firm of any size, the PM or CM must supply the field-personnel structure, the extent of the physical facilities, and the scope or cost of field operations may fall to the estimators if effective pricing in the estimate is to be achieved.

Construction companies operating in a given market environment often develop a percentage of total project cost for estimating the indirect costs in the field. If good job-cost records are maintained for a number of projects, it is possible to develop a reliable percentage figure. This can at least be used as a check number to evaluate the efficacy of a detailed estimate of indirect costs in the field.

Temporary Site Facilities

The size and complexity of the temporary site facilities vary greatly from site to site. For small modification jobs, temporary site facilities are usually minimal because existing facilities and utilities are available to satisfy most of the field operating needs.

For large greenfield projects, temporary site facilities such as field offices, warehouses, roads, drainage, fencing, and the like can be quite extensive and costly. The best solution to that cost problem is to schedule the work so that the applicable new facilities can be used as temporary facilities under construction. This is especially true for site improvements, utilities, office space, and materials storage.

On large, complex projects involving extensive temporary facilities, a sound management approach from the construction-management staff is required. The layout of temporary facilities in relation to the work must be studied to maximize productivity and to reduce the cost of the work site. It is important that any temporary structure not be located in an area on the site where it has to be moved to accommodate the new construction, roads, or underground utilities.

Requirements of a Good Estimate

We can summarize the discussion of project-cost estimating with a checklist of the basic requirements for a good estimate:

- A thorough and detailed review of the bid documents and accompanying plans and specifications
- A realistic project-execution plan
- Good estimating methods and accurate data
- Neatly documented detail
- A reasonable estimating budget with appropriate contingencies applied
- A knowledgeable and experienced estimator

Budget Format

Budget formats can vary widely depending upon company standards. A few things, however, are required to make any budget work. A typical summary of detailed account breakdowns and their contribution to the overall budget will look something like this:

1. Home-office services (3 to 5 percent)
   a. Procurement services
   b. Support-services salaries and expenses
   c. Project accounting
2. General construction trades (65 to 70 percent)
   a. Site development
   b. Subsurface structures (foundations)
   c. Structural framing
   d. Architectural trades
3. Mechanical trades (10 to 15 percent)
   a. Plumbing
   b. Heating, ventilating, air-conditioning (HVAC)
   c. Fire protection

4. Electrical trades (10 to 15 percent)

   a. Power supply

   b. Power distribution

   c. Lighting

   d. Communications and security

5. Field indirect costs (5 to 7 percent)

6. Contingency

7. Overhead and profit added to these total costs

The percentages shown in the parentheses are order-of-magnitude numbers provided only to indicate the relative portion of the whole project cost that each section might represent. The percentages will vary somewhat from those listed depending upon the type of project being built. Obviously a project with considerable electrical requirements or stringent HVAC loads and tolerances will skew these percentages.

The budget format will be converted into another form: the monthly request for payment. This may combine several cost-estimate breakdowns into one figure. The common consolidation is the general-conditions costs, which are billed as a total and requisitioned on the basis of monthly payments derived by dividing the total by the number of months the project is expected to take from beginning to completion. For example, a general-conditions cost which will incorporate home- and field-office expenses totaling $1,000,000 for a project scheduled for 10 months will bill not as a percentage complete but as the same amount each month, in this case $100,000.

Another consolidation of costs is site work and site utilities, lumped together in one dollar amount in the request for payment. This and all other cost items will be billed on a percentage-completion basis each month.

Front-End Loading

Many contractors will add disproportionate amounts to those activities that occur early in the project. For example, if the actual site-work costs are $125,000 to be dispensed as a percentage complete each month, the contractor may increase that cost to $150,000 so that he or she can actually receive more in payment than the actual amount of work performed in the early stages of that operation. There are any number of reasons why some contractors do this, but none are really justified.

Summary

Accurate estimating is the lifeblood of any construction company. The objective of any estimate is to assure that all costs are included, all bases are covered, and the company will be successful in being awarded a contract for the project at hand. Accurate estimating depends upon accumulating a database of costs to fill various needs; an order-of-magnitude request from a client; assistance to an architect or owner pursuing the viability of a project by requiring design-development costs; and on-the-money cost estimates for competitive hard-bid projects.

The estimating department needs to work hand in hand with project management and field personnel who can add their knowledge and experience to the estimator's database. When all of these protocols, relationships, and just plan hard work meld together, successful estimates are most likely to emerge.

CHAPTER **6**

Project Resources Planning

U p to this point we have prepared a project-execution plan, a time plan, and a money plan. Now we are ready to turn to planning the human and physical resources needed to implement the execution of the construction project.

There are seven key project resources that require early evaluation and forward planning on the part of the construction manager (CM) or project manager (PM). Simpler projects may involve only some of these areas, whereas complex projects may involve them all. The seven key areas are:

- Human resources
- Engineered equipment and materials
- On-site facilities
- Construction equipment
- Project services and systems
- Transportation arrangements
- Project financing

Obviously, project financing and funding availability will have a major influence on performance early in the project since procurement of all of the other components depends upon having the resources to acquire them when needed.

Human-Resources Planning

At this stage we are discussing only the types and numbers of people required, not the way in which we will organize them into a team. Human resources for construction-project planning breaks down into three major categories:

- Home-office personnel
- Construction personnel (field supervisors and labor)
- Construction subcontractors and their supervisors

The personnel segment of the construction industry can best be described as fluid, mobile, and demographically changing. We are talking about a major industry segment that hires large numbers of talented and highly skilled people, largely on a temporary basis, to perform dangerous work geared to strict schedules on sites that are rarely the same. Some liken this to an outdoor factory producing a one-off product.

The necessity of effectively staffing the world's construction projects makes construction contractors' personnel departments a key to successful contracting.

The Changing Face of the Construction Workforce

The U.S. Bureau of Labor Statistics (BLS), in the 2010–2011 *Career Guide to Industries*, painted a detailed picture of construction-industry personnel that numbered 7,268,000 paid employees. There were four significant points:

- Job opportunities are expected to be good, especially for skilled and experience construction-trades workers.

- Workers have relatively high hourly wages.

- About 68 percent of establishments employ fewer than five people.

- Construction includes a very large number of self-employed workers.

The construction workforce is aging. It is estimated by the Association of General Contractors that by 2030, people 55 and older will make up 40 percent of the adult population. Recruiting and training new workers for the construction industry will help to alleviate the worker shortage that occurs as more and more older workers retire. Older workers tend to have fewer accidents, but their accidents are generally more severe, so employers need to look at changing their disability-management plans.

More members of minorities have been entering the construction workforce. According to the BLS, in 2010 the number of foreign-born workers rose while the number of U.S.-born workers declined. Latinos accounted for 49.9 percent of the foreign-born labor force that year and workers of Asian descent accounted for 21.8 percent.

This 2010 survey also revealed that 26.5 percent of the foreign-born labor force age 25 and over had not completed high school, as compared with 5.4 percent of U.S.-born workers. These figures regarding the aging population and the percentage of foreign-born workers lacking a high-school education require construction managers and general contractors to rethink their training programs in order to maintain an efficient workforce.

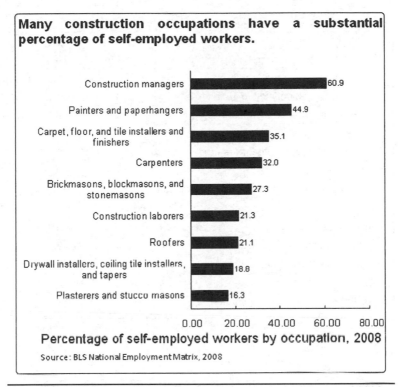

FIGURE 6.1   Percentage of self-employed workers by occupation as of 2008.

Other BLS figures provide another insight into the makeup of the industry:

Figure 6.1 shows percentages of self-employed workers by occupation as of 2008.

Table 6.1 shows percentages of workers by craft who were wage and salary workers as of 2008.

Table 6.2 shows employment of wage and salary workers in construction as of 2008 and projected percent change extended to 2018.

Table 6.3 shows median hourly wages of the largest occupations in construction as of May 2008.

Table 6.4 shows entry-level education and 2010 median pay for construction and extraction occupations.

Practical Personnel-Loading Curves

Figure 6.2 shows a theoretically ideal bell curve for personnel loading as a dashed line and the front- and back-loaded curves as solid lines.

Occupation	Percent
Insulation workers	91.7
Cement masons, concrete finishers, and terrazzo workers	89.4
Structural iron and steel workers	84.6
Drywall installers, ceiling tile installers, and tapers	80.2
Plasterers and stucco masons	79.9
Roofers	76.2
Pipelayers, plumbers, pipefitters, and steamfitters	71.6
Electricians	69.7
Brickmasons, blockmasons, and stonemasons	69.0
Glaziers	67.5
Carpenters	56.1
Carpet, floor, and tile installers and finishers	49.5
Painters and paperhangers	43.9

Source: BLS National Employment Matrix, 2008-18.

TABLE 6.1   Percentage of Wage and Salary Workers by Craft Occupation as of 2008

Occupation	Employment, 2008		Percent Change, 2008–18
	Number	Percent	
All occupations	7,214.9	100.0	18.5
Management, business, and financial occupations	571.4	7.9	21.6
General and operations managers	121.2	1.7	7.8
Construction managers	176.9	2.5	26.1
Cost estimators	128.0	1.8	32.6
Office and administrative support occupations	699.6	9.7	15.8
Bookkeeping, accounting, and auditing clerks	141.0	2.0	19.4
Executive secretaries and administrative assistants	75.9	1.1	17.5
Secretaries, except legal, medical, and executive	151.9	2.1	8.2
Office clerks, general	159.2	2.2	19.7

TABLE 6.2   Employment of Wage and Salary Workers in Construction as of 2008 and Projected Change to 2018

Occupation	Employment, 2008		Percent Change, 2008–18
	Number	Percent	
First-line supervisors/ managers of construction trades and extraction workers	442.1	6.1	22.7
Brickmasons, blockmasons, and stonemasons	110.5	1.5	14.3
Carpenters	721.0	10.0	15.2
Carpet, floor, and tile installers and finishers	79.4	1.1	13.3
Cement masons, concrete finishers, and terrazzo workers	184.7	2.6	13.9
Construction laborers	771.0	10.7	26.0
Construction equipment operators	297.5	4.1	18.1
Drywall installers, ceiling tile installers, and tapers	151.3	2.1	15.5
Electricians	484.0	6.7	15.3
Painters and paperhangers	197.6	2.7	8.2
Pipelayers, plumbers, pipefitters, and steamfitters	398.0	5.5	21.6
Roofers	113.5	1.6	6.5
Sheet metal workers	107.9	1.5	10.1
Helpers, construction trades	349.2	4.8	21.0
Installation, maintenance, and repair occupations	545.8	7.6	28.6
Heating, air conditioning, and refrigeration mechanics and installers	178.6	2.5	42.8
Line installers and repairers	83.5	1.2	21.4
Transportation and material moving occupations	251.8	3.5	12.6
Truck drivers, heavy and tractor-trailer	104.0	1.4	15.6

Note: Columns may not add to total due to omission of occupations with small employment.
Source: BLS National Employment Matrix, 2008–18.

TABLE 6.2 Employment of Wage and Salary Workers in Construction as of 2008 and Projected Change to 2018 (*Continued*)

Occupation	Construction of Buildings	Heavy and Civil Engineering Construction	Specialty Trade Contractors	All Industries
Construction managers	$37.45	$39.87	$38.34	$38.39
First-line supervisors/ managers of construction trades and extraction workers	28.49	28.10	27.49	27.95
Plumbers pipefitters, and steamfitters	22.83	21.27	21.78	21.94
Electricians	21.26	22.85	21.69	22. 32
Operating engineers and other construction equipment operators	20.48	20.02	18.98	18.88
Carpenters	19.17	19.42	18.50	18.72
Cement masons and concrete finishers	17.41	17.13	16.85	16.87
Painters, construction and maintenance	15.41	16.84	15.46	15.85
Construction laborers	14.35	14.29	13.57	13.71
Heating, air conditioning, and refrigeration mechanics and installers	**	18.54	18. 25	19.08

** Data not available.
Source: BLS Occupaional Employment Statistics, May 2008.

TABLE 6.3   Median Hourly Wages of Largest Occupations in Construction as of May 2008

U.S. Bureau of Labor Statistics			
Occupation	Job Summary	Entry-Level Education	2010 Median Pay
Boilermakers	Boilermakers assemble, install, and repair boilers, closed vats, and other large vessels or containers that hold liquids and gases.		$54,640
Brickmasons, Blockmasons, and Stonemasons	Brickmasons, blockmasons, and stonemasons (or, simply, masons) use bricks, concrete blocks, and natural stones to build fences, walkways, walls, and other structures.	High school diploma or equivalent	$45,410
Carpenters	Carpenters construct and repair building frameworks and structures—such as stairways, doorframes, partitions, and rafters—made from wood and other materials. They also may install kitchen cabinets, siding, and drywall.	High school diploma or equivalent	$39,530
Carpet Installers	Carpet installers lay carpet in homes, offices, restaurants, and many other types of buildings.	Less than high school	$36,090
Cement Masons and Terrazzo Workers	Cement masons pour, smooth, and finish concrete floors, sidewalks, roads, and curbs. Using a cement mixture, terrazzo workers create durable and decorative surfaces for floors and stairways.	See How to Become One	$35,530
Construction and Building Inspectors	Construction and building inspectors ensure that new construction, changes, or repairs comply with local and national building codes and ordinances, zoning regulations, and contract specifications.	High school diploma or equivalent	$52,360
Construction Equipment Operators	Construction equipment operators drive, maneuver, or control the heavy machinery used to construct roads, bridges, buildings, and other structures.	High school diploma or equivalent	$39,460

TABLE **6.4**   Construction and Extraction Occupations—Entry-Level Education and 2010 Median Pay

U.S. Bureau of Labor Statistics			
Occupation	**Job Summary**	**Entry-Level Education**	**2010 Median Pay**
Construction Laborers and Helpers	Construction laborers and helpers do many basic tasks that require physical labor on construction sites.	See How to Become One	$28,410
Drywall and Ceiling Tile Installers and Topers	Drywall and ceiling tile installers hang wallboards to walls and ceilings inside buildings. Tapers prepare the wallboards for painting, using tape and other materials. Many workers do both installing and taping.	Less than high school	$38,290
Electricians	Electricians install and maintain electrical systems in homes, businesses, and factories.	High school diploma or equivalent	$48,250
Elevator Installers and Repairers	Elevator installers and repairers install, fix, and maintain elevators, escalators, moving walkways, and other lifts.	High school diploma or equivalent	$70,910
Glaziers	Glaziers install glass in windows, skylights, storefronts, and display cases to create distinctive designs or reduce the need for artificial lighting.	High school diploma or equivalent	$36,640
Hazardous Materials Removal Workers	Hazardous materials (hazmat) removal workers identify and dispose of asbestos, radioactive and nuclear waste, arsenic, lead, and other hazardous materials. They also clean up materials that are flammable, corrosive, reactive, or toxic.		$37,600
Insulation Workers	Insulation workers install and replace the materials used to insulate buildings and their mechanical systems to help control and maintain temperature.	See How to Become One	$35,110
Oil and Gas Workers	Oil and gas workers carry out the plans for drilling that petroleum engineers have designed. They operate the equipment that digs the well and that removes the oil or gas.	Less than high school	$37,640

TABLE 6.4   Construction and Extraction Occupations—Entry-Level Education and 2010 Median Pay (*Continued*)

U.S. Bureau of Labor Statistics			
Occupation	**Job Summary**	**Entry-Level Education**	**2010 Median Pay**
Painters, Construction and Maintenance	Painters apply paint, stain, and coatings to walls, buildings, bridges, and other structures.	Less than high school	$34,280
Plasterers and Stucco Masons	Plasterers and stucco masons apply coats of plaster or stucco to walls, ceilings, or partitions for functional and decorative purposes. Some workers apply ornamental plaster.	Less than high school	$37,210
Plumbers, Pipefitters, and Steamfitters	Plumbers, pipefitters, and steamfitters install and repair pipes that carry water, steam, air, or other liquids or gases to and in businesses, homes, and factories.	High school diploma or equivalent	$46,660
Reinforcing Iron and Rebar Workers	Reinforcing iron and rebar workers install mesh, steel bars (rebar), or cables to reinforce concrete.	High school diploma or equivalent	$38,430
Roofers	Roofers repair and install the roofs of buildings using a variety of materials, including shingles, asphalt, and metal.	Less than high school	$34,220
Sheet Metal Workers	Sheet metal workers fabricate or install products that are made from thin metal sheets, such as ducts used for heating and airconditioning.	High school diploma or equivalent	$41,710
Structural Iron and Steel Workers	Structural iron and steel workers install iron or steel beams, girders, and columns to form buildings, bridges, and other structures. They are often referred to as ironworkers.	High school diploma or equivalent	$44,540
Tile and Marble Setters	Tile and marble setters apply hard tile, marble, and wood tiles to walls, floors, and other surfaces.	Less than high school	$38,110

TABLE 6.4 Construction and Extraction Occupations—Entry-Level Education and 2010 Median Pay (*Continued*)

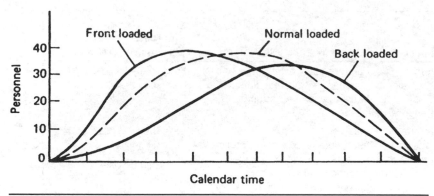

Figure 6.2   Practical personnel-loading curves.

The latter two result when the personnel loading occurs earlier or later than planned on a project.

The significance of these conditions becomes apparent when we look at the set of S curves resulting from plotting the percentage of hours expended against the schedule time, as shown in Fig. 6.3.

The S curve for the ideally loaded project has a gradual start and finish, which indicate smooth starting and finishing conditions. The front-loaded curve shows a rapid project start-up and a phaseout at the end that is even more gradual than normal. The back-loaded curve indicates a more relaxed start and a very steep finish slope. The steep finish leads to such problems as inefficient use of personnel and

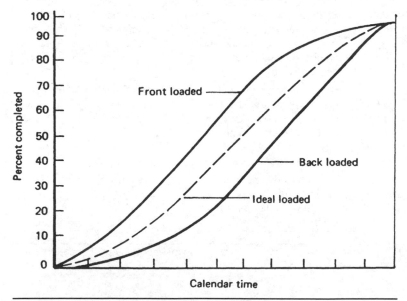

Figure 6.3   S curves from bell curves.

overrunning of the budget. The inefficiency results from having too many people working on only a few remaining tasks.

Normally, projects cannot phase personnel off the job too quickly, and delays in doing so may overrun the schedule, finish late, and likely overrun the budget. The simple lesson to be learned here is that front-loaded projects may slip to a normally loaded mode and still finish on time. There is little or no hope that a back-loaded project will finish on time. Any slippage during execution further exacerbates the phaseout problems and makes the project finish still later and more over budget.

CMs and general contractors (GCs) must remember, however, that front-loaded projects do not just happen to draw the personnel-loading curves that way. All necessary start-up requirements of design documents, facilities, personnel or materials, and equipment must be available to support an early labor pool.

The classic forms of bell and S curves introduced here are used throughout the project, as we shall see in succeeding chapters.

Planning the Construction-Project Personnel

Personnel planning for field work involves detailed craft-loading curves and includes the field supervisory team. Although the home-office support personnel play an important role in the project effort, their numbers are usually relatively small, so loading curves for them are not practical. They may be assigned for a duration of time or even on a part-time basis. Also, the home-office people may not be under the direct administration of the CM or GC; they need to be supervised by the construction or project manager whenever their activities bring them out to the field.

Among noncraft personnel assigned to the field may be a secretary to answer phone calls and receive e-mails; distribute incoming correspondence from and to the owner, architect, engineer, and various subcontractors; and receive and distribute revised plans or specifications to the appropriate parties. Some personnel may be assigned to review and process requests for information generated by the field supervisors or subcontractors. Processing daily activity reports, either written or generated by computer, may be a full-time job for another person, depending again upon the size and complexity of the project.

Planning Field Supervisory and Staff Personnel

The field-personnel group breaks down into two groups: field supervisors and craft labor. Percentagewise, the field-supervision labor hours in toto will be less than the aggregate of the craft labor hours, merely by virtue of the number of workers. The quality of the field supervisors are critical in contributing to the success of any project. Although directly under the supervision of the CM or PM, they are more or less left to

Contracting Basis	Number of People		Types of People
	Process	**Nonprocess**	
Self perform (Direct-hire craft labor)	30–50	15–25	Managers, craft supervisors, foremen, administration
Construction management (All subcontracted)	10–20	5–10	Manager, supervisors, control people, etc.
Third-party constructor	25–40	12–30	Managers, craft supervisors, foremen, administration

TABLE 6.5   Typical Sizes of Field Supervisory Staff

operate on their own, with periodic meetings with their immediate boss to discuss job progress and problems with subcontractor performance or responses from the design team needed to maintain job progress.

The breakdown of self-performed work, subcontracted work when a CM, is engaged, or a mixture of the two when a GC is employed will vary depending upon whether the project is a process-type project or a nonprocess, vertical-construction-type project. Table 6.5 shows an approximation of the range of supervisory staff, depending upon the type of project.

The construction-management approach requires less supervision because all of the work is subcontracted, whereas the general-contracting approach may introduce some self-performed operations. And of course, in a project where all work is self-performed, the need for more supervision is quite evident.

Planning the Completion of the Project as Construction Is Just Beginning

Project closeout—commissioning, preparing the punch list, and completing the project—if not planned beforehand, has been the downfall of many a project, contractor, and supervisory staff.

Some projects have rather simple closeout and commissioning and—with diligent supervision during construction—a limited punch list to be completed; while others may have sophisticated HVAC equipment and other complex commissioning requirements. Whether simple or complex, the time to plan for a successful project closeout is at the beginning of the job.

Closeout requirements and procedures are generally contained in the project's specification manuals; the CM or PM would be well advised to make a separate, in-depth review of the specs for the sole purpose of creating a planned closeout procedure that will be addressed with subcontractors and materials and equipment suppliers at the first project meeting. Being aware of what is required of these subcontractors and suppliers during the project closeout process before they even start work will uncover some items that may have escaped their attention:

- Attic stock. In specifications sections dealing with painting, suspended ceiling work, or floor covering, the painter is often required to provide X gallons of paint to the owner for each type and color used in the project. The acoustical-ceiling contractor may be required to provide the owner with X boxes of each type of ceiling tile used in the project and maybe even X lineal feet of each type of suspension system. The flooring contractor, whether using resilient or ceramic tile or both, may be required to leave the owner X boxes of each type of tile plus resilient baseboard or, in the case of carpeting, X square yards of each type of carpet. The subcontractors will need to be reminded to increase their orders for material to accommodate the attic-stock requirements; they will be grateful that it was called to their attention prior to their placing orders with their suppliers.

- Spare tools or spare parts. Special equipment required in the project may have a requirement to leave the owner's maintenance people with certain special tools required for replacement of parts or maintenance of that equipment. It is a common requirement in the fire-protection section of the specifications to leave the owner spare sprinkler heads and even a storage cabinet for them. Extra filters, extra HVAC diffusers, or extra variable-air-volume components are often stipulated in their related specification sections.

These are only a few of the requirements that must be fulfilled as part of the project-closeout procedures. After a careful review of the specifications, the project manager would be wise to type a list and distribute it to each of the affected suppliers and subcontractors.

Commissioning—the start-up, testing, balancing, and verification of the electrical and mechanical equipment—can be a complicated process that can actually start prior to the end of the project. It often requires the presence of the equipment manufacturer's representative on-site to participate and observe that the equipment is operating as required and within the tolerances set forth in the specifications. (This is dealt with in more detail in a later chapter.)

The Numerous Inspections, Tests, and Accompanying Reports Also Required as Part of the Closeout Documents

There are a number of inspections, tests, and certifications required to be submitted at the end of the project, and many of them are generated at the beginning of the project. It is helpful if separate files are created when the field office is being set up; as these inspections and reports are generated, they can be placed in the appropriate file for easy access at the end of the project.

What follows here is a partial list of these inspections and tests, but once again, the specifications and the owner's contract should be carefully reviewed and a report prepared listing all of these requirements; this report should be distributed to the concerned parties and supervised by the project superintendent:

- Earth-compaction inspections and tests
- Concrete-compression tests
- Infiltration and exfiltration tests for underground utilities—storm and sanitary systems
- Mill reports from the structural-steel supplier
- Weld, bolt-up steel-connection tests, shear stud testing, and accompanying reports
- Mortar-cube tests
- HVAC and plumbing tests, not only by the design engineers but also as required by local building officials
- Inspections of various substrates and approval before encapsulation or encasement
- Tests of fire-protection systems, including underground piping, fire pumps, and pressure testing of filled lines, with an official sign off by the local fire marshal
- Operation and maintenance manuals in the form of a three-ring binder, usually, and the number of each

Building Commissioning

Along with end-of-project procedures, commissioning of the various mechanical and electrical components of the building should be an important part of the project-resources planning. The CM or PM should remember that until the building is officially accepted by the owner and the owner's design consultants, the costs to maintain its systems fall to the contractor. A 250,000-square-foot building's heating or cooling system, for example, can generate some very substantial utility costs.

The commissioning process for new construction is abbreviated Cx, and the recommissioning of an existing structure is abbreviated as ReCx. If the work requiring commissioning is to be subcontracted, then as part of the initial project planning, the subcontractors engaged for the various systems need to be made aware, up front, that their project planning must include provisions for what is typically included in this process:

- HVAC, including ducting (specified flow and volume) and accessories
- Piping systems—pipes, valves, valve operators, instrumentation, and related work
- Plumbing—fixtures and piping
- Direct digital controls—both software and hardware
- Electrical systems, including switchgears, transformers, grounding, bonding, lighting, and lighting controls
- Building envelope, including roofing, sheet metal, and air infiltration through envelope penetrations
- Fire protection, pumps, standpipes, and hoses
- Specialty systems—security and fire alarm and voice or data communication systems
- Emergency power systems

Attention to these items early in the project-planning stage will alert all subcontractors and suppliers to the importance of having the necessary personnel and equipment standing by as the project nears completion.

Construction Subcontracting

Trades as electrical, mechanical, plumbing, roofing, site work, and site utilities, as well as finishing trades, are generally subcontracted on most projects. When the subcontractors commence work, the problems of communication, cost control (in the case of extra work), and quality all fall to the field supervisors and the construction manager or project manager. The prime contractor's project-control people will oversee the supervision of the subcontractors for adherence to scheduling, progress payments, and change orders, if they arise.

The subcontracting plan is initially formulated during the estimating process, since proposals will be solicited from several subcontractors in each trade. These bids need to be scrutinized by the estimator and the PM or CM assigned to the project to insure that the best price is, in fact, all-inclusive. As with any subcontract, prices should be obtained from only those subcontractors who will be selected if their price and scope of work meet the project's requirements. It does not make much sense to obtain pricing from subcontractors with whom the company has had a bad experience or from those who may not be qualified to perform the work for any number of reasons—not enough properly trained trade workers, a record of financial problems, or a history of inadequately qualified supervision.

The potential low bidder's proposal should be reviewed to ascertain that the scope of work is all-inclusive, that no areas have fallen through the cracks. Anything that is included in the prime contract but left out of the subcontract falls to the prime contractor's account. Make sure that the ancillary contractual requirements of schedule, warranties and guarantees, quality, payment schedules, change-order procedures, and the like are included in the subcontract agreement. If materials are being furnished as part of the subcontract, ensure that the quality standards are equal to the prime contract's specifications.

A common practice in the architect's writing of the specifications is to use the word *contractor* in, say, the electrical section of the specifications, stipulating that the contractor shall perform all cutting and patching as required. The question arises of who is the contractor— the electrical contractor or the general contractor? Fortunately, most specifications also include a definitions section, and looking up the word *contractor* to determine how it is defined in the specifications will provide the answer to that question.

Most subcontractors are major players on the construction team, so be sure to foster mutual respect and goal participation among the field staff and the subcontractors. If problems arise, do not hesitate to discuss them with the subcontractors' management and to make personnel changes if necessary.

Construction-Material Resources Planning

The basis for construction-material resources planning is in the project's material plan. That document is created by the CM or PM in conjunction with the estimating and purchasing departments. The report resulting from making the materials resource plan should include the division of procurement responsibility, current delivery data, bulk materials, and subcontracts. This is in addition to a survey of current market conditions, pricing trends, bulk-material availability, and vendor lists. These data are invaluable in formulating the project's material resources plan and the project schedule.

The division of procurement responsibility is defined by the project manager or the construction manager. In most industrial or process-engineering work, the engineered equipment is purchased by the design team because of the lengthy delivery times involved. In projects like this, the contractor is involved only in buying the bulk materials, such as concrete, masonry, bulk piping, and electrical materials.

In commercial and institutional projects, procurement of all materials via purchase orders or the award of a subcontract is performed according to the purchasing plan devised by management. Bulk orders for concrete, lumber, and miscellaneous materials are obtained via purchase orders with instructions to ship as released by the project manager or the project superintendent. It is vitally important that these materials be purchased from established sources from whom required delivery dates are assured in order to maintain job progress.

Today's materials plan is usually a detailed document listing, by account code or Construction Specifications Institute code, the quantities of required materials and equipment, a description, and the date the materials are required in the field, either as a specific date or on an as-released basis. Either a series of purchase orders or a computerized spreadsheet can be used to track the placement of orders and their receipt on the jobsite.

Many construction managers and project managers have had input into material purchases, taking specific or basic product information from the specifications and assisting the estimating department with quantities and qualities required, even commenting on the need to purchase some extra quantities based on to their experience in predicting the amount of waste that should be factored in.

Long-Lead Materials

Ensuring that the material resources for the project arrive on time involves planning for the following important areas:

- Long lead time for delivery of equipment, some of which requires the submission, review, and approval of shop drawings
- Special materials and alloys
- Common materials in short supply (this may change from project to project depending upon demand)
- Heavy construction equipment and special tools
- Services and system requirements
- Financial resources

Placement of required equipment or materials will be highlighted in the Critical Path Method (CPM) schedule. This will dictate when orders need to be placed and when delivery is required.

Special Materials and Alloys

It is a good idea to review all materials specified for the project, whether they be purchased by the contractor or supplied by a subcontractor. Special alloys such as Hastelloy G, and—to a lesser degree— Monel, Inconel, and some special stainless steels may be required for specific applications in the project. If these materials are included in a subcontractor agreement, it is important to have the subcontractor provide documentation that the material has been ordered, the anticipated delivery date, and periodic updates indicating that delivery is still on track. Among these special materials that require periodic checking to ensure that delivery will be as specified are special aggregate for precast concrete panels, special window systems, and special quarried stone, from either local or foreign quarries.

Common Materials in Short Supply

Common materials often in short supply depend upon many different market conditions. A boom in residential construction may create a shortage of wood studs and Sheetrock; if your project specifies wood studs, it may be wise to switch to steel studs. Structural steel for high-rise commercial buildings will generally be subcontracted to a structural-steel fabricator and erector. The steel will usually be ordered directly from the mill rather than from a warehouse, where the price would be higher. The mill will provide the steel subcontractor with a rolling schedule, but if special sizes and weights are required but not available in a rolling schedule that coincides with the project schedule, some warehouse steel might be required.

Cement used in the production of concrete has from time to time been in short supply because of world demand, so a project requiring large amounts of ready-mix concrete needs to be monitored closely by the CM or PM to obtain a source that will be able to meet delivery needs.

Special Construction Equipment

Special construction equipment is more prevalent in process and chemical or petrochemical construction; on vertical-construction projects, its use is generally relegated to the construction process, not incorporated into the building itself. Tower cranes and mobile or track-mounted large cranes would fit into this category. Depending upon the physical location of the project, say in an isolated area away from major metropolitan areas, procurement of special lifting equipment may require some searching.

Services and Systems

Strictly speaking, project services and systems are not physical resources, but they must be planned for at about the same time as the physical ones. Planning for them is particularly critical on large projects.

The project scheduling, accounting and cost-control, and administrative systems must be decided at this time so that the necessary manual or computer operations can be effectively implemented, such as payroll, project control systems, and review and approval procedures for materials and subcontractor payment requests.

Even such ordinary resources as the office's service and site facilities for construction have to be planned early. What form of security system will be employed at the site? How much space is needed for parking, materials storage, and access around the site? What design documents are needed, and how many of each? Are reduced sets of drawings needed for the project superintendent? These are just a few typical questions that must be considered and answered by the CM or PM. Some items will fall on the critical path and some will not. The CM or PM must be familiar with those that will and those that will not. Formulation of the various priority lists will begin at this stage.

Transportation Systems

With the global economy that we live in today, it will not be unusual for some specialized equipment, products, or materials to be manufactured abroad. In that case, the method of transportation may become a factor. Smaller items needed rather quickly can be delivered via air freight, whereas larger items will most likely be containerized and shipped by boat. Logistics may play a large part in insuring that items of foreign origin are firmly in the transportation system and are being tracked by the supplier.

Prefabrication has become more prevalent today, increasing productivity but creating new concerns, such as the shipping of oversize over-the-road items via truck that may only be able to travel during certain hours and may require escort services. Some prefabricated products such as mechanical-equipment rooms or prefabricated HVAC modules may require large-capacity cranes, a clear access path for the truck delivering them, and a designated area for the crane to unload and either store or lift and set it in place in the building.

Financial Resources

It has always been the owner's responsibility to arrange for the project financing, to a degree that almost wholly excludes this aspect of the project from the CM or PM's responsibility—but not quite. There may come a time when the owner's payment is late, and the old "pay when paid" clause in many subcontract agreements has been deemed nonenforceable by a number of court decisions. So the CM or GC may not be able to fall back on that tactic when an owner's payment is late and a particular subcontractor is operating on a hand-to-mouth basis and cannot wait another two or three weeks for payment. In the case of the CM who is merely monitoring subcontractor and supplier payments and approving them for payment, the owner will need to be advised that some interim financing may be required to satisfy the contractual payment obligations for a subcontractor or supplier.

For a general contractor, if such an instance arises, the GC should not take money from one project to pay another—a practice known as commingling of funds—so the GC will need temporary access to additional funds, if the regular requisition request to the lender is delayed an unreasonable length of time.

Meeting financial obligations is a very important part of the construction process, and the inability of a CM or GC to meet financial commitments does little to create the spirit of cooperation and trust so necessary for a successful project.

Summary

This chapter wraps up the planning portion of project's activities. Sound planning forms the foundation for everything that comes later. Effective CMs and GCs must train themselves to plan all facets of the project, as well as their day-to-day work activities.

We must remember, however, that a plan is only a proposed baseline for the execution of the project. Any plan is subject to changes along the way. Although it is often necessary to change plans, we do not recommend making any radical changes to the original plan. If your plans were well throughout, you should resist pressures to change them.

CHAPTER 7

Project Organization

With planning out of the way, we are now ready to get into the organizational part of our project management and-execution philosophy of *planning, organizing, and controlling*. We are sure you understand that these three activities are not performed in a compartmentalized fashion as we present them in this book; all the activities in fact overlap and proceed concurrently. Continuing the review of all three management functions simultaneously and modifying them as necessary to meet day-to-day operating conditions should be an ongoing activity. As in previous chapters, when we use the term *construction manager* (CM) it refers to a manager working under a construction-management contract; when we use the term *project manager* (PM) it refers to a general contracting firm, generally involved in third-party contractual relationships (the other two parties are the owner and design consultants).

Organization Overview

Organizing can be defined as *the function of creating, in advance of execution, the basic conditions that are required for successful achievement of objectives*. The operative phrase here is "achievement of objectives"—the objectives being the *project goals*. Never allow the organizational structure to get in the way of meeting the project goals!

The first law of organizing is: To meet the objectives, design the organization around the work to be done, not the people available. This law is based upon the assumption that you have access to a bank of skilled personnel to perform the necessary activities. However, we know that that rarely happens, and that some compromises have to be made in selecting a project team.

The general goal of any organizational structure is to establish the proper relationship among

- the work to be done,
- the people doing the work, and
- the workplace(s).

Later, we will work our way into building some typical project organization charts based on these principles.

Organizational Design

By way of building some background in organizational design, it will be valuable to review a bit of basic theory and practice used to design organizations. Most project organizations tend toward the functional or military style, using a combination of vertical and horizontal structures. As with the military, we also use line and staff positions.

In recent years other forms or organization charts have been put forward, but they do not seem to have caught on in the construction of capital projects. A vertical structure, as the name implies, places one position (block) over the other, as shown in Fig. 7.1. The vertical dimension establishes the number of layers in the organization. Keep in mind that every layer introduces a communication filter into your organization, and each filter is a potential checkpoint for the necessary flow of vital project information.

The horizontal dimension is shown in Fig. 7.2, with a number of horizontally arranged positions reporting to a single block above. The horizontal dimension introduces the term *span of control, i.e., the number of workers that can reasonably be controlled by the number of managers assigned to control them.* In the vertical organization shown in Fig. 7.1,

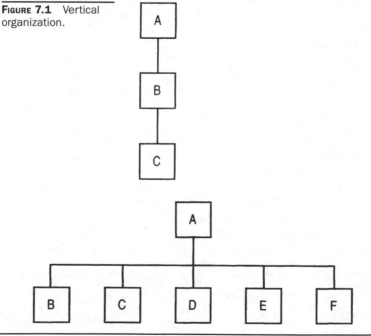

FIGURE 7.1 Vertical organization.

FIGURE 7.2 Horizontal organization.

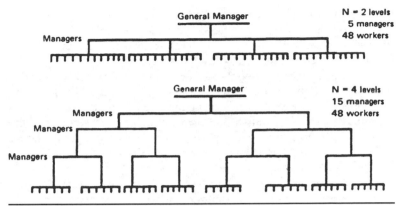

FIGURE 7.3   Vertical and horizontal organizations.

the contacts are one-to-one; in a horizontal one, they are n-to-one (with n being the number of levels). If n becomes too great, supervisor A will not have enough time to devote to each of the supervised blocks and will require additional assistance to do so. There is no fixed maximum for n, since the ability to supervise depends on the nature and complexity of the contract. The number six is often mentioned in management circles as a normal maximum span of control, and it is one we would not exceed without giving the matter a lot of thought. Remember too that communication filters are added in the horizontal direction as well as the vertical.

Figure 7.3 compares vertical and horizontal organizations and the ratio of levels, managers and workers in each type.

Obviously, an organization should have an optimum balance of managers and workers to give the most effective control of the project work. We prefer to have our project supervisors slightly overloaded, as this encourages them to extend their capabilities and to have an opportunity to grow with the job. Remember that some people exceed their capacity of control because they are not organizing their time effectively. We do not consider that a sound reason for revising the organizational structure to eliminate the problem. Some on-the-job training or a shift in personnel is definitely a better solution to such an apparent span-of-control problem.

These organizational charts may be considered "text book" theory, but real-life situations depend upon a number of events.

Corporate Organizational Structure

The corporate structure of a construction company depends upon a number of things: the size of the construction company, number of employees, and annual volume. The Bureau of Labor Statistics indicates that most construction companies are small enterprises; about

68 percent employ fewer than five employees, so the discussion about organizational structure may be a moot point, but it is still worthy of discussion since someday you may be offered an opportunity to work for one of America's top 400 contractors.

Corporate organizational structures and policies vary greatly, depending upon the size and type of market the firm serves. Differences in size tend to reflect themselves in the number of duties assigned to an individual. Small firms, which comprise most of the construction industry, use multifunctional people, whereas larger firms have enough volume to afford specialists or departments to handle the work.

In this chapter, we will take a look at the organizational structure of a design-build firm and a large general contractor and provide some detail relating to the various departments in both types of organizations. Both line and staff home-office construction people often move into the organization from service in the field. A little experience in field activities, problems, solutions, and day-to-day dealing with sometimes difficult problems is always helpful when one moves from field to office operations.

Figure 7.4 shows a typical corporate organizational structure for a large design-build firm. The structure may vary, depending upon whether the firm is design driven or construction driven. As you may recall, our discussion of design-build in an earlier chapter described a couple ways in which a design-build team can be structured: a contractor bringing a design team on board to fulfill a client's requirement for that project delivery system and, conversely, a design firm joining with a construction company to satisfy a client's request to proceed with a project as design-build. One or the other—either builder or architect—will be the leader of the team.

There are some construction firms that employ a full complement of designers and engineers in order to perform design-build projects. Figure 7.4 describes one of these firms. Although it appears to be a design-build firm involved with engineering-driven projects, its engineering-manager position could just as easily become an architectural-design manager and the staff below would be various designers in lieu of project and design engineers.

Figure 7.5 reflects the organization of a large contracting company. This organizational structure would be similar to that of a construction-management firm. The home-office staffing has a direct relation to the general and administrative expense; this must be included in the company's overall overhead and profit that is included in bids, whether for lump-sum, design-build, or cost-plus guaranteed-maximum-price contracts. Therefore, these large contractors are at a disadvantage when bidding against small shops with 15 or so office personnel, since those smaller firms can afford to have a lower percentage of overhead and profit. In some cases that is the difference between being a successful bidder or an also-ran.

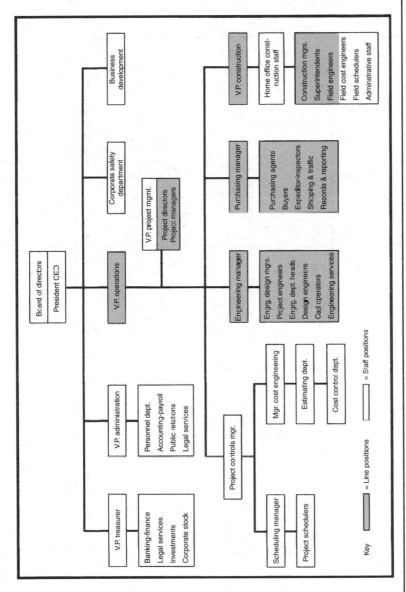

FIGURE 7.4 A design-build firm's corporate structure.

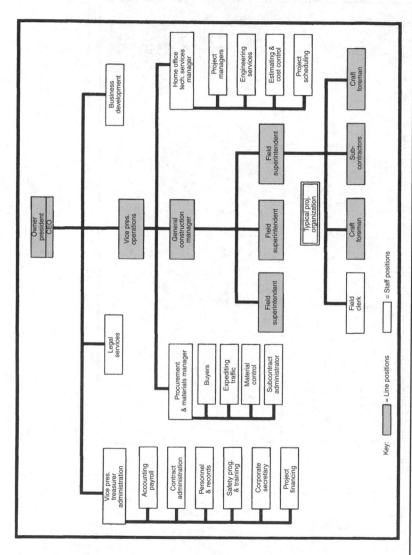

Figure 7.5 A construction-only firm's corporate structure.

Constructing the Project Organization Chart

Any operating project is worthy of a organizational chart showing positions, lines of authority, titles, group relationships, and even the names of current group leaders. This is especially true in large organizations; it may not be of so much importance to a small contractor whose cousin is the bookkeeper, whose niece is the office manager, and that's about the extent of the office structure.

Depending upon the size of the project, the project organization chart may vary from a single letter-size sheet of paper to something much more complex. The main advantage of the organization chart is that it shows in an easily understood format the key project functions and the related players. This makes it easier for new hires, in the field or in the office, to become familiar with the company and the lines of authority and responsibility.

One concern as to the effectiveness of an organization chart is the feeling of compartmentalization and *empire building* that can occur and may endanger the efficiency and morale of the organization. Any chart tends to set up psychological and communication barriers that may detract from the team effort everyone is anxious to develop in order to have a successful project.

The best way to overcome those obvious disadvantages is to have superimposed on the functional structure a network for handling interpersonal contacts and information flow. A good analogy is to consider the organization chart as the skeleton of the project organization, with the unseen network acting as a circulatory system through which the lifeblood of the project flows. A good CM or PM must work hard at breaking down barrier-building tendencies by fostering the unseen network, which is accomplished by exerting strong and charismatic leadership on the project team. If a cooperative team spirit is not created across all blocks of the chart, project performance will suffer.

Typical Project Organization Charts

Since the number and types of project organizations used in the capital-projects business is virtually limitless, it is impossible to produce a sample chart for each. The purpose of the organization chart is to define the work to be done, the people doing the work, and their location—field or office. The work to be done is a function of the project scope and the contract format. A turn-key project involves design, procurement, and construction activities in one organization. A third-party organization chart shows another staffing posture, and a construction-management organization a third, slightly different type of staffing.

Remembering that the overall project goal is to produce a quality facility that meets the owner's needs and provides the constructor with a reasonable profit, it might be a good idea to start at the bottom of the chart, with the human resources required to reach that goal effectively. The project-execution plan goes a long way toward shaping the overall organization chart because it defines how the job will be done.

Construction Organization Charts

The CM or PM interfaces that exist in the various construction-project settings differ widely and are often confusing. As projects became larger and more complex, the construction manager or project manager finds him- or herself dealing with more sophisticated owners who may have construction or design professionals on their staff, and these people become more intimately involved in the progression of design and construction activities than the CM or PM may have experienced in other projects.

The project manager will generally appear as a subsection of the construction company's management organizational chart. The office staff, accounting, sales-development, and top management people will retain their place filling out the balance of the chart. Estimating, purchasing, and accounting workers may appear as staff members.

Figure 7.6 shows the organizational chart for a process-type design-build firm providing engineering, architectural, and construction services. Although the caption calls it a typical design-build contractor's project organization chart, it represents a fairly large and substantial construction company, certainly in excess of the five-employee maximum that the Bureau of Labor Statistics says characterizes 68 percent of the country's construction companies. Figure 7.7 is a more typical general-contracting organizational chart.

Field Superintendent

One of the most important line functions is that of the field superintendent, the person responsible for the day-to-day on-site construction activities. He or she manages the largest commitment of human and physical resources, being responsible for supervising any self-performing labor force and all subcontracted work as well as for keeping the project stocked with materials and equipment in an ever-changing environment.

With all of that responsibility, the field superintendent may—again, depending on the size and complexity of the project—require a staff in the field office, composed of forepersons to supervise specific operations and staff to keep accurate daily field records of operations being performed, subcontractors on the job, the size of their crews, and the operations being undertaken that day by those subcontractors. There may be a need for a secretary to answer the phone; send, receive, and print out e-mails for documentation; keep the set of drawings up to date as changes are made and individual drawings are updated; and perform a myriad of other bookkeeping-type activities.

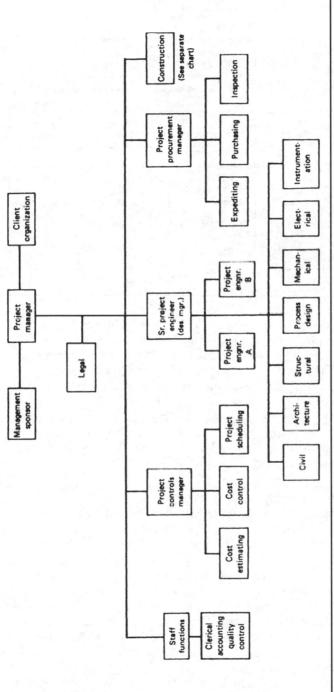

Figure 7.6 A typical design-build contractor's project organization chart.

171

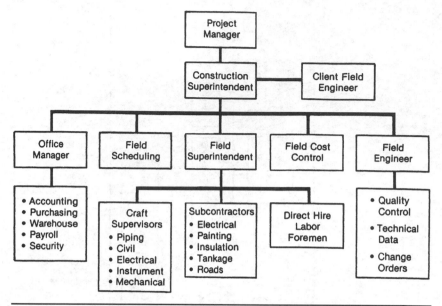

Figure 7.7   A typical contractor's field project organization chart.

The field superintendent, working closely with the project manager, tracks progress, discusses any problems that may arise that should be included in the agenda of the next progress meeting, and, in general, plans and coordinate the activities that lie ahead.

The position of field superintendent is quite often the beginning of a path to positions in upper management, possibly being followed up first by a promotion to project manager on a small project and eventually, after the person has gained experience, being given the responsibility for a larger project and advancing into higher management positions.

Field Staff

The field scheduling group, specifically, the scheduler assigned to the project in question, reports to the construction manager or project manager and is responsible for maintaining the construction schedule. The key role of the field scheduler is to lay out the work to be accomplished in the field for the next week or two. This is often referred to as a one- or two-week look-ahead schedule. The CPM schedule can be used to create a simpler bar chart containing the subcontractor activities that need to take place during that week or two. If the ability to meet any of the dates is in doubt, then the field superintendent, the project manager, and the subcontractor(s) involved must resolve those doubts at a meeting.

After the look-ahead schedule is reviewed—usually at the weekly progress meeting—to verify that the activities on it are feasible and will be completed by the subcontractors concerned, the field supervisor's job is to ensure that these activities occur.

The field cost-control group will include the field superintendent, since supervision to meet preapproved goals and costs are interrelated. Chief among the responsibilities of the cost-control group is monitoring the costs versus the budget; we will discuss this in further detail in the next chapter.

If change orders increasing or decreasing the scope and cost of work have been authorized and approved, the field cost-control group will include that analysis in their study. Through balancing the percentage completion of the field work against the cost, the calculation of productivity of the field forces will be produced. If productivity lags, it is extremely important to get it back on track as soon as possible and, equally important, to investigate the reasons for the lag. Was it due to lack of proper-size subcontractor crews, delays in obtaining required materials or equipment, inclement weather, or slow decisions from the owner, the design consultants, or the company's own upper management?

Field Engineering Group

Depending upon the nature of the project, field engineering may play a large or a minimal part in daily operations. The field engineering group receives the technical documentation from the design group and distributes it to the field crew. All field revisions and as-built drawings are handled by this group, along with any technical interpretation of the plans or specifications that is required by the construction people. The field engineer is the liaison between the field and designers for design clarifications and changes. The group also handles the field-survey work needed at the site and sets the lines and grades for site utilities, ancillary structures, and roads and parking areas.

Involvement in construction-quality control with the field engineering group is a potential conflict-of-interest area. In the absence of an owner's or outside inspection team, the field organization is placed in the position of inspecting its own work. Pressure from field-organization peers to meet the field schedule can often result in a temptation to cut corners. Instances like this place a strain on the ethical practices of the CM or PM and the field engineer and must be resisted.

If the contract or the specifications are vague with respect to inspections, the CM or PM should meet with the owner and convince him or her of the need to have an impartial inspection service, which may be cause for a change order. The scope of field-inspection services ranges from planning the inspection program to completing the final facility checkout and acceptance. In between, it includes checking incoming materials and equipment, erected equipment and systems, field welding and bolt-ups, steel alignment, civil work, and laboratory-inspection services. Proper planning of the inspection work and its associated budget is critical to meeting the allocated field indirect costs. The inspection program must be designed to catch any errors early, before costly corrections are needed.

Field change orders are an important part of the work handled by the field engineering and cost-control groups. When the need to present a change order arises, it is important that no such extra work be started until the owner or design consultants agree to the scope of work and the cost to perform that work. Many field superintendents have a tendency to proceed with the extra work without written authorization, so as not to slow down the schedule. However, a verbal OK is not sufficient to commence that extra work. A brief description of the change and either a lump-sum or Time and Material (T&M) figure should be prepared by the field engineer and signed off on by the owner or the owner's representative. This can then be submitted to the CM or PM so that a formal change order can be prepared and sent to the owner.

The position of field engineer is another possible springboard into the construction-manager position, with some additional training in basic construction techniques. The field engineer gets good all-around exposure to construction technology and management. The CM or PM should keep this in mind and be on the look-out for field supervisors or engineers with the potential to fill positions of higher authority.

The Integrated Project Organization Chart

If we assemble all the examples that we have discussed in this chapter, it is readily apparent that even a small project can involve a large number of people. On a relatively small project of $10 million, there could be a design team of 25 which would cover civil, structural, architectural, HVAC, plumbing, and electrical. Combine that with a construction team of 75, which would include supervisors, managers, and subcontractors, and there is a total of 100 people contributing to the project. On a project of $50 million, the number could be as high as 500. Each person's project activities must be channeled into effective work if project expectations are to be met. This points up the importance of designing an effective project organization chart as the first step in building a quality project team. Designing the chart is, however, only the first step in building the team.

Selecting and Motivating the Project Team

We said earlier that we first design the organization to suit the work to be done, after which we fill the organization from the pool of people available. In the ideal situation, we would like to have two or three candidates presented to us for each key position. Unfortunately, that does not happen very often, except in periods of low workload. Even in those periods, it does not happen often, because the staff has already been reduced to meet the existing low-workload situation.

Selecting the Project Team

The owner of the project may have gone to the contractor's shop to get a good look at the contractor's operations and personnel. The owner may have met two or three top people during the contractor-selection process and worked closely with these people during the contract-negotiation process. The owner may have also met the project manager and field superintendent during one of those contractor-interviewing sessions and been impressed with their knowledge and attitude. The owner may have also met some supervisors or managers that he or she did not have good feeling about and therefore would not like to work with.

Most contracts, particularly the ones from the American Institute of Architects, include a clause allowing the owner to request a change of supervisory personnel at any time during the construction process if the owner is not happy with a supervisor's performance; the initial assignment of personnel that will be in a position to interact with the owner ought to be made with this in mind. This does not mean that just because an owner likes a particular supervisor or project manager who may not be fully capable of running the project, that person should be assigned the job. But having a construction team that the owner can relate to is an important consideration in assembling the project team.

After establishing a candidate's ability to do the job, you must look at how the person will fit into the team. Does the candidate subscribe to your management method of operation? Will he or she fit in with the other players and the client? Do not select anyone who you feel has a stronger loyalty to a department head than to you or the project team. Do not under any circumstances fill a key slot with anyone who is not a team player! You will only have to change the person later, at some cost to the organization.

Having said all of the foregoing about selecting only quality people, we also know that pulling it off is very difficult and even a matter of luck. You will have to make some compromises and accept some lesser-quality performers. If possible, it is best to blend them in to less critical assignments within the organization, where they are least likely to hurt project performance.

If, politically, you are forced into taking marginally qualified people at key levels, back them up with strong people who will require less supervision. Also, it may be possible to place them in positions reporting directly to you, which gives you an opportunity to develop them on the job. If certain people show good potential, do not be afraid to take a chance on them. There are also qualified people who have been assigned to the wrong operation and who, when they are transferred to another operation where their skills can be exploited, become a plus to the team.

Motivating the Project Team

The first motivation tool is to establish the project goals and install them into the minds of the key players. In addition to the general goals of quality, budget, and schedule, some project-specific goals need to be formulated and written down. They may involve key milestone dates or budget targets that must be met to earn a bonus, if in fact the company policy includes performance bonuses. The client may have set some unusually tough aesthetic designs and construction standards, or there may be some tough environmental goals that have to be met within a tight schedule and budget. If these things are thrown out as challenges, many key personnel will take them as personal challenges and motivate themselves to meet them.

The early project goal-setting activity is probably the most overlooked and underrated activity in project-goal initiatives and team building. Without setting project goals and continuing their reinforcement, the likelihood of meeting those goals is indeed slim. We urge you to remember that this goal-setting philosophy applies equally to every facet of the project as well as to the major areas of design, procurement, and construction.

Writing Job Descriptions

One theory on job descriptions is that if the person does not know what is expected of him or her, he or she is in the wrong position to being with. However, someone new to a position may need a job description to alert him or her to the duties and obligations of that position. Another approach to writing job descriptions is to have the team member write what he or she believes the job entails. This will not only help the person crystallize his or her duties but present them to the person's immediate supervisor for affirmation.

Project Mobilization

Project mobilization is a critical time in the birth of any project. Everyone is gung ho to get the project started. The client, company management, and just about everyone wants to see some dirt fly. There are any number of procedures that need to be defined and established before the project gets into high gear, and to accomplish this a few key people—such as the field superintendent, a foreperson or two, the CM or PM, and possibly a secretary—need to meet at the site and establish some ground rules, the procedure for the flow of important documents, and a list of dos and don'ts.

A review of the initial section of the specifications manual in the presence of all field supervisors will be in order. Specification section 1 is usually devoted to project-management and coordination procedures and generally contains the following directives that the owner and design consultants expect the contractor to follow:

- Coordinate the tie-in of all underground utilities to the appropriate local government agencies.

- Prepare coordination drawings for work where close coordination is required to install products and materials in the space allotted to them. These must be passed on to the appropriate subcontractors so that they can be submitted to the design consultants for review, comment, and approval.

- Prepare memoranda for distribution to each party involved, including notices, reports, and attendance at meetings.

- Hold any preinstallation or prefabrication conferences with the design consultants and appropriate subcontractors or vendors.

- Run progress meetings and provide the schedule and proposed agenda for each.

- Submittal procedures.

- Work documentation and periodic site observations required of the contractor.

- Regulatory requirements and the methods by which they will be handled.

- Procedures the contractor is to follow for dust and noise control.

- Product substitution requirements if the contractor plans to submit materials or equipment not specified in the contract, but for which the contractor considers of equal or higher quality.

- Close-out procedures.

- Close-out submittals.

Procedures for receiving, logging, and distributing shop drawings must be set in place. The method by which requests for information are received, logged in, and logged out needs to be addressed. The change-order process should be reviewed. The receipt, logging in, and logging out of samples to be submitted to the design consultants must be addressed, as must simple tasks such as creating daily reports—who should be given that task? Are there forms at the jobsite? What is to be included? and so forth.

It is not unusual for the design consultants to issue revised drawings or revisions to certain pages in the specification manuals. Often these changes and revisions are extensive; a system must be set in place to log these changes in, insert the revised drawing or specification page in the manual, and distribute the changed documents to all interested subcontractors and vendors.

Once the project activity begins, there will be lots of questions about contract interpretation and about construction details as represented on the drawings or in the specification manual. There will not be much time to devote to the mechanics of paper flow, so the time to establish these procedures is before the first shovel of dirt begins to fly.

Shop Drawings

A review of the specifications will establish the procedures for the submission, review, and distribution of shop drawings. The time allotted to the architect or engineer for review will probably be included in the section on shop drawings, but if not, that subject should be addressed at the first project meeting. There are many other requirements for this critical activity:

- Is a special stamp required for the subcontractor or general contractor to apply to each shop drawing?
- How many copies are to be submitted for submissions? Are sepias required?
- How are samples to be handled, packaged, identified, etc.?
- Are all shop drawings to be submitted to the architect or can drawings pertaining to the mechanical, electrical, and plumbing engineers be submitted directly to them with an information copy to the architect?

The CM or PM needs to create a log to track all of the outgoing and incoming shop drawings, the date received, the date submitted to the architect, the date returned, the action taken (approved, approved as noted, disapproved), the date of resubmission if required or distribution if approved, and the distribution recipients. Someone in the field or in the office must be designated to periodically review this log to ensure that shop drawings are flowing properly and not being held up—or if they are being held up, by whom. This should be made a part of the weekly or biweekly job meetings.

Change Orders

Very rarely is there a project that does not generate change orders, either to include work required but not indicated in the contract or to make adjustments based on owner-generated additions or deletions. The proper preparation of change orders is a topic unto itself, but here we are only concerned with their tracking. Change orders are initially submitted as change-order proposals or proposed change orders. They are not officially classified as change orders until they have been approved as such by the owner. Each proposed change order will be assigned a number. A log must include that number, the date the proposed change order was submitted, the date it was approved or rejected, and any changes made in either the scope or the cost of the initial submission.

Requests for Information (RFIs)

Again, it will be a rare project where there are no questions raised about items in the drawings or specifications which will require clarification and interpretation by the design consultants. Such RFIs may be generated by the contractor, the subcontractor, or a vendor. They will be transmitted to the architect for clarification, and as a result may also be cause to prepare a change order, if the response warrants extra work and extra costs. A log for RFIs will include the assigned RFI number, the date the RFI was received and the person from whom it was received, a brief comment as to its content, the person to whom was it sent (the owner via the architect or the design consultants), the date it was sent, and the number of days required for a response. This later item should be spelled out in the section of the specifications dealing with the RFI process or, again, addressed at the first project meeting. Someone must review the RFI log religiously to insure that RFIs have not been outstanding and unanswered beyond the time allotted in the specifications.

Daily Reports

The importance of accurate and precise daily reports cannot be over emphasized. They may form the basis for approval of a claim or, if improperly prepared, a reason for denial of a claim that may cost the contractor a ton of money. The daily report should include, as a minimum, the following:

- The date and weather conditions—clear, sunny, cloudy, rainy, snowy, etc. (This information is important if weather delays are justified by patterns that vary from the norm—e.g., a 100-year storm—so be specific as to the amount of snow or inches of rain and the intensity.)

- Temperature readings—three per day is preferable: one at 7:00 A.M., one at noon, and one at 4:00 P.M. (These will be critical in case of a claim based on the need for additional winter protection, if not included in the contract, or unusual weather patterns that have caused more costs to be incurred.)

- The subcontractors on site that day, the number of workers in their crews, a brief statement as to where they were working in the building or on the site and what work tasks they were performing, and the number of hours they were on-site.

- Any of the company's workers on-site—the numbers, even their names, and the tasks they were assigned and the number of hours they worked that day.

- Any visitors to the site—various inspectors, the owner or the owner's consultants, or office personnel from your company, including their names, the times when they were on-site, and the reasons for their site visits.

- Comments on any unusual occurrences, such as an accident on-site, involving either vehicles or personnel—if the latter, including names, type of injury, degree of injury (minor or major requiring hospitalization), and any OSHA notification filed.

Drawing and Specification Revisions

The requirement for coordination drawings frequently uncovers problems where mechanical, engineering, and plumbing systems do not fit in the space allotted. Ceiling heights may need to be lowered and chases may need to be increased in size or relocated; and these changes should be accompanied by a drawing change, which might be only on an 8½ × 11 page. Other, more serious architectural revisions may require full drawing revisions. Change orders often result in reissued drawings, as do RFIs that result in more than minor changes. Errors and omissions, if they occur, will also be cause for reissuance of drawings or specifications. Sometimes there are only a few such changes, but we have seen projects where 50 or more drawing revisions were made and both old and new drawings needed to be incorporated into the contract set and distributed to the concerned subcontractors or vendors.

When a flood of 8½ × 11 changes or clarifications to the design or specifications are issued by the architect or engineer, they should be placed in a binder so they do not get misplaced. A log should be included indicating where information copies were sent—i.e., to a subcontractor or vendor—and on what date. A response should be requested from the recipient, either accepting the document with no change in scope or price or providing a proposal to be sent on to the owner and indicating a change in scope or price.

Once the field supervisory organization has been staffed, it is difficult to make major shifts in loading during the execution phase. People who are released temporarily in the middle of a project often will not be available when they are needed again. If the people are working on out-of-town sites, it is not feasible to relocate them to another site and return them later. That is a nasty problem that occurs during unforeseen project suspensions or slowdowns, and it can have disastrous side effects on project budgets, productivity, and morale. There is virtually no way to manage slowdowns out of existence; at best, the damage can be minimized with an intelligent approach to reducing staff and restaffing when the need arises.

When a major destaffing problem occurs in the middle of a project, created by the owner for any number of reasons, all major players—the owner, contractors, and design consultants—must participate in finding the best solution to the problems.

The project and construction managers must inform interested parties of the short- and long-term effects of early destaffing on the execution of the project's master plan. An agreement whereby the project can be reorganized with minimum damage to all participating project partners must be reached. One major problem is dealing with subcontractors. They will reduce staff and workers and shift them onto other projects. If and when the current projects staffs up again, the contractor may have some serious problems getting the subcontractors to staff up to previous levels, and the owner should be aware of the impact this will have on the overall schedule and cost.

Organizational Procedures—The Field Procedure Manual

A key CM or GC responsibility in organizing any construction project is the development and issuance of the field-procedure manual (FPM). This document lays the ground rules under which the field organization will function in executing the work. Each company will have a different standard to which the field procedures are prepared. The procedures should spell out the minimum regulations under which your company management wants its CM or PMs to operate. They can vary from being too simple to offer good project control to being too ponderous to allow efficient execution. A good FPM gives the CM or PM a sole source for the procedures in effect on the project and the relations of the various field functions to each other. Please remember the KISS principle for this task; the FPM should be as simple as possible to suit the work being controlled.

The Typical Contents of a Field-Procedure Manual

The heart of the operating procedures for any construction project is the field-procedure manual. The CM or PM has the prime duty of seeing that the FPM is produced on time and that it works effectively for the life of the project.

Almost every company has a standard form of an FPM that is geared to its type of work. In the event that your firm does not have such a standard, we have included a typical table of contents for one for a design-and-construct project in Fig. 7.8. This sample can be expanded or contracted to suit the size and complexity of your particular project. The manual for a small, uncomplicated project can be just a few pages, whereas one for a large project usually runs into one or more loose-leaf notebooks.

Introduction

The introduction should contain a statement of purpose for the project. What is the owner hoping to accomplish with the project? What needs

Table of Contents

1.0 Introduction
  1.1 Statement of purpose
  1.2 Contract controlling statement

2.0 Project description
  2.1 Brief description of project
  2.2 Location and site description
  2.3 Scope of company services
  2.4 Work by others
  2.5 Owner's responsibilities
  2.6 Owner's project objectives
  2.7 Contractor's project objectives

3.0 Contractual matters
  3.1 Type of contract
  3.2 Secrecy requirements
  3.3 Checklist of reimbursable and nonreimbursable charges
  3.4 Subcontracting procedures
  3.5 Special contractual requirements
  3.6 Statement of guarantees and/or warranties
  3.7 Applicable codes and government regulations

4.0 Project field organization
  4.1 Project organization chart
  4.2 Brief project job descriptions
  4.3 Client project organization chart
  4.4 Brief client job descriptions
  4.5 Organization chart of companies or divisions
  4.6 Client's resident field team (if applicable)

5.0 Project personnel policies
  5.1 Statement of craft labor posture
  5.2 Site labor agreement
  5.3 Project hiring practices
  5.4 Site personnel records
  5.5 Travel and expense account policies
  5.6 Labor relations procedures

6.0 Project coordination
  6.1 Communications procedures
  6.2 Communications systems
  6.3 Key personnel names and addresses
  6.4 Correspondence logging procedures
  6.5 Document distribution schedule and transmittals
  6.6 Document approval procedure
  6.7 Meetings and preparation of meeting notes
  6.8 Telephone and verbal information confirmations
  6.9 Project language (international projects)
  6.10 Project filing system
  6.11 Reproduction procedures

7.0 Planning and scheduling
  7.1 Contracting strategy
  7.2 Project execution plan

FIGURE 7.8   A sample table of contents for a field-procedure manual.

Table of Contents, Cont'd.

7.3 Project scheduling procedures
7.4 Preliminary—final schedules
7.5 Schedule control
7.6 Earned value and progress status reports
7.7 Special scheduling requirements

8.0 Project procurement procedures
8.1 Materials management master plan
8.2 Approved vendors list
8.3 Buying procedure
8.4 Expediting and inspection services
8.5 Shipping traffic control procedures
8.6 Vendor invoice review, approval, and payment
8.7 Subcontracts administration

9.0 Field warehousing procedures
9.1 Warehousing plan and inventory control
9.2 Material receiving and storage
9.3 Over, short, and damage reports
9.4 Material issue procedure
9.5 Warehouse security and weather protection
9.6 Disposal of surplus property

10.0 Heavy construction equipment and small tools
10.1 Heavy-lift and equipment studies
10.2 Project heavy-equipment policy (rent versus own)
10.3 Heavy-equipment schedule
10.4 Operation of equipment pool
10.5 Small-tools policy and control

11.0 Project estimating
11.1 Project cost estimating plan (accuracy and methods)
11.2 Appropriations estimate
11.3 Cost-trending and -reporting procedure
11.4 Definitive estimate
11.5 Project change order estimating

12.0 Project control and reporting
12.1 Code of accounts
12.2 Project budgets
12.3 Cost-control procedures
12.4 Field productivity measurement
12.5 Project reporting procedures
12.6 Project cost reports
12.7 Project progress reports
12.8 Cashflow plan and report
12.9 Project accounting and auditing procedures
12.10 Schedule of reporting dates
12.11 Invoicing and payment procedures
12.12 Computer charges

13.0 Site safety and security procedures
13.1 Owner's safety requirements
13.2 Site safety operating plan and enforcement

FIGURE 7.8 (Continued)

Table of Contents

Figure 7.8   *(Continued)*

does the owner anticipate the project is going to fill in the community, industry, or market? The people involved with the work need to know what the owner's goals are. Also, it is always appropriate to include a statement to the effect that the FPM does not replace the contract and that any conflict between it and the contract will be resolved by the provisions contained in the contract.

Project Description

A project description gives the location of the project, a site description, an overview of any processes involved, and any other outstanding features of the project. An outline of the scope of work and the services offered is important to the general knowledge of the team members. It can be neatly tied into the project objectives, which form the basis of the project program. All the goal-oriented groups involved on the project should be covered in this section, including any project-team performance incentives. Any work by others involved on the project, including major subcontractors or licensors, along with their contributions to the project, should be mentioned here.

Contractual Matters

Since the contract is a quasiconfidential document, it should be treated as such, but key areas affecting project performance should be included in this section. The people who are working on the project but will not have access to the contract need to know how the contract can affect their work. For example, it makes a difference to the project team's performance whether the contract is on a lump-sum or a cost-reimbursable basis.

Any requirements for project secrecy or confidentiality must be addressed in the FPM. All members of the team need to conform to the regulations for security and secrecy agreements, including the handling of confidential documents and equipment. This need for secrecy or confidentiality might be present on a project where a new product is being manufactured and competitors would want to know the special features of the construction.

Project Organization

This section covers the project organization charts, work descriptions, and any information pertaining to organizations involved with the project. If there are any special organizational interfaces—for example, the design-construction interface—they should be described in this section. You may also want to include the key project-personnel job descriptions here.

Project Personnel Policies

This section covers the project's labor policy and the handling of the related personnel policies for the project's craft and supervisory people. The hiring practices can go into such highly sensitive subjects as prehiring interviews, on-the-job drug testing and substance abuse, and past criminal records, all of which are necessary information for a safe project.

Project Coordination

The main part of the project-coordination section covers the communication procedures for the project. The key names, e-mail addresses, and phone numbers (both landline and cell) should be provided, and any correspondence logs should be described that are to be set up for expediting the handling of project communications. Properly logging volumes of letters, memos, transmittals, and meeting minutes generated during the project expedites the location and retrieval of correspondence when it is needed quickly later on. A system to handle this work should be made clear to all field members.

Minutes of meetings and confirmations of project information that has been transmitted verbally are critical to maintaining control over the project. Any verbal instructions should be followed up by a written statement of that instruction and the action being taken.

The document-distribution schedule, which sets up who gets copies of correspondence, drawings, specifications, and so on, plays a key role in controlling the project. It establishes the budget for project reproduction costs, which can be substantial on most jobs.

Any e-mail system(s) should include a method by which older e-mails are preserved for the length of the project. Historical files can be set up to keep e-mails beyond the normal period of six months or so.

Document-approval procedures are the key to controlling project progress. They should be set up with reasonable but fixed time limits for the approval process. This list of documents needs to be reviewed on a weekly basis to determine if approvals are proceeding according to the prescribed period of time set up for review. Agreement must be obtained from all parties to the review-and-approval process that the time allotted for this process is acceptable.

Having a standard project filing system is a big help in organizing the work in the field office. Having one throughout the company makes for easier access to project information by key team members as they move from project to project.

Planning and Scheduling

Planning and scheduling comprise a key area that has to be decided on early in the construction project. Quite a bit of generalization has probably gone into it up to this point. Now is the time to crystallize all the prior thinking about scheduling the construction of this project. Agreement with the client also is critical in this area.

Project Procurement Procedures

The procurement-procedures section lays out the work plan for the procurement and delivery of the physical resources for the project. We are speaking of a procedure to control about 30 to 40 percent of the total project budget, so this area deserves a good deal of management attention. The starting point is an approved-vendors list—an often overlooked item. If inquiries are sent to ill-chosen vendors, the whole procurement chain will suffer.

As we will state several times in this book, do not for any reason slight the procurement effort on your project, because it plays such an important part in attaining your project goals!

Field Warehousing Procedures

This section covers that portion of the materials-management master plan that begins when the materials start to arrive in the field. It needs to cover the physical storage facilities' size and location and the procedures for controlling the materials passing through them. The cost of the materials passing through the field storage facilities can constitute significant costs, so establishing a system for that purpose can cut down on the project budget.

However, keeping costs in mind, there are other things to consider. It may be advantageous to spend a little more money on the procurement of certain items, such as lumber and drywall, in order to get partial shipments instead of the entire amount of the purchase order at one time. If a purchase order is prepared with a statement such as "To be shipped as released by the project superintendent" and the minimum amount of each shipment has been negotiated with the vendor, the usually small increase in price may be well worth the added cost. Materials can be delivered to the exact location required instead of having to be moved from the primary location where the entire shipment is delivered. Many of these materials mysteriously disappear from the site because of theft. It is easier to spot a loss in a small inventory than a larger one.

Heavy Equipment and Small Tools

Not including the heavy equipment required for site grading, both rough and fine, heavy equipment such as cranes and excavators will usually be needed at the construction site. It is important to think ahead and determine where this equipment will be needed and what path it will take during its operation. Placing large stockpiles of materials in an area that will be in the path of the excavator installing underground utilities will cause a disruption to many other operations while this stockpile is moved out of the way of the excavator. Placing a field office or storage trailers where future on-site roadways or parking areas are to be built is another consideration. Moving a field office requires relocating all power, communication, and other utilities installed when the field office was first set up.

Small-tools requirements may be difficult to predict as the job commences, but it is important to have a source for either rental or purchase when the need arises.

Project Estimating

Estimating is the foundation of the project financial plan, so it must be well conceived if the project is to be a successful one. The field personnel should be aware of the various estimate components that fall within their purview. This serves as the budget against which related costs will be applied. Unless monitored carefully, costs to date plus costs to complete as opposed to the budget, need to be reviewed with the field personnel so they can determine whether they are hitting their goal, exceeding the budget, or effecting savings in the accounts under their control.

Project Control and Reporting

Project control and reporting is generally the largest section in the FPM because there is a lot of ground to cover. It is also a pivotal section, because failure here can cause a loss of project control, which is sure to result in unmet project expectations. The CM or PM plays a key role in all the activities listed, in the Field Procedure Manual but he or she can also delegate a great deal of the work to project team specialists. In those instances, the CM or PM becomes the editor of the material generated by the specialists. It is important to read and check the procedures for content, writing systems, conflicts, and project-goal criteria before releasing them to the key players. It will be your first chance to evaluate the ability of your key people to perform effectively. Particularly important items in the section are the cost-control procedures, the project budget, project reporting, and project accounting. They, at a minimum, will appear in the FPM for most projects.

Site Safety and Security

This section spells out the safety program and procedures to be followed for the specific project, including the statement that all OSHA regulations are part and parcel of the safety plan. It starts with a statement from top management of their concern that safety be given top priority in the conduct of the project.

Change Orders

The change-order procedure can be included in the FPM or, on occasion, included in the Specification Manual prepared by the project owner's architect. Change orders on a project will occur, with more frequency on some than on others, and there are legal and procedural measures that are to be followed in the initiation of a proposed change order, the documentation and response that follows, the determination of whether the proposed change order involves increased or decreased scope and related costs, and the designation of which party is to receive the proceeds of the change order when it is officially executed. Although some change-order work may originate via verbal authorization, a primary rule to follow is "No work will be started on the change until the parties agree on the scope, schedule, and cost of the additional work—in writing."

Design Procedures

The owner and the owner's design consultants are obliged to prepare contract documents that comply with applicable local, state, and federal rules and regulations. Quality-control procedures are established in the specification manual, and if new regulations are enacted, the owner and the design consultants have the option of incorporating them into the contract documents after the CM or PM assesses whether costs or completion dates are affected.

Issuing the FPM

A key factor to remember when issuing your field procedures is to get them published as early in the project as possible. An issue date more than four or six weeks after project kickoff is too late. Issuing the FPM with holds to be cleared up in later issues as the information is finalized is quite normal, so do not be made late by trying to perfect the first issue.

Early issuance of the FPM is an excellent project-personnel indoctrination tool: It gets the new team members up to speed in a hurry. It is essential that they learn the who, what, when, where, how, and why of the project without any false starts. This is especially true when a particular group of people many not have worked together as a team before.

Interoffice Coordination Procedures

We are including in this section interoffice procedures because at some time you may be involved with a construction project that is being run part-time in the field and part-time in the office. The potential for an increase in problems can occur when a split basis is being used. Confusion can arise over who has responsibility for a specific task or tasks. The only way to minimize these problems is to put in place a good coordination procedure to organize and harmonize the work at both sites.

A good starting point for the interoffice procedure is to use the FPM and tailor it to suit the interoffice operations. The satellite office has to perform the same functions as the primary office, so the systems should be made compatible at the outset. In areas in which the same systems will not fit for some special reason, a workable adaptation must be developed. The difference must be minimized to the highest degree possible in order to maximize the opportunities for meeting the project goals.

A typical case for an interoffice procedure occurs when another specialty construction company has a contract directly with the owner, say for installation of office workstations, that must be coordinated with the electrical requirements contained in the contract. The division of work for the handling of materials and equipment, design modifications, drawing interpretations, responsibility for testing for completion, and associated costs must be resolved early in the project.

Summary

This chapter has given the CM and PM an insight into the organization of the human resources, materials, and procedures crucial to meeting the project goals. The section dealing with building the project organization involves many human factors that sometimes do not come easily to the CM or PM. Leadership in those areas cannot be delegated to subordinates, so these managers must train to overcome any weaknesses in that area.

The area of the many detailed project procedures that need to be developed is an excellent one for delegation, but strong input from the CM or PM is needed to guarantee a uniform, effective set of working rules to make the project run smoothly.

CHAPTER 8

Project Control

Project control is the pivotal activity that ties together the project-management techniques of planning and organizing that we have already discussed. Project control is certainly important in leading us toward meeting our project goals, but it is absolutely essential that it be effective project control. A manager might be a little off target on planning and organizing and get away with it, but he or she cannot fail even a little bit at control and hope to come through in one piece.

A definition of control that we think is particularly appropriate for construction project work is the *work of constraining, coordinating, and regulating action in accordance with the plans to meet specific objectives*. We have set our objectives of safely building a *quality project, on time, and within budget*. We have made our time and financial plans and created an organization to execute them. Now all we have to do is proceed with the constraining, coordinating, and regulating activities that will deliver the desired results.

The Control Process

The basic mechanism of the control function is shown in Fig. 8.1. It is necessary to always measure "Actual Performance" from "Desired Performance" taking the appropriate measures to identify and analyze the deviations between the two, developing a program for corrective action and implementing that program.

Basically we start the cycle in the upper right-hand corner, measuring actual performance, which is then compared against planned performance. In Ch. 10 we will discuss how building information modeling can include software to graphically display the status of construction as indicated in the approved CPM schedule, so that at a project meeting we can just look at the window, see where we *actually* are, then project the scheduled model on our laptop to determine if we are on, behind, or ahead of schedule.

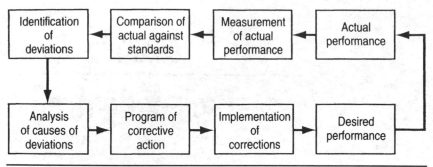

Figure 8.1 The control process.

Areas of Control

Which key project activities must we *constrain, coordinate, and regulate* to reach our project goals? The primary control areas are the ones developed in the project-planning phase, namely

- The money plan (budget)
- The time plan (schedule)
- Quality standards
- Labor supply and productivity
- Material resources and delivery
- Cash-flow projections

By concentrating on control of these six key areas, we should be able to meet our project goals successfully. Of course there are a myriad of lesser areas of control, but most of them, with the exception of the *human factors*, (labor supply and productivity) are related to these six areas.

Controlling the Money Plan

The cost-control system lies at the heart of controlling the money plan—the budget. Many well-designed cost-control systems have been developed over the years to suit the broad spectrum of capital projects. Regardless of the system used, it will take your single-minded devotion throughout the project to make any system work!

Cost-Control Definitions

Cost control, despite its simple name, means a lot of different things to different people. Some often-heard synonyms are *cost engineering*, *cost reporting*, *value engineering*, and *cost reduction*. None of them alone, though, is equivalent to *cost control*. Let us define these phrases in order to understand the differences in meaning:

- Cost engineering—a generic term that covers the total field of cost estimating, budgeting, and cost control. It is too general a term to use for real cost containment.

- Cost reporting—the gathering of cost data and reporting of the actual versus planned results without mentioning the operative word *control*.

- Value engineering—getting closer to cost control, since it looks at ways to reduce costs on specific items or activities while still maintaining project quality. However, it does not look at the total project picture or check the daily performance. Value engineering focuses only on specific items in the area of design, procurement, or construction.

- Cost reduction—also getting closer to cost control; it would be fine if it included cost reporting. The result would then be evaluation and containment of costs on the complete project.

As it turns out, true cost control for capital projects involves all of these activities at various times. We feel that cost control means *the purposeful control of all project costs in every way possible*. That means that every member of the project team has a part in controlling costs. The construction/project managers are the leaders of the cost-containment program; they must constantly reinforce this philosophy throughout the life of the project!

Cost-Control Philosophy

A comprehensive philosophy for cost control that we have developed over the years is based upon three building blocks:

1. The encouragement and promotion of cost-consciousness in the performance of all phases of the work.

2. The provision of accurate and timely data on cost status and outlook, and the highlighting of any unfavorable cost conditions or trends

3. The taking of prompt and effective action to correct problems and to provide positive feedback for continuous evaluation of those problem areas

A major problem area in most cost-control systems arises under the second point: the provision of accurate and timely data on cost status and outlook. Most good project and construction managers are willing to accept minor differences in the accuracy of the cost data if they can get those data in a timely fashion. But they must be forewarned by the staff member presenting the cost data that one or two figures may be off by X percent.

Effective managers are looking for trends in the control of the financial plans, so an accuracy of plus or minus 2 to 3 percent is close enough—percentages that the accounting department could not tolerate when compiling costs to date and the final project cost.

Cost reports take time to prepare. The figures from the field may need to pass through several hands for review and modification before being presented to the construction or project manager. And the cost data from other field and office sources, which must be compiled, take some time to obtain, verify, and convert into the format established for cost reporting.

Cost-Control System Requirements

If the cost-control philosophy is to be successfully implemented, the control systems should include these basic features:

- A simple but comprehensive code of accounts (per the Construction Specifications Institute system)

- Assignments of specific responsibilities for controlling costs within the field organization

- Standard forms and formulas based on the standard code of accounts throughout the estimating, procurement, design, construction, and cost-control groups

- A sound budget based upon a sound estimate

- A mechanized system for handling the data on medium and large-sized projects

By now everyone in the field and office should be using the Construction Specifications Institute's MasterFormat division numbers to assign numbers to the various project costs for identification purposes. The estimate was most likely prepared using this numbering system, the architect will use this system when issuing instructions or responding to queries about sections of the specifications or drawings, and subcontractor awards will be segregated by subject and their respective division numbers. This numbering system will be carried over into the field when reporting costs related to carpentry (Division 06, Wood, Plastics, and Composites) or concrete (Division 03) or masonry (Division 04) are assembled and reviewed by the project superintendent before being passed onto the construction or project manager.

The assignment of specific responsibilities should be delineated in the project cost-control section of the field-procedure manual. The description should include the forms to be used, the formats for any reports, the assignment of duties, the liaison with home-office personnel, and other matters needed to create an effective field cost-control system.

If the cost-control systems are to be effective, they must be based on a sound cost estimate and project budget. If the cost figures being controlled are inaccurate to start with, no amount of control can make them right; if that situation occurs, the heart will go out of the cost-control program.

The estimating department will be closely connected to the field cost reporting. As more data become available, adjustments will be made in the estimating database of costs. A continuing report, from the field, of per square foot of concrete foundation forming, labor to install rebar, and per-cubic-yard placement of ready-mix concrete flows back to the estimator, who incorporates it into the database. After tens or hundreds of these unit costs are received by the estimating department, the estimators will adjust their database accordingly and develop representative costs for each of these items. So the cost-reporting system becomes not only an analysis of project costs (costs to date plus projected costs) versus budget but a feed of unit costs to the estimating department to refine their database.

If a bad budget is discovered early enough in the job, the project manager and construction managers should press for a new estimate and a revised construction budget. The project manager and the field superintendent need to review the operations that fail to meet the budget in order to determine whether the operation is in fact proceeding efficiently and the estimate is wrong or, conversely, the estimate was proper but the operation is being performed inefficiently and a change in foreperson may be required along with a discussion of the inefficient work and what needs to be done to improve efficiency.

With today's easy access to laptops and smartphones, the monitoring and reporting of ongoing costs can be accomplished quickly. Weekly updates should be a standard procedure so that the various supervisors can remain current on their costs versus their budget responsibilities.

A Typical Cost-Control System

Figure 8.2 shows how the work flows through the major elements of a typical cost-control system. The idea here is to have all the necessary cost information flowing to the cost-control group on a routine basis so that it can be assimilated and organized into accurate and timely cost reports. The diagram includes the design function as it would occur on a design-build project. It also relates to an engineering-type project, and the concept is basically the same for design-bid-build projects and commercial or institutional projects.

The process starts with the project estimate passing through the client and management reviews before it is converted into the project

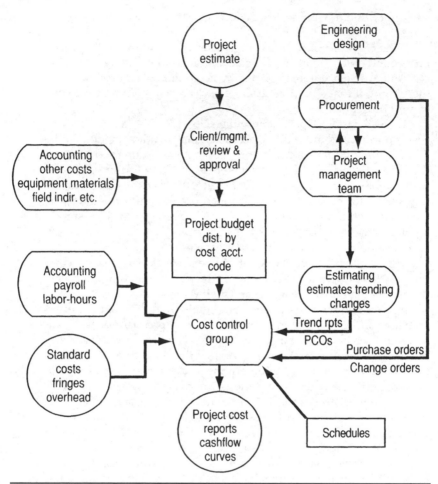

FIGURE **8.2**   Cost-control flow diagram.

budget (this is the process for design-build, but it does not apply to design-bid-build, where builders are in a competition and their estimates for the work are a closely guarded secret).

On the left side of the diagram are the normal staff departments that feed routine construction costs and expenditures to the cost-control group on a regular basis. On the right side are the design, procurement, construction, and estimating functions feeding commitment, change-order, and estimating data to the cost-control group. At the lower right is the schedule, which is regularly issued to the cost-control group so that the effects of schedule changes on project costs can be evaluated. At least monthly, the cost-control center issues the project cost report and cash-flow curves.

The idea behind this flow diagram is to have the large volume of routine data flowing in standardized, normal channels to ensure that nothing affecting construction cost is overlooked. Missing data can result in inaccurate reports, which lead to loss of control of the project's money plan. This in turn results in unmet expectations for finishing the project on or under budget.

Staff Functions

The staff functions on the left side of Fig. 8.2 are recording the monies paid out for the human and physical resources that flow into the project. These are mainly accounting functions that are paying for the commitments made by the operational groups on the right side of the diagram. And of course, in a design-bid-build project neither engineering nor architectural design costs would be part of the contractor's budget or their responsibility.

In cost control, it is important to differentiate between *commitments* and *expenditures*. Commitments are made when material and equipment are ordered and subcontract agreements issued, and they include the cost of personnel who charge their time to the project. Expenditures occur when the bills for goods and services are paid and payrolls are met. Payrolls convert from commitments to expenditures fairly rapidly, usually in a matter of days or weeks. The time lag between commitment and expenditure for materials, equipment, and subcontracted work can stretch for a month or more.

Line Functions

The activities of line functions on the right side of Fig. 8.2 are more variable and much less routine. The progress of the job can affect schedules, which in turn affects costs. Project-change orders, with their impact on budgets and schedules, will always arise. The procurement program is the chief source of the longer-term project commitments that show up in the cost report.

The project and construction teams are the center of the cost-control activities on the project. As we said earlier, the PM or CM, as the project leader, must generate the project's cost-control philosophy. The project and field report groups provide the coordination needed to make the other groups on the right side of Fig. 8.2 perform effectively. All the corrective cost-control actions and feedback are processed through the project and construction teams.

In addition to coordinating the procurement efforts, and the design efforts if the project is design-build, the project supervisory teams initiate and process all construction-change orders. They are responsible for getting the change-order estimates, submitting them, and obtaining approval so the orders can be fed into the cost-control center for use in revising the *current* budget, and possibly the *current* schedule if the change order has impacted the completion date of the project.

The project and construction teams are continuously monitoring the cost-trending reports that are maintained between the various project estimates as construction progresses. In the case of design-build, this process is ongoing from conception to completion of the plans and specifications.

The visible products of the systems shown in Fig. 8.2 are the monthly cost reports and the cash-flow curves. These are usually very detailed reports that account for all commitments and expenditures on the project. Any off-target trends in costs and cash flow must be brought to the attention of the PM or CM very quickly on an informal basis either in memo form or at a meeting to explain and discuss the impact, if any, of these trends.

It is always wise to get these types of problems out in the open early, along with some suggested corrective action, rather than to bury them in the hope that they will go away. They usually get worse, not better, unless some corrective action is taken.

Another key activity that does not show up on the flow diagram is the execution supervisor's *estimate to complete* either the labor or materials for the project. This key value, added to the commitments and expenditures, allows the *estimated costs to complete* to be compared with the baseline budget. Because these values are still estimates, their generation must be pragmatic rather than overly optimistic.

How a Cost-Control System Really Works

Let us look into some of the details of just how a typical cost-control system can work on a design-build project and, with some minor adjustments, a design-bid-build project. Assume the project has already gone through the estimating and budgeting phase; now it is in the early stages of execution.

The budget, or the contract sum, is the baseline of the cost-control system, so we must refresh our memories as to what it contains. We can expect the major cost accounts for a design-build and a design-bid-build project to be as follows (all items represent a percentage of the budget or contract sum):

	Design-Build	Design-Bid-Build
Design services	6–10	0
Major equipment accounts	25–30	30–40
Bulk materials	15–20	15–20
Construction and field costs	45–50	45–50
Contingency and inflation	Variable	Already included in each item

Controlling Labor Costs

Variances in labor budgets can stem from three possible sources:

- an original hourly-takeoff error (underestimating the amount of labor required for an operation)
- a variation in the assumed labor rates (not taking into account expiring union contracts, where labor and fringes generally increase, or the normal annual pay raise for open-shop workers)
- a variation between standard and actual labor productivity

Figure 8.3 is an example of a computer-generated printout used for controlling field-office and related indirect costs. It reflects the budget for each of these activities, many of which will be for field-office operation and personnel. It should be noted that the initial budget was modified, for any number of reasons—accommodating change orders or reflecting a rethinking of the initial budgeted items.

This form is actually the perfect format to create a cost projection, because it contains not only the cost to the period of the report but an estimate to complete each of the items displayed. This cost-projection exercise would have been created by the job superintendent, the CM or PM, and several key field-office personnel. They would also have received some assistance from the home office. Faced with the costs to the date of the report, all of these people would sit down and discuss what their best estimate is of each item for which they have control. Those costs would show up in the "Estimate to Complete" column. Adding those costs to actual costs to date, they would get the final cost for each item. If some are over budget, the reasons for the overages would be discussed, along with whether it is possible to bring those items back onto budget. If some items are project to completion at considerably below budget, some of the savings can possibly be applied to an overage to zero out two accounts.

This same process should be used for all construction-related costs, including self-performed work, material and equipment purchases, and subcontract agreements. One might ask, "Why project the total cost of a subcontract agreement when a lump sum based upon plans and specifications has been executed?" It is not uncommon for the purchasing people to forget to include, or to simply overlook, some incidental work in a subcontract agreement, only to be faced with having to issue a small change order to include that work whose costs are not reimbursable by the owner.

Changes in Labor Rates

Changes in labor rates for salaried or hourly supervisory or office personnel usually are not a problem because they are fairly predictable. Home-office and field supervisory personnel will generally

DIAMOND SHARP CONSTRUCTION COMPANY
FIELD INDIRECTS COST REPORT
LABOR HOURS & FIELD COSTS

CONTRACT NO: S-3030
CLIENT: ABC CHEMICALS, INC.
REPORT NO: 7
TYPE: LUMP SUM CONSTRUCTION

CUTOFF DATE: OCT. 24, 1993
RUN DATE: OCT. 25, 1993
PHYSICAL % COMPL.: 55%

PERSONNEL HOURS

COST CODE	DESCRIPTION	ORIGINAL BUDGET	CHANGES	CURRENT BUDGET	HOURS EXPENDED TO DATE LAST PER.	THIS PERIOD	TO DATE THIS PER.	ESTIMATE TO COMPLETE	PROJECT FINAL COST	CURRENT PROJECT VARIANCE	PERCENT OF TOTAL	PERCENT EXPENDED
1000	CONSTRUCTION MANAGEMENT	2500	80	2580	1259	172	1431	1275	2706	-126	4.82%	55.47%
1100	FIELD SUPERVISION	8575	350	8925	3499	860	4359	4200	8559	366	16.68%	48.84%
1200	PROCUREMENT - EXPEDITING	7450	175	7625	3684	675	4359	3140	7499	126	14.25%	57.17%
1300	PERSONNEL ADMINISTRATION	4000	-135	3865	2010	234	2244	1745	3989	-124	7.22%	58.06%
1400	ACCOUNTING - PAYROLL	3690	0	3690	1945	165	2110	1700	3810	-120	6.90%	57.18%
1500	FIELD SCHEDULING	3200	80	3280	1456	185	1641	1620	3261	19	6.13%	50.03%
1600	COST CONTROL - ESTIMATING	3500	150	3650	1687	180	1867	1750	3617	33	6.82%	51.15%
1700	FIELD ENGINEERING	5400	-235	5165	2679	243	2922	2130	5052	113	9.65%	56.57%
1800	STENO - CLERICAL	4500	80	4580	2359	344	2703	1876	4579	1	8.56%	59.02%
1900	FIELD MAINTENANCE LABOR	3500	-25	3475	2243	195	2438	1123	3561	-86	6.49%	70.16%
2000	SAFETY PROGRAM	6550	120	6670	3546	350	3896	2675	6571	99	12.47%	58.41%
2900	CONTINGENCY	2500	0	2500	0	0	0	0	0	0	0.00%	0.00%
	TOTAL FIELD HOURS	2854710	560	53505	26367	3603	29970	23234	53204	301	100.00%	51.84%

FIGURE 8.3 Computer-generated printout to aid in monitoring and controlling costs.

FIELD EXPENSES - DOLLARS

COST CODE	DESCRIPTION	ORIGINAL BUDGET	CHANGES	CURRENT BUDGET	DOLLARS EXPENDED TO DATE LAST PER.	THIS PERIOD	TO DATE THIS PER.	ESTIMATE TO COMPLETE	PROJECT FINAL	CURRENT PROJECT VARIANCE	PERCENT OF TOTAL	PERCENT EXPENDED
3000	TEMPORARY FIELD OFFICES	20000	9600	29500	27850	278	28128	1000	29126	374	8.41%	95.34%
3100	TEMP. FIELD WAREHOUSE	23450	4500	27950	25875	775	26450	2100	28550	-600	7.97%	94.63%
3200	FIELD UTILITY INSTALLATION	17890	2345	20235	18775	321	19096	785	19881	354	5.77%	94.37%
3300	TEMP. SITE IMPROVEMENTS	26750	-2350	24400	23070	1254	24324	1688	26012	-1612	6.96%	99.69%
4000	UTILITY COSTS	12500	1235	13735	6555	897	7452	6500	13952	-217	3.92%	54.26%
4100	OFFICE MACHINE RENTAL	12450	0	12450	6233	875	7108	5788	12896	-446	3.55%	57.09%
4200	OFFICE SUPPLIES	7500	576	8076	3465	234	3699	3976	7675	401	2.30%	45.80%
4300	LAB & TESTING SERVICES	22540	0	22540	16567	1085	17652	3800	21452	1088	6.43%	78.31%
4400	COMPUTER & SOFTWARE	40000	-1500	38500	32988	1254	34242	4500	38742	-242	10.96%	88.94%
5100	HEAVY EQUIPMENT RENTAL	88900	1200	90100	45688	7659	53347	37500	90847	-747	25.70%	59.71%
5200	SMALL TOOLS	15760	575	16335	14555	875	15430	980	16410	-75	4.66%	94.46%
5300	CONSUMABLES & SERVICES	18000	-1230	16770	8975	1474	10449	6500	16949	-179	4.78%	62.31%
5400	SAFETY EQUIPMENT	9500	0	9500	7566	743	8309	1950	10259	-759	2.71%	87.46%
9000	CONTINGENCY	25000	-4500	20500	13545	750	14295	5560	19855	645	5.85%	69.73%
	TOTAL INDIRECT EXPENSES	340240	10351	350591	251507	18472	269979	82627	352606	-2015	100.00%	77.26%

FIGURE 8.3 (Continued)

receive an annual raise in pay and perhaps a bonus, depending upon their performance and the overall profitability of the company.

Craft rates are somewhat predictable. Nonunion craft rates are often market driven; when there are lots of construction projects in the geographic area where the company operates, demand and supply work hand in hand, and last year's hourly rate may need to be increased by 10, 15, or 20 percent in order to obtain skilled workers. Fringe benefits are usually not excessive and can be reasonably predicted.

Craft rates for union workers are another matter. Although the term of the current hourly rates is known by each subcontractor hiring union labor, it is up to the general contractor to ensure that the negotiations being conducted with the various subcontractors will include some cushion or contingency to take into account a wage and fringe-benefit increase if the project in question will extend into a new union-contract time period.

Under full employment, a general-contractor team faced with the task of negotiating a statewide contract with a union trade delegation may not have much leeway in the proposition presented by the union delegates. Next year's total hourly rate, including an increase in the workers' base pay, may be accompanied by a fringe-benefit package that has the effect of significantly increasing the base pay rate of $25 per hour. With all of the union fringe benefits, and including state and federal taxes and the subcontractor's overhead and profit, a $25 base rate can rapidly escalate to $100 per hour or more.

Variations in Productivity

Worker productivity is the most likely area in which labor-hour budgets will go astray. Most standard construction unit labor costs are based upon average hours expended on similar tasks in previous projects. These standard hours are deemed sufficient for bidding purposes and for use in comparing actual costs versus estimated costs. Productivity can vary from project to project based upon the intricacy of the task, efficiency of the worker, weather conditions, and number of hours worked.

Extended workweeks prompted by the need to have trade workers put in long hours can be due to a lack of qualified workers in the area or a need to get back on schedule or compensate for slowdowns due to extreme weather conditions—either very high ambient temperatures or subzero winter conditions. Working more than an eight-hour day for extended periods of time will significantly reduce productivity. Working seven- to ten-hour days for a period of four weeks reduces workers' efficiency to below 65 percent.

Although some premium or overtime work may be necessary from time to time, the impact on labor costs when extended periods of overtime are required are dramatic.

Productivity is largely a result of good supervision and a management culture that views it as an essential part of a successful construction company. According to a 2012 survey conducted by FMI (a management and consulting firm for the engineering and construction industries that is headquartered in Raleigh, North Carolina) the construction industry as a whole has not seen a significant increase in productivity during the last five decades when compared to other industries. Since the construction industry does not have any industry-wide productivity standards, this makes the industry-wide study of productivity difficult.

Improving Construction Productivity

A study by Scott Kimpland and Phil Warner for FMI in 2009 asked respondents to comment on productivity by asking them a simple question: "Of the initiatives you've taken to improve productivity, which have given you the best results?"

The answers to this question all point to better involvement of managers in construction activities and better training of these managers:

- Provide weekly status reports versus budget incentives to field personnel to beat the labor budgets.
- Employ all methods to improve communications between all parties.
- Conduct an improved explanation of project expectations prior to the start of the project, followed by regular inspection and reporting.
- Institute cost coding and productivity monitoring.
- Establish baseline productivity numbers and then measure changes against them.
- Have supervisors measure productivity on a daily basis and report their findings to their project manager.
- Establish key performance indicators and report them weekly to their field managers.
- Become selective in building teams that work well together.
- Conduct preconstruction coordination meetings and gain commitments of coordination for team members.
- Perform close planning of day-to-day activities.
- Formalize the planning and scheduling process and incorporate these activities into site-specific safety and quality programs.
- Plan the work that is to be measured, establish goals, communicate those goals, and measure and provide feedback of results in a timely fashion.

- Develop personnel schedules for each project and follow through with these schedules throughout the project.
- Increase training at the project-management and field supervisory levels.

When FMI asked respondents which of the actions they had taken to improve productivity worked the best, the answers they received were as follows:

- Continuous improvement with fabrication.
- Developing tracking metrics that are simple while providing for a high degree of accuracy.
- Fabricating everything possible.
- Getting project managers to take responsibility for productivity.
- Greater use of automatic welding.
- Greater focus on quality.
- More in-depth planning up front.
- Hiring better field managers.
- Labor forecast reports e-mailed to field managers on a weekly basis after payroll is updated for the week.
- Lean-construction techniques that require seven- to ten-day and six-week look-aheads.
- Making productivity an integral and specific part of work planning. As work is planned, the contractor's engineers not only think about how to build it, they must also plan for safety and productivity, which includes considering access, lay-down areas, material handling, etc.
- Measuring field productivity and sharing the results with supervisors.
- Moving to totally integrated design-build teams.
- Putting together a complete operational plan before starting an activity, and monitoring and making follow-up revisions to the plan as the work progresses.
- Eliminating marginal workers.

FMI also created some bar charts and graphs to visually reflect the responses they received during their survey:

Figure 8.4 reflects how contractors perceived their productivity trend over the past two years.

Figure 8.5 reveals the largest internal challenges to improving productivity.

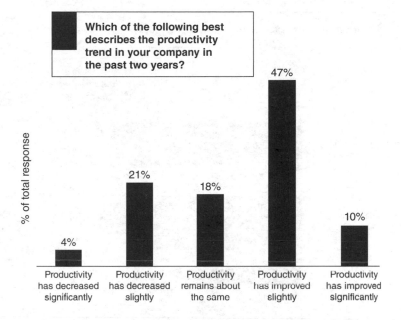

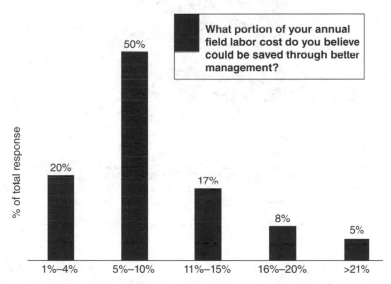

FIGURE 8.4 Perceptions of productivity and the trends. (*Reprinted with the permission from FMI Corp. For more information visit www.fminet.com*)

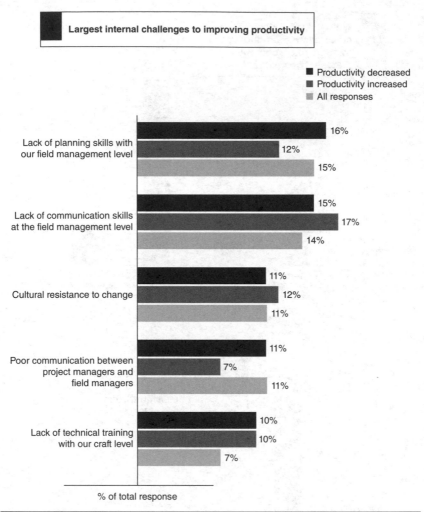

Largest internal challenges to improving productivity

■ Productivity decreased
■ Productivity increased
▨ All responses

Lack of planning skills with
our field management level
- 16%
- 12%
- 15%

Lack of communication skills
at the field management level
- 15%
- 17%
- 14%

Cultural resistance to change
- 11%
- 12%
- 11%

Poor communication between
project managers and
field managers
- 11%
- 7%
- 11%

Lack of technical training
with our craft level
- 10%
- 10%
- 7%

% of total response

FIGURE 8.5 Internal controllable variables' effect on productivity. (*Reprinted with permission from FMI Corp. For more information visit www.fminet.com.*)

Figure 8.6 shows the largest external challenges to improving productivity.

Prefabrication of piping, ducts, and framing systems has always intrigued contractors and subcontractors, and it appears from the FMI study that the majority of the respondents are using prefabrication as a method to improve productivity. Figure 8.7 shows the percentage of contractors using prefabrication and their estimate of the impact it has on productivity.

In Ch. 10 we discuss the design process known as building information modeling (BIM), whereby the virtual building is created in 3-D,

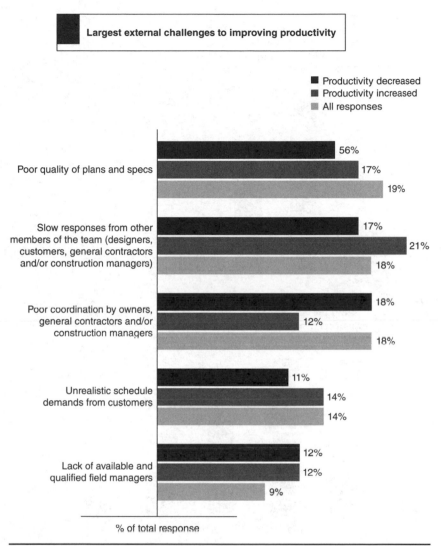

Largest external challenges to improving productivity

- ■ Productivity decreased
- ■ Productivity increased
- ▨ All responses

Poor quality of plans and specs
- 56%
- 17%
- 19%

Slow responses from other members of the team (designers, customers, general contractors and/or construction managers)
- 17%
- 21%
- 18%

Poor coordination by owners, general contractors and/or construction managers
- 18%
- 12%
- 18%

Unrealistic schedule demands from customers
- 11%
- 14%
- 14%

Lack of available and qualified field managers
- 12%
- 12%
- 9%

% of total response

FIGURE 8.6 Largest external challenges to improving productivity. (*Reprinted with permission from FMI Corp. For more information visit www.fminet.com.*)

a process that has revolutionized the industry. Figure 8.8 shows the increase in productivity attributed to the use of BIM, according to FMI's survey respondents. It also shows in what capacity BIM aided the projects. Clashes and conflicts of various systems designed to fit into a particular horizontal or vertical space have always presented a problem to construction managers and project managers. The drawing-coordination process required in most contract specifications provides for subcontractors and vendors to meet with the contractor

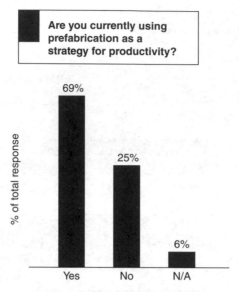

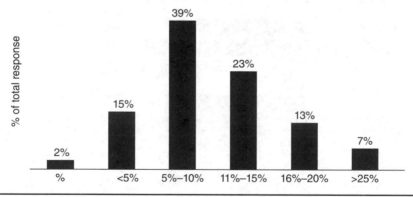

FIGURE 8.7   Prefabrication and productivity. (*Reprinted with permission from FMI Corp. For more information visit www.fminet.com.*)

and review their systems installations to determine if there are any conflicts between, say, a fire-protection main and an HVAC duct both occupying the same space above the ceiling or in a chase. As you can see from Fig. 8.8, clash detection and interference management represent 31 percent of the usefulness of BIM.

In Ch. 2, as you may recall, we discussed the new forms of standard contracts issued by the American Institute of Architects (Integrated Project Delivery), the Associated General Contractors of America (ConsensusDocs), and the Lean Construction Institute.

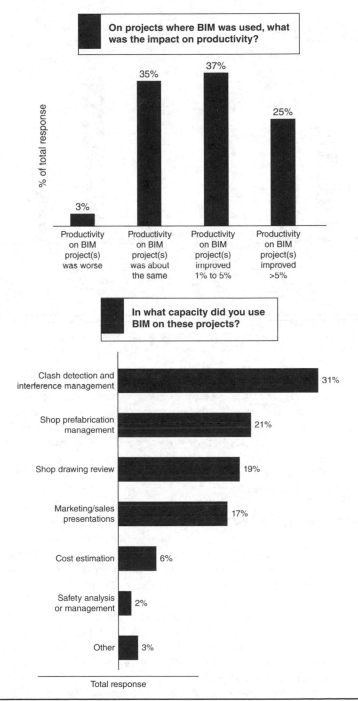

FIGURE 8.8 BIM—it's impact on productivity. (*Reprinted with permission from FMI Corp. For more information visit www.fminet.com.*)

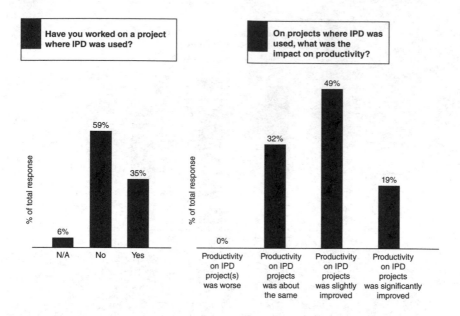

FIGURE 8.9 Integrated project delivery and it's impact on productivity. (*Reprinted with permission from FMI Corp. For more information visit www.fminet.com.*)

Figure 8.9 reveals the impact on productivity for those contractors using the Integrated Project Delivery (IPD) system. Figure 8.10 shows contractors' familiarity with and opinions of lean construction.

Emerging techniques and the concepts of collaboration and risk sharing all offer the potential for greater productivity, but we cannot dismiss the basics: developing a comprehensive plan incorporating time and resources, committing to adhere to these plans, and monitoring them frequently.

Under-running the Labor Hours

Some people do not consider it a problem when the fortuitous circumstance of underrunning the labor hours arises, but some comment on how to handle it does seem to be called for. Remember Parkinson's law: The amount of money (or time) expended always rises to the amount allotted.

When actual labor unit costs are consistently less than the budgeted amount, this may of course be due to a very efficient field supervisory crew; but it needs further investigation. One example we experienced will illustrate this need to delve into the reason for costs that are lower than budgeted.

On a high-rise apartment project that was in the substructure, foundation stages, the actual unit costs reported weekly for stripping concrete foundation walls were well below the budgeted cost.

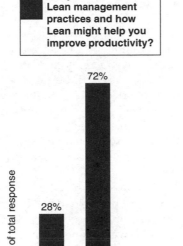

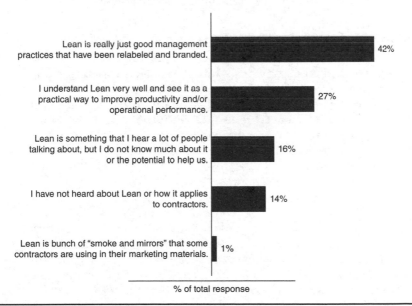

Figure 8.10   Lean management and it's impact on productivity. (*Reprinted with permission from FMI Corp. For more information visit www.fminet.com.*)

One day the writer visited the project after all workers had left the job and only the field superintendent was there. As the writer approached the site, he saw the wooden foundation forms come flying out of the excavation; the superintendent was stripping and removing the forms. He asked the superintendent, "What are you doing down there, doing physical work? Are you applying your hourly rate to this task?" The answer was, "No, I needed the exercise so I decided to strip forms." The writer responded, "So, Billy, that's why your form-stripping unit costs are so low—you are not applying your costs to this task. Why are you doing this?" His answer was, "job security." The writer replied, "You are a top-notch super, you don't have to worry about your job. But you sure are screwing up the unit costs for form stripping, and I want it to stop."

Actual unit costs that are lower than budgeted costs, when they occur with some consistency, need to be reviewed with the estimating department. It may be possible that the company lost a previous bid because some of the unit costs were higher than they should be. The lesson of underrunning actual costs is: Investigate the reason for the occurrence. Is it a one-off event, or has the supervisor experienced similar underruns on previous projects?

The CM or PM must impress upon the field supervisors that they need to monitor all unit costs and that a savings in one cannot be used to counter an overrun in another. Each element of the budget stands on its own and must be treated as such. The maintenance of accurate labor-cost reports and a weekly review should become part of the field superintendent's weekly meeting with the staff of forepersons.

Controlling Material-Resource Costs

The physical resources on a given project can run from 40 to 60 percent of total installed costs. Obviously this percentage varies depending upon the nature and physical composition of the project, but even at only half of this percentage, it is a budgetary force to be reckoned with. The human element is present to a much lesser degree here than in the self-performing-labor portion of the budget. A CM or PM does not have to deal with labor productivity factors when dealing with equipment and materials. Prices are controlled by market forces and negotiated costs with the vendors.

The overall philosophy for controlling the physical-resources budget is much the same as for controlling human resources, in that we start with an estimate and convert it to a budget, which becomes the baseline against which all future costs will be compared. As the project progresses, we check the actual delivered cost against the estimated cost for each item. The estimate of the cost to complete the delivery of each physical resource occurs in the same way as the estimates for human resources. Any variance between budgeted or planned and actual will be reflected in the job cost report.

Unfortunately, jobsite theft of materials occurs on many projects, so security of stored materials takes on dual importance. Not only is the loss of material via theft costly in that the material must be replaced, but if a close watch is not kept over inventory, a crew may need to cease work when they run out of materials and must await the replacement shipment.

Reviewing a Cost Report

The best way to get an overview of construction control of the materials budget is to review a typical capital-project budget, like the one in Fig. 8.11. This material-status report relates to a process-engineering project, as can be seen from the items in the "Description" column—vessels, towers and reactors, pumps and drivers. So for a commercial or institutional project, one can substitute plywood, framing lumber, steel studs of various sizes and metal gauges. Although the items may vary, the concept remains the same: The description of the item is listed in the first column, followed by the original budget, followed by any changes to that budget (possibly by the addition of change-order work). Also included are a "Commitment to Date" column—i.e., the buyout amount—the costs to date, and the estimated costs to complete, which then result in the total cost for the items that is to be compared with the budget.

Is There a Cost-Control Problem in Figure 8.11?

This cost report contains one element that is often overlooked—just going to the bottom line to determine whether the project is on budget, over budget, or under budget. The bottom line shows a savings of $353,000, which indicates that the total material-and-equipment account is under budget and therefore in good shape. In looking at the numbers in the "Projected Variance" column, however, it is obvious that something is wrong. The equipment account shows an overrun of $1,719,000, certainly not a good showing. The bulk-material account shows a savings of $2,072,000, which by itself is fine. But how can that be, when the costs for aboveground piping are being exceeded by $2,400,000?

Account number 0199 for a steam pipeline shows a savings of $5,040,000, in the Project Variance column, which turns out to be the culprit. The item was recently deleted from the project in Costs to Date and Estimate to Complete columns as unnecessary, but it is still carried as a savings instead of being shown as a negative change order in column 3. Carrying the account in the wrong column has created a false sense of well-being by turning a $4,687,000 overrun into a $353,000 savings. We would surely have a very strong conversation with the field or office personnel who caused the need for this obvious correction to the cost report.

CONTRACT:
CLIENT:
PROJECT:
LOCATION:

*** COST PROGRESS REPORT ***
MATERIAL STATUS REPORT
SUMMARY ALL AREAS

REPORT DATE: JULY 27,1989
PERIOD ENDING: JULY 24, 1989
REPORT NO.: 005
ALL FIGURES IN THOUSANDS
PAGE   OF

DESCRIPTION	ORIGINAL BUDGET	TRANS	APPR EXTRA	CURRENT BUDGET	COMMIT PERIOD	COMMIT TO DATE	COST TO DATE	EST TO COMPLETE	PROJ COST	PROJ VAR	% COMMIT
MAJOR EQUIPMENT	2582	0	0	2582	0	0	0	2998	2996	414	0.
0111 TANKS	138	0	0	138	101	172	0	198	370	232	46.5
0112 VESSELS AND DRUMS	1363	0	0	1363	857	1447	0	773	2220	857	65.2
0113 TOWERS AND REACTORS	6646	0	0	6646	0	4030	0	3641	7671	1025	52.5
0114 INTERNALS	1435	0	0	1435	0	0	0	1519	1519	84	0.
0115 HEAT EXCHANGES	4460	0	0	4460	-25	1072	0	4018	5090	630	21.1
0117 COMPRESSORS/DRIVERS	3490	0	0	3490	0	2235	0	356	2591	-899	86.3
0118 PUMPS AND DRIVERS	1674	0	0	1674	4	4	0	1745	1749	75	.2
0119 SPECIAL EQUIPMENT	4958	0	0	4958	634	636	0	3871	4507	-451	14.1
MAJOR EQUIP ESCAL	1246	0	0	1246	0	0	0	998	998	-248	0.
*TOTAL EQUIPMENT	27992	0	0	27992	1571	9596	0	20115	29711	1719	32.3
0121 A/G PIPE	2285	0	0	2285	0	0	0	4685	4685	2400	0.
0122 A/G FLANGES	1281	0	0	1281	0	0	0	1226	1226	-55	0.
0123 A/G FITTINGS	1715	0	0	1715	0	0	0	1695	1695	-20	0.
0124 U/G PIPING MATERIAL	957	0	0	.957	0	0	0	957	957	0	0.
0125 SHOP FABRICATION	2379	0	0	2379	0	0	0	2379	2379	0	0.
0126 A/G VALVES	2661	0	0	2661	0	0	0	2137	2137	-524	0.
0127 PIPING SPECIALTY	1326	0	0	1326	0	0	0	1340	1340	14	0.
0131 FDNS/STRUCT/PVG MAT	1016	0	0	1016	0	0	0	1095	1095	79	0.
0138 STRUCTURAL STEEL	2095	0	0	2095	0	0	0	2131	2131	36	0.
0140 MAJ ELECT EQUIP	3286	0	0	3286	0	314	0	3033	3347	61	9.4
0141 ELECT MATL/DEVICES	1583	0	0	1583	0	0	0	1822	1822	239	0.
0150 INSTR/CONTROL DEV	4020	0	0	4020	0	0	0	4249	4249	229	0.
0151 INSTR VALVES	1242	0	0	1242	0	0	0	1242	1242	0	0.
0152 INSTR/BULK MATL	160	0	0	160	0	0	0	160	160	0	0.
0165 BULK FREIGHT	390	0	0	390	0	0	0	390	390	0	0.
0166 VENDOR REP	192	0	0	192	0	0	0	192	192	0	0.
0199 STEAM PIPELINE	5040	0	0	5040	0	0	0	0	0	-5040	0.
BULK MATERIAL	6957	0	0	6957	0	0	0	7360	7360	403	0.
BULK MATL ESCAL	3659	0	0	3659	0	0	0	3765	3765	106	0.
*TOTAL BULKS	42244	0	0	42244	0	314	0	39858	40172	-2072	.8
***TOTAL DIRECT MATERIAL	70236	0	0	70236	1571	9910	0	59973	69883	-353	14.2

FIGURE 8.11 Cost progress report—equipment and materials.

Although this is an obvious screwup on the part of the project's supervisors, such failures to update committed costs in the cost report can be more subtle and can slip by more easily. This can occur, for example, when a change order of some dollar significance has been approved by the owner but either accounting staff or field supervisors fail to make the required adjustments to the budget.

Cost-Control Summary—All Areas

Figure 8.12 shows a typical budget summary sheet for the same project as in Fig. 8.11. Let us examine the major accounts to see if we can spot any problem areas from the numbers shown. From the "Total Pro-Services" numbers (home-office costs), we can see that the design is about 17.6 percent completed, so the design is just starting to build to a peak. The total home-office service is the sum of pro-services, overhead, and profit, totaling $21,826,000. That number divided by the Grand Total $187,836,000 gives a ratio of 11.6 percent, which is in line for a petrochemical project. Figure 8.13 is a typical cash flow chart for a project such as this.

On the other hand, the pro-services account is already showing a projected overrun of $2,172,000, or a 17 percent increase. The higher value is still within the 11.6 percent rule-of-thumb number relating to the increase in design cost that early in the project, without any change orders listed, does look suspicious. It could indicate that the design cost was underestimated or that change orders are not being processed properly, because it has surfaced so early in the project. In any event, a thorough review of the home-office services accounts seem to be in order.

We have already discussed the false indication of a savings in the direct materials caused by the deleted steam pipeline. Another suspicious savings is in the construction-subcontracts account in Fig. 8.12, which looks peculiar at this stage of the project. The account shows a projected savings of $11,185,000 on a base budget of $37,633,000; that is 30 percent. Since no subcontracts have been committed as yet, it is difficult to imagine that such a prediction could be made without having a reduction in the scope of work. The same comment goes for the field-labor account, which shows a savings of $5,168,000, or 23.2 percent, before one field labor hour has been expended. All of these add up to a nifty savings of $16,706,000, when only 14.2 percent of one account has been committed. If we drop down to the total line, only $13,702,000 has been committed, which is really only 8 percent of the budgeted funds.

Those are only a few points about the cost report that we have developed from analyzing just two summary pages. There may be more suspicious areas hidden away in the backup to the summary pages. Having wide swings of that sort in some of the accounts so

CONTRACT
CLIENT
PROJECT
LOCATION

** COST PROGRESS REPORT **
MONTHLY JOB PROGRESS REPORT
SUMMARY ALL AREAS

REPORT DATE: JULY 27, 1989
PERIOD ENDING: JULY 24, 1989
REPORT NO.: 005
ALL FIGURES IN THOUSANDS
PAGE OF

DESCRIPTION	ORIGINAL BUDGET	TRANS	APPR EXTRA	CURRENT BUDGET	COMMIT PERIOD	COMMIT TO DATE	COST TO DATE	EST TO COMPLETE	PROJ COST	PROJ VAR	% COMMIT
**TOTAL DIRECT MATERIAL	70236	0	0	70236	1571	9910	0	59973	69883	-353	14.2
**TOTAL SUBCONTRACTS	37633	0	0	37633	0	0	0	26448	26448	-11185	0.
**TOTAL DIRECT LABOR	22203	0	0	22203	0	0	0	17035	17035	-5168	0.
***TOTAL DIRECT COST	130072	0	0	130072	1571	9910	0	103456	113366	-16706	8.7
**TOTAL FIELD INDIRECTS	18797	0	0	18797	0	0	0	16146	16146	-2651	0.
**TOTAL PRO-SERVICES	12933	0	461	13394	551	2733	2733	12833	15566	2172	17.6
**TOTAL OTHER COSTS	6099	0	0	6099	0	0	0	5618	5618	-481	0.
***TOTAL INDIRECT COST	37829	0	461	38290	551	2733	2733	34597	37330	-960	7.3
**TOTAL OVERHEAD	5417	0	173	5590	271	1059	1059	5136	6195	650	17.1
***TOTAL	173835	0	643	174469	2339	13702	3792	143831	157533	-16936	8.7
**TOTAL ESCALATION	0	0	0	0	0	0	0	0	0	0	0.
**TOTAL CONTINGENCY	14001	0	0	14001	0	0	0	13194	13194	-807	0.
***GRAND TOTAL	187836	0	634	188470	2339	13702	3792	157025	170727	-17743	8.0

FIGURE 8.12 Cost progress report—summary.

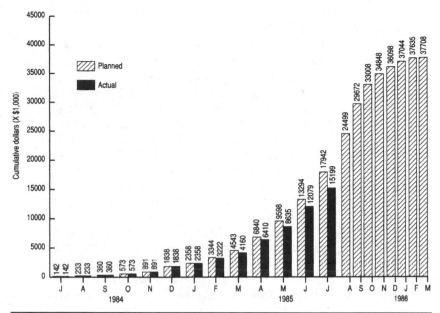

Figure 8.13 Cash-flow chart.

early in the project indicates that some person or group of people has done a horrible job of handling the budget accounts on this project.

Escalation and Contingency

Since we are already dealing with a project team that has done a terrible job in handling the cost progress report, we view the escalation and contingency accounts with equal suspicion.

We previously discussed holding escalation and contingencies in separate reserve accounts until needed. Assigning fixed amounts to the major accounts is probably a good practice, since it allows more flexibility in putting money where it is most needed.

It is normal to reduce the main contingency and escalation funds as the commitment percentage increases and the exposed portion of the unspent budget becomes smaller. Any changes in those accounts require the approval of the company management. For project and construction managers, it is best to manage these funds with a lot of thought before recommending any revisions. The complexion of a budget can change overnight, so maintaining a conservative balance as long as you can makes sense. You will get lots of credit if the healthy escalation-and-contingency balance shows up at the end of the project.

Specific Areas for Cost Control

The amount of control a project manager or construction manager has over a project depends upon the type of contract being considered by the owner. In a design-bid-build project, all bidders are provided with complete plans and specifications. When the bidding is for a public-works project, as we have stated, any changes to the scope of work outlined in the bid documents, if submitted with the contractor's bid, may be cause for disqualification. Once awarded the public project, the contractor then has the flexibility of meeting with the owner and the design consultants to discuss and offer changes to the scope of work and some potential savings.

In a design-build project, one of the prime advantages to the owner is that design and costs are being monitored constantly. Designs or construction details that add costs to the owner's initial budget are reviewed by the owner and the design-build team and resolved before the design moves on. The expectation of the owner is that once construction drawings have been completed, the design and budget coincide. If, in fact, the budget has been increased, the owner has been aware of the increase and the reason for it and agreed to it.

When a construction-management contract is negotiated with an owner and the owner elects to have the CM work with the design consultants, then similar to the design-build scenario, cost control will go hand in glove with design.

And then there is the negotiated contract, which can take two routes: As one possibility, an owner, having worked successfully with the contractor on a previous project or two, invites the contractor to participate in the project at an early design stage, providing estimating knowledge. Alternatively, after a design-bid-build bidding session, if the project comes in over budget, the owner may select the low bidder or another bidder that the owner feels comfortable with and ask that person to work with the design team to effect changes that lower the cost of the project.

All of these scenarios require one important factor: The contractor has an up-to-date database of costs that he or she will stand behind while working with the owner to achieve the project goals.

Procurement-Cost Control

Buying the large amount of physical resources and subcontracts required for a capital project requires careful analysis and a search for competitive bids from reliable vendors and subcontractors. Depending upon the company's purchasing setup, the CM or PM may have primary responsibility for purchasing, may share responsibilities with the purchasing department, or may merely act as a final check on the contents of the purchase order or subcontract agreement.

Effective cost control in procurement is founded upon good procurement procedures, which should comprise the following factors:

- A comprehensive list of approved vendors and subcontractors
- Ethical bidding practices
- Sound negotiating techniques
- Change-order controls
- Control of open-ended orders and subcontracts
- Control of procurement, expediting, and inspection of goods and services
- Maintenance of purchasing-status reports

Procurement's goal is to buy the specified amount of goods and services for the project at the best possible price. The first four items just listed contribute the most to maintaining cost control early in the project; the remaining three deal with later procedures.

The suppliers on the vendor list must be capable of supplying the proper kind of goods at competitive prices to meet the schedule. When you are operating in the geographic area where most of your work is performed, it is presumed that you have an established list of materials vendors that you have dealt with over the years. For out-of-town operations, new vendors must be sought and investigated as to their financial strength, reliability, and merchandise quality, by either the purchasing department or the CM or PM. A request for a statement from a vendor's primary bank as to the vendor's financial stability is in order, and checking with other contractors in the area may be helpful in verifying the vendor's reliability. A trip to the short-listed vendors' warehouses will allow the CM or PM to quickly inspect their materials and obtain a feeling as to whether they are organized and run a tight ship.

Ethical bidding practices are essential in establishing relationships with vendors and subcontractors of a like mind. Word of unethical bidding practices by a construction company or construction manager gets around rather quickly, and will create an environment where the CM or PM may receive a good price initially but not in the long run. If bidders know their price will be "shopped," they will not provide the best price initially; the result of such a practice will be considerable time added to the procurement process—time better spent on planning and organizing.

Sound negotiating techniques make it easier to ferret out respectable vendors and contractors. The first step in procurement is to have complete knowledge of what exactly is to be purchased or subcontracted. Obtaining materials quantities from the estimating department is the first step in dealing with materials suppliers. The specifications for the materials will be found in the specification manuals or possibly on the drawings. Unless a prior discussion with the design consultants allows a change in these specifications, the CM or PM should provide the vendor with a copy of the appropriate specification and indicate that this spec will be included in any purchase order issued.

As we discussed previously, paying a slightly higher price for split shipments of certain materials may be well worth the price to have them delivered to the location where they will be consumed rather than to a central storage area, which will require the field superintendent to have them redistributed as construction progresses.

Spot inspections of such items as wood or metal studs, plywood, Sheetrock, and flooring materials may reveal excessive damage requiring replacement by the vendor. Along with any such demand for replacement should come a cautionary statement that any excessive damaged goods in future deliveries will be cause to cancel the vendor's purchase order.

A typical purchase order should include the following:

- The vendor name; address; phone or fax number or e-mail address; and contact person for product deliveries
- An instruction that the project name and number are to appear on all delivery receipts and vendor billings, along with the cost code assigned by the PM or CM (this information will aid the accounting department in assigning these costs to the proper categories and projects, which in turn will be of great assistance in tracking these costs on the cost-control or cost-projection form)
- The quantity of material and equipment ordered, a brief description, and the specification section confirming the type and quality standards to be adhered to
- Trade discounts, delivery charges (if any), the tax-exempt number (if the purchase is tax-exempt, for nonprofit work), an indication whether partial releases are to be made, and the minimum quantity for each item
- A preprinted statement that signed receipts for all deliveries are to be attached to the vendor's invoice and include the project name, project number, and product cost code
- Another statement on the purchase order that prices indicated on the purchase order are not subject to change without prior notification and presentation before that delivery is scheduled for shipment
- Any special delivery requirements, such as one that all shipments must be made from a certain location, like "All deliveries are to be made at the Elm Street entrance to the site" (this may be the area where you have an antitracking driveway so that trucks leaving the site will not carry dirt onto the public street)

Subcontract agreements require extra scrutiny to ensure that all required items are included in the scope and price of the subcontract agreement. With respect to subcontract awards, there are a number of procedures that ought to be followed to ensure that all terms and conditions are met. Your company's attorney has probably assisted in the preparation of the contract boilerplate, but it remains for the purchasing agent or the CM or PM to fill in lots of blanks regarding scope and price:

- A review of division 1 of the specifications may reveal a requirement for a particular subcontractor to participate in the preparation of "coordination drawings."
- Preinstallation or prefabrication conferences may be required by the architect for certain items of work.
- Attendance at scheduled progress meetings may be required for each subcontractor's project manager.
- Shop-drawing submission procedures and sample submissions will most likely be included in this section, of the subcontract agreement as well as in the specific section devoted to the relevant product or material. Each contractor may be required to provide a special stamp for each submission.
- Product-substitution procedures must be spelled out in detail.
- Responsibility for cutting and patching should be assigned, and by the project manager.
- Closeout procedures are specified for the general contractor and his or her subcontractors.

A review of the applicable division in the specification manual for each subcontractor may include some unique provisions that should be included in the subcontract agreement:

- Specific quality standards, accompanied by manufacturer's certification
- Cold- or warm-weather precautions, as applicable to weather-sensitive trades
- Requirements relating to testing and inspection by the design consultants
- Mock-ups required for approval before work can proceed
- Procedures for product substitution
- Requirement of inspection by the design consultant before the contractor's work can be enclosed
- Tolerances allowed as acceptable by the design consultants

And then there are the specific requirements of the construction manager or the contractor:

- On a daily basis, the subcontractor shall clean up the work in the area where he or she was operating.

- The subcontractor shall not sub-subcontract any work to a second- or third-tier contractor without the knowledge and approval of the CM or PM.

- The subcontractor is responsible for repairing damage to other subcontractors' work.

- The subcontractor is bound to the same conditions as are included in the contract between the owner and the contractor. (This may already be included in the contract boilerplate.)

- If a subcontractor falls behind schedule based upon the subcontractor's performance, that subcontractor will be required to work additional hours at no added cost to the owner or the contractor in order to get back on schedule.

- Change-order work, if required, shall not proceed without written agreement between the CM or PM and the subcontractor.

This is only a partial list of items to consider. The primary goal in negotiating a purchase order or a subcontract agreement is to include all materials or an inclusive scope of work, to avoid extra charges down the road which could have been negotiated in the purchase order or subcontract agreement at little or no extra cost.

Construction Costs and the Argument for Negotiated or Design-Build Work

Since construction costs absorb the lion's share of the project budget even in the design-build mode, we expect them to be a fertile ground for cost control and cost reduction. The following procedures to review for savings all make the case for construction management, design-build, and negotiated work; they can even serve as a sales-development tool in seeking new work.

- Conducting a constructability analysis
- Developing a contracting strategy
- Planning temporary facilities and utilities
- Planning heavy-lifting equipment and associated equipment
- Assigning lay-down and working areas on-site
- Organizing the field supervisory staff
- Establishing the construction philosophy

- Setting field-personnel mobilization and demobilization procedures. When it is possible to get the constructor's involvement early through any of the types of owner–contractor relationships such as Cost Plus a Fee with a Guaranteed Maximum (GMP) or design-build, the task of cost reduction becomes much easier. Owners may also consider hiring a construction expert to assist in reviewing constructability issues before the final set of plans and specifications is prepared.

As we look at these areas, we can see that the best time to save is early in the project.

Conducting a Constructability Analysis

A constructability review is a review of the plans and specifications to check for buildability issues. A check for incompleteness of design, missing details, and dimensional discrepancies is in order, as is a review of notes on the drawings, completeness of wall-section details, window-framing and water-infiltration issues, and door and window schedules that match the door and window types and sizes noted on the various floor plans. With respect to site work, invert elevations for underground utilities as designated on the building's structure should match the inverts on the appropriate mechanical, plumbing, and electrical drawings. In a complete constructability review, every detail, every section, every floor plan, and every page of the specifications manual is reviewed. The time for such a review is when the plans, specifications, and bidding instructions have been completed and before the project is submitted for bid.

Once an experienced constructability reviewer has been selected, the method by which he or she is to proceed on the review must be established, along with his or her method of reporting. A kickoff meeting should be held with the constructability expert and the design group to develop the standard procedure for marking up drawings and noting questions in the specifications manual. This information can be placed on a spreadsheet for ease of review, comment, and correction.

The purpose of the constructability review is to reduce the number of questions that may arise during the bidding process, if the project is to be competitively bid, and reduce the number of requests for information or clarification and the potential for change orders that can have an impact on the project's budget.

Developing a Contracting Strategy

Owners have multiple choices in their method of contracting; if they are experienced in previous capital projects, they may have knowledge of these choices. Early involvement with construction experts is the best way for owners to control costs. They can go the route of design-bid-build and employ a constructability expert to reduce the potential for bids to come in over budget. They can go the route of construction management, whereby they utilize the experience and knowledge of a construction specialist during the design stage. They can decide to invite one or several design-build firms to present conceptual drawings and budget proposals in a competition or, via competitive bidding (for owners that are not government agencies), select a contractor who may not have had the low bid but with whom they will negotiate both scope and price to achieve their end goal.

Planning Temporary Facilities and Utilities

Since costs for temporary facilities and utilities become part of the overall construction estimate or bid, they need to be reviewed to make sure the facilities provided are adequate but not gold plated. General conditions can assume a significant portion of the budget. If the project is of moderate size, a full-time field engineer may not be required; the field superintendent can assume those functions. Reproduction costs can be high, and it is not unusual for the contractor to provide, free of charge, one complete set of plans and specifications to each major subcontractor. If subcontractors need more copies of either or both, they will have to purchase them from the reproduction company. As stated previously, the location of the temporary field office, storage sheds, and subcontractor field offices should take into consideration the possibility of a future need to move the facilities to excavate for underground utilities or allow passage for materials and equipment to be stored.

Planning Heavy-Lifting Equipment and Associated Equipment

In high-rise structures, it is common practice for the general contractor to provide a man lift for personnel and small tools, but for any lifting of heavy equipment such as rooftop mechanical equipment, the subcontractor may need to provide his or her own lifting equipment; or if the general contractor or construction manager has a crane on-site, the subcontractor may be charged for the use of that crane. Pathways for this equipment need to be planned.

Assigning Lay-Down and Work Areas On-Site

The same considerations that apply to the planning of temporary structures and heavy-lifting equipment should apply to assigning lay-down and work areas on-site. At some stage in the building's construction, the general contractor's or subcontractor's field offices may be able to be relocated inside the building, freeing up space on the site for pad-mounted mechanical, engineering, and plumbing equipment or for grading for roadways or parking and landscaping areas. The main consideration to keep in mind is avoiding the expense of moving sheds, offices, materials, equipment, etc., once they have been placed on-site.

Organizing the Field Supervisory Staff

Organizing the field supervisory staff is probably the largest single cost item in the field indirect-cost budget. Since many of these people are highly skilled and trained, their salaries will reflect that skill and training. A good philosophy to adopt is to be slightly understaffed for best efficiency. Avoiding idle hands is not only a cost saver but a reflection on the ability of the CM or PM to plan effectively. Monitoring and improving staff efficiency is an ongoing activity; at times a supervisor may need to be temporarily borrowed from another project, such as when a backlog occurs or there is a temporary period of intense construction activity.

Establishing the Construction Philosophy

This activity will have occurred in the planning stages of the project, going all the way back to the way in which the project estimate was prepared. Should the job be worked as a union project or does the construction environment in which the project is to be built allow the work to be performed as a merit-shop contractor, using either union or nonunion personnel based upon which subcontractor is most competitive and qualified? Once that philosophy has been established, the ground rules for operational planning will have also been established.

If it will be a union-shop project, there may be a glitch down the road if the owner is preparing to perform some of the work with its own contractors. Two such processes come to mind. One possibility is the owner's purchase and installation of demountable partitions requiring electrical connections to rough in provided under the general contract. If the owner has selected a nonunion contractor for the installation and electrical connections, the matter of jurisdiction may arise, as the union contractor may claim that those final connections fall within the union's contractual jurisdiction. The same may apply for owner communication and data systems installations where final electrical connections are required. It is best to have the owner advise the general contractor or CM what work the owner plans to install, how the owner plans to select an installer. If the installer is to be a nonunion company, the Project Manager needs to arrange a meeting before the work begins to settle the issue and avoid a labor dispute and potential work stoppage.

Setting Field-Personnel Mobilization and Demobilization Procedures

Mobilization can be effected by waiting for the preferred staff to complete a current project that is due to finish shortly after the present project requires staffing, so some interim supervisors may be required. Conversely, the superintendent for the new project may have completed his or her prior project and be currently doing menial work somewhere else until the new project starts. The CM or PM ought to take this slack time to set up the field office on the new job and have the superintendent spend time reviewing the plans and specifications in a leisurely but effective manner that is not often afforded when a professional has to rush from one project to another. As far as demobilization is concerned, the danger there lies in removing staff too early and not having the proper, high-quality personnel on-site to insure a swift, efficient closeout of the project.

Field Personnel Leveling and Field Productivity

Field-personnel changes and productivity are ongoing activities, and the tendency of a CM or PM to back off supervision when things are going well is a mistake common to many. But to keep the efficient momentum going, one should not let up on weekly reviews of actual versus budget costs and costs to complete, constant supervision of subcontractor performance, and frequent inspection walks through the ongoing construction areas to let everyone know that it's not over until it's over.

Cash Flow

Two parties are very much interested in cash flow: the owner and the contractor. The contractor must watch cash flow and ensure that requisitions are prepared and submitted promptly as specified in the contract with the owner. There are bills to be paid to vendors, generally on a 30-day basis from receipt of invoice, and there are payments to subcontractors, due and payable based upon the terms of the subcontract agreements. When payments from the owner are late, the contractor will need to tap his or her own letter of credit with their bank, effectively lowering his or her profit margin on the project by the amount of interest that has to be paid until the requisition is paid.

The contractor's accounting department not only needs to keep tabs on the timely preparation and submission of the monthly requisition, but they must also track payment. The CM or PM plays a role in this process. Some contracts allow the contractor to include costs projected to the end of the billing period. This takes into account the time lag between submission and payment of requisitions, when construction still continues, and puts more value in place than would be represented if the percentage complete were capped at the date of requisition submission.

The CM or PM must review the subcontractor's request for payment, as projected, to determine if in fact the amount to be completed during that lag date is accurate and is not more than the subcontractor can reasonably be expected to complete. This must be done quickly, as the clock starts ticking as of the date the contract indicates that the requisition can be submitted.

From the owner's viewpoint, construction-financing interest is much more expensive than permanent financing, so the owner wants to hold borrowing to a minimum and will be scrutinizing the requisition with the design consultants when the requisition is received. Any questions posed by the owner and the owner's team need to be addressed quickly to keep the process in motion. At times it may be the more practical approach to reduce one or two costs that are sticking points, concurrently convincing the subcontractor or vendor that a slight reduction in the request for payment will speed up the processing of the requisition and get funds more quickly.

Cash-Flow Curves

A typical cash-flow curve follows the traditional S shape, with a slow start and finish and a straight center section. Cash flow is sometimes shown in the form of a table or as a vertical bar chart.

The cash-flow projection is made up of three major cost components:

- Progress payments for home-office services
- Payments for goods and services made through procurement
- Direct and indirect construction costs such as labor, material, subcontracts, and field costs

Cash Flow on Lump-Sum Projects

The cash flow on lump-sum projects requires the CM or PM's special attention. It is vital that he or she ensure that progress payments are keeping up with the cash outflow. On most lump-sum projects, progress payments are set on a schedule, and it is not unusual for a contractor to *front-load*—place slightly more value on the initial work items than has actually been accomplished. One argument for such a practice is that the owner will most likely withhold a certain percentage of funds referred to as retainage. This does not ring entirely true, because the contractor's agreement with the subcontractors usually includes the same percentage of retainage as does the contract with the owner. However, payments to vendors for supplies require full payment, so there is some validity to the front-loading practice that allows the contractor to get the funds needed to pay suppliers 100 percent of their bill.

The normal retainage percentage is usually 10 percent, but many contracts allow a reduction to 0 percent when the project is adjudged 50 percent complete, effectively creating a project-wide retainage of 5 percent.

Change-Order Work and Cash Flow

Most proposed change orders address the total cost of the change order, hopefully fully detailed to allow the owner and the architect to review, comment on, revise, accept, and sign off on the extra work. Unless stated elsewhere in the contract, the work will be included on the requisition when it has been completed, inspected, and accepted by the design consultant. On a change order of significant value, the subcontractor may expend a considerable amount of money but get the work to only, say, 50 percent by the time the next requisition is prepared; this subcontractor will therefore be out of pocket a few bucks. Unless the contract calls for partial payment when change-order work is performed, there is not much the contractor can do if the subcontractor indicates that he or she needs a partial payment, except pay the subcontractor without reimbursement from the owner.

So Mr. or Ms. CM or PM, remember this when the next contractor–owner contract is being negotiated. Insert a clause that allows for partial payment of change-order work upon review and acceptance by the architect. It will improve subcontractor relations, you can be sure.

Schedule Control

We sometimes hear schedule control referred to as *time control* in relation to making the *time plan* for the project work. Actually time cannot be controlled, since it marches on relentlessly in fixed amounts and without any known means yet of controlling it. Time is, in fact, quite inelastic.

The project schedule is a plan of work per unit time, so it is somewhat flexible. By speeding up the task (doing it in less time),we can make up for past or future lost time and still gain the predetermined schedule completion date. Therefore, work is *elastic* and must be conformed to the *inelastic* time scale. We realize that this discussion sounds very basic, but you would be surprised how many managers think they can create more time to make the schedule.

Monitoring the Schedule

If we are to control the schedule, we must monitor progress against the time scale. We break the work down into specific tasks as a convenient way to check elements of our work against time. The average completion status of those work activities is our measure of overall physical progress of the total project.

We must also consider that each work activity has to be *weighted* to arrive at its percentage of the total project. Each weight is determined by the value of the human and physical resources expended to accomplish the task. The weighted value of the activity multiplied by the physical percentage complete tells us how much *earned value* that activity is contributing to our percentage complete. Conversely, the earned value divided by the total budget for that activity gives us the physical percentage complete for that activity.

Measuring Physical Progress

Again, it is necessary to point out the fallacy of using labor hours expended to calculate the physical percentage of completion. If you are not operating at exactly 100 percent efficiency, the percentage completion will be in error. That is why we stress the necessity of estimating physical progress and not expended progress.

The physical percentage complete must be tied to the *earned value* of the work accomplished to date. The earned value of the work activity is measured by breaking the task down into a logical system of checkpoints and assigning a percentage of completion of that task up to that point.

To calculate earned value, it is necessary to break down the various construction tasks into each trade's contribution to the work. A ready example is the installation of foundations that is generally required on all capital projects. A typical breakdown can be made as follows:

Operation	Percentage Progress
Layout	5
Excavation	20
Forming	50
Setting of rebar and anchoring of bolts	75
Pouring of concrete	90
Stripping of forms	100

So one can see from the above that when layout is complete, 5 percent of the total cost to excavate, form, install rebar, place concrete, and strip forms has been attained.

The cost-effectiveness of any earned-value systems must be evaluated in relation to the benefits involved. If the process of estimating foundation walls uses software that breaks these operations down during the estimating process, it may not involve too many extra steps to have the estimator also assign a percentage and even a cost to each of these components.

Planned Versus Actual Physical Progress

Once the physical progress for the period has been calculated, it becomes a simple matter to plot the new value on the planned-progress S curve as show in Fig. 8.14. The curve is derived from the original project schedule by plotting planned percentage complete and personnel hours over the elapsed time for the project.

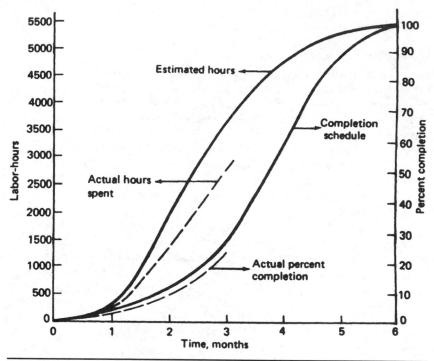

Figure 8.14   Curve of planned versus actual physical progress.

The resulting S curves are shown as the solid lines in Fig. 8.14. The actual physical progress and hours are then plotted monthly as dashed lines on the same graph.

If the plot for the actual completion falls on the solid line, the progress is on schedule. If the actual-progress curve falls below the solid line, the project is running behind schedule. That could be due to an understaffed project, because the curve of actual hours expended is also below the planned-progress curve.

There may be extenuating circumstances, such as unrecorded change orders, lack of owner's performance, or force majeure forcing the actual-progress curve below the planned-progress curve. The failure to maintain the schedule then lies at the door of the construction team and the CM or PM in particular, and they must take urgent steps to rectify the problem.

First, determine the cause of the problem. Is it due to a general cause such as staffing problems, poor productivity, low morale, and lack of leadership? Or is it due to the failure of a specific group to perform up to standard? In most cases, the problem will involve more than one cause, so the suspected areas must be prioritized in order to attack the worst ones first.

Ways to Improve the Schedule

We can control the future, as we said, by speeding up those activities that are now expected to finish late. This discussion refers to all applicable phases of the project. There are a few ways to improve your rate of progress when you are behind schedule in any of the project phases:

- Improve productivity.
- Increase staff.
- Work overtime.
- Reduce the workload.
- Subcontract part of the work.

Improving productivity is the best and cheapest way to increase the speed of doing the work, either by using better qualified workers or by improving management and work methods. Because it takes time, improving productivity may not work when a quick fix is needed, but it is a good cure for the longer term.

Overtime work is the next option and it is the one most commonly used. It can immediately increase the available staff hours and these added hours come from people already trained in the work. But, productivity drops off sharply after just one week.

Reducing the workload to regain schedule is another route we might consider using. Subcontracting some work, and thus freeing up workers to switch over to the areas where we are falling behind, will create extra costs, but in the long run it may be a more cost-effective approach than overtime.

Monitoring Procurement Commitments

Procurement procedures are closely tied to job progress, for without the proper materials on hand when they are needed, work tasks can come to a halt. At times, it will appear that it is impossible to obtain a subcontract agreement to meet the budget, so valuable time is spent interviewing still another and another subcontractor in the hope of negotiating a contract that falls within the budget. There are dangers in pursuing such a strategy and negotiating a contract with a subcontractor of questionable ethics or financial stability: The subcontractor may take shortcuts that fail to meet the contract requirements, or delete items or substitute lower-quality items, or staff the job with fewer and less-qualified trade workers, or go out of business halfway through the job.

Finally coming to the realization that the budget is wrong, the CM or PM ends the time wasted in chasing shadows by hiring a competent subcontractor at the market price—but wasting several weeks in doing so. This puts that subcontractor's particular trade behind schedule and may also impact other trades that follow it. The lesson to be learned is that after seeking prices from a group of qualified subcontractors, if none can agree to do the work for the amount of the budget, a conversation with the estimating department is in order to alert them to your dilemma so that they can review their estimate to determine where their error occurred.

Monitoring Field Productivity

There are as many ways of monitoring field productivity as there are contractors. The important thing is to select a process that works for your environment. Basically all productivity measures use some format that compares physical progress against labor hours expended.

One system for calculating the productivity rate is represented by the following formula:

Productivity rate = budgeted hours divided by actual hours

This calculation is made weekly for each craft performing a measurable task. Any unexplained adverse trends must be evaluated and acted upon quickly to stave off time and money overruns. The productivity rate does vary over the life of the project, as shown in the productivity curves in Fig. 8.15, so do not be too upset because of job start-up problems and the learning curve of the crew as they begin to get the feel of the work task and the natural rhythm that will be developed as that learning curve turns into a productive flow of work. The objective of a good field performance is to finish the field work at a cumulative productivity rate of 1.0 or better.

CMs and PMs must monitor the productivity of the individual field crews to keep their productivity at least as planned. Some units will stay on track and others will slip. Never be complacent about productivity in any group, because it can fall off quickly and recovery is always difficult.

Keep a sharp eye on operations that fail to meet their productivity goal over a period of several weeks, especially after they have passed the learning-curve portion. Are they getting the materials, supplies, and tools that they need to be productive? Are they working efficiently and is their foreperson or supervisor planning today's and tomorrow's activities in advance and in the proper manner? If the answer to all of these questions is yes, then pay a visit to the estimating department to review the budgeted amount for that operation. It may be that the estimate did not accurately reflect the working conditions of the new project or is incorrect in some other way and needs to be revised. Possibly invite the estimator to the jobsite to observe firsthand how that work task is being performed.

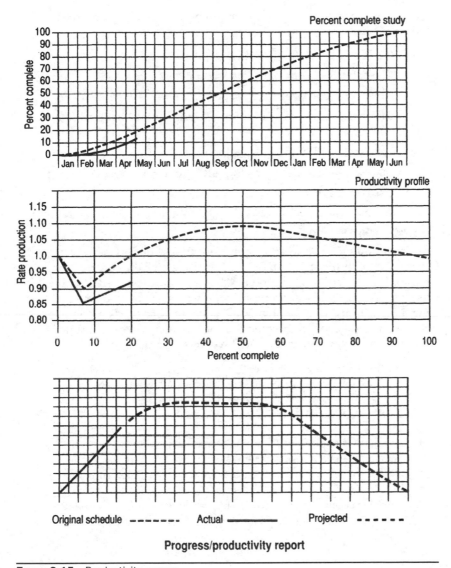

FIGURE 8.15 Productivity curves.

Good productivity is not solely a matter of being lucky in getting trained and experienced trade workers; it is rather a combination of sound management, good work methods, and a motivated workforce. The following checklist gives the combination of management methods and human factors that the CM or PM must skillfully manage to maximize field productivity on every project level. Virtually every facet of the total construction project management approach espoused in this book must come into play if the goal of top project productivity is to be reached.

- Cycle and check the schedule regularly.
- Look ahead to control the future; the past is history.
- Check physical progress, not just labor hours expended.
- Check the schedule by exception, using the CPM critical-item and look-ahead sorts.
- Take early action to correct slippage.
- Make corrective decisions logically; do not overreact.
- Use overtime judiciously.

Controlling Quality

There is a difference between quality control and quality assurance. Quality control is the standard or standards to which the construction or assembly of a building component has been incorporated into the project's design. Each section of the contract's specifications manual generally includes one section titled "Quality Control." A sample quality-control section in the specifications manual can be illustrated by looking at a typical masonry quality-control paragraph:

Preconstruction testing: Owner will employ a qualified laboratory to perform the following preconstruction testing indicated, as well as inspecting and testing services required by referenced unit masonry standards or indicated herein for source and quality control:

1. Clay masonry unit tests: For each different clay masonry unit indicated, units shall be tested per ASTM C67. Test each type and grade of brick for compression, water absorption, and efflorescence per ASTM C67.
2. Concrete masonry unit tests: For each different concrete masonry unit indicated, units will be tested for strength, absorption, and moisture content per ASTM C140.
3. Mortar properties will be tested per property specifications of ASTM C270.
4. Mortar efflorescence: Test each mortar type per ASTM C780. Test each mortar type which will be exposed to weather for efflorescence in accordance with the "Wick Test" procedure in BIA Research Report Number 15, *The Causes and Control of Efflorescence in Brickwork*. Mortar mixes which show efflorescence shall not be used in the work.

Quality assurance is the process of ensuring and verifying that the bricks and concrete masonry units meet the ASTM International specifications and the mortar meets the ASTM International requirements.

Quality assurance produces a project that everyone can be proud of and marks your company as one devoted to producing high-quality projects, thereby providing your sales-development group with a valuable tool.

The University of Wisconsin in 2008 offered a course in engineering professional development and included a survey to develop a list of deficiencies in construction documents. Without a quality set of contract documents, a quality building cannot be built unless the contractor's professional staff can uncover these deficiencies and convince the project owner that changes need to be made, probably at a price, to allow them to create a project everyone can be proud of.

The respondents to the survey at the University of Wisconsin produced a list of recommended actions an owner and design consultants should consider:

1. Conduct an in-house peer review of the plans and specifications. Coordination concerns should be at the top of that review.

2. Scrutinize the drawings as they are being developed; outside reviewers may be invited to provide a more objective opinion of the quality of the design drawings.

3. Organize a meeting of key participants in advance, and if a contractor has been selected for a negotiated contract, invite the contractor. If construction management is the contract format, the CM should be immersed in this process.

4. Keep communications open during construction with ongoing, on-site project meetings to review progress and look for potential problems down the road.

5. Involve a quality-assurance or quality-control consultant.

6. Review and revise key documents, if necessary, during the entire life of the project; the owner, design consultants, and contractor should all participate in this exercise.

7. Create a quality-assurance section in the contract to streamline standard details, drawings, and specifications.

8. Create a series of functional checklists to act as quality-assurance reviews, distribute them to the appropriate parties, and *use* them.

9. Participate in workshops and training related to improving document quality. Invite your contractor and his or her major subcontractors to get their opinion on how to improve the quality of the design documents.

10. Establish accountability for everyone at every stage of document development.

11. Update any system that stands in the way of quality document development.

12. Conduct a lessons-learned meeting when the project has been completed.

Benchmarking

Benchmarking is a process whereby a company seeks out and studies best practices in order to produce superior performance. Benchmarking can be an interior company exercise, where the investigation of practices and performances will isolate a specific component of construction for analysis. Let us say the company self-performs work and it is confined to carpentry or concrete work. Look at a project where the costs per unit were the lowest and investigate the practices that created this quality, cost-effective work. If high productivity coupled with strong supervision produced these results, this particular operation could become the *benchmark* against which all similar future operations will be tested.

If the concrete work was subcontracted and that particular subcontractor produced a quality product at a competitive price, then not only should that subcontractor's operations be scrutinized by one of your field engineers, but the process whereby the purchasing department selected the subcontractor should be reviewed and that process used as a benchmark for future awards.

External benchmarking attempts to compare your company's practices against those of superior firms in your field. The Construction Financial Management Association in Princeton, New Jersey, often produces financial data provided by its members annually. It then collects and analyzes the data and posts best-in-class statistics. There are other construction-related organizations, such as the Construction Industry Institute in Texas, that can also be sources for benchmarking procedures and related information.

International Organization for Standardization (ISO)

More and more you will see processes and products with an ISO number. ISO was founded in 1946 to promote voluntary manufacturing and trade communication standards.

ISO 9000 and its variants ISO 0991, 0992, 9003, and 9004 relate to construction and involve basic quality-management principles, including the following:

- Management and responsibilities
- Quality-systems principles
- Internal audits
- Contract review
- Design control

- Process control
- Product identification and traceability
- Inspections and testing
- Corrective actions
- Handling, storage, packaging, and delivery
- After-sales servicing
- Document control
- Quality records
- Training
- Purchaser-supplied products

Several construction companies have sought ISO certification; by 2009 there were 20,000 such contractors across the globe. The U.S. Department of Defense has adopted ISO 9000 in lieu of its own MIL-Q-9858A specification wherever possible.

One of ISO's involvements has been to work with world organizations to create a common computer language for the design and construction industries so that vendors and manufacturers can communicate more clearly with design consultants and builders.

The Six Sigma Approach to Quality Control

In the 1970s, Motorola sold its Quasar television business to a Japanese owned company located in the United States. Much to Motorola's surprise, the Japanese company began producing the heretofore lackluster Quasar TV with only 5 percent as many defects as Motorola. Motorola's investigators found that the new owners were using the same labor force, the same technology, and the same designs. Over the following 10 years, Motorola devoted a considerable effort to improving the quality of its own products; the system was called Six Sigma, named after the Greek letter used by statisticians to measure the variability in any process. Motorola introduced an extensive training program, and those completing the 160 hours of classroom training were known as Black Belts. The Master Black Belts, selected from the top ranked Black Belts were responsible for the technical leadership of the Six Sigma approach.

The process of improving a particular operation is known by the acronym DMAIC:

D—define the goals.

M—Measure the existing systems to establish a metric by which progress towards the goal can be tracked; in other words, create a baseline against which to measure progress.

A—Analyze the system to identify ways to eliminate the gap between existing performance and the desired goal.

I—Improve the system.

C—Control the system.

It is not difficult to see how the Six Sigma system could be adapted to the construction industry.

Operating the Controls

As a means of summarizing the far-flung requirements of project control, we have prepared some recommended guidelines for handling the items under each of the headings on the next few pages. The best-designed control system in the world is not going to lead to a successful project unless we operate the controls properly.

Establish Priorities

The number-one requirement for effective construction or project management is the establishment of priorities for your work. The most urgent activities should go to the head of your "Things to Do" list. Do not handle the activities in the order in which they arrive in your *in-box*. Remember, the project organization is functioning on the basis of what comes from your *out-box*.

Sort the whole contents of your in-box at least once daily in addition to handling the day-to-day operations of your project. Make sure that items related to project quality, cost, and schedule receive top priority. Also, items related to project and construction/client relations should receive high priority. Be responsive to people problems.

All of the priority items have to be worked in with the routinely scheduled project activities such as project reviews, monthly progress reports, and trips, which also make a heavy demand on your time. We have found that keeping a written or electronic priority list or daily things-to-do list is a valuable tool when it comes to managing your time. It is a simple way to keep yourself from overlooking or forgetting a high-priority item now and then.

Control by Exception

Controlling by exception is the only way to stay on top of the plethora of daily project activities. This is true even if you have properly delegated as many of them as possible to other project team members. Many control systems such as CPM schedules, control budgets, and quality- and material-control reports, are designed to highlight exceptional conditions. Make sure to check the exceptional items first, then go into the normal items as time permits.

Keep Looking Ahead

Running a project is like piloting a ship. It is nice to know the depth right under your keel, but it is more important to know what dangers lie immediately ahead. Keep your radar turned on! Take advantage of the control systems that have look-ahead capability, such as periodic CPM look-ahead sorts and estimated-to-complete projections. Keep an eye on the applicable project-performance curves for design, procurement, and productivity to note any trends away from the projected norms.

Check the actual work of the groups performing the work—whether it is design, procurement, or construction—that is charted on the curves. Has there been a sudden drop in personnel just when the group should be at maximum production? How is productivity holding up? Checks should be made by visiting the groups performing the work and physically observing the work in progress. Also, talk to the group leaders; will you be amazed at what they will tell you verbally that they would not dream of putting into their written reports. Take a supervisor to lunch, visit the craft managers, and keep tuned in to the project grapevine to get a real feel for how things are going. Check all of the intelligence you have gathered to ensure that it is correct, then act on it accordingly.

Have a Quick Reaction Time

The function of any control system is to have a rapid response time to off-target items. Corrective action should be initiated immediately. The longer you delay, the worse the condition is likely to become and the longer it will take to get back on target! Once time has been lost from a schedule or money from a budget, it is very difficult to get even again. Also, you will have lost one of your opportunities to beat the schedule or budget. Do not postpone taking needed action to correct a situation just because it may result in unpleasantness to someone else or yourself. When you feel that the need for action is real, get right to it.

Be Single-Minded

A lot of people along the way will try to blunt your single-mindedness and divert you from your goals. Some may not share your goals; they may even have a personal dislike for you. Others may be incompetent or just plain lazy. All of these factors must be brushed aside and not allowed to deter you from reaching your intended goals. There may be times when a suitable compromise is the only solution, but never sell out completely.

If you have reached an impasse when going through your normal channels, do not be afraid to go over to the opposition's head to meet your project's goals. Remember, top management gave you a mission to accomplish—the project—and you accepted it. We have found that most top managers are very supportive of their project and construction managers when an impasse arises on a project.

A final word of caution: Single-mindedness is essential, but the way you present it to get your project needs satisfied is equally important.

Summary

We have found that some mistakes in the operation of project controls occur fairly often:

- Failing to plan, organize, and control the project

- Failing to maintain good client/subcontractor relations

- Underestimating the construction cost and thereby making it impossible to meet the profit goals

- Overestimating the construction cost and thereby making the project uneconomical

- Failing to get a good procurement program organized

- Failing to properly document meetings, project changes, voice mails, e-mails, and other important communications (if the job runs into trouble later on, you will need the documentation to defend your position against a variety of possible criticisms)

- Failing to hold construction scheduling and review meetings with the client and management to properly check progress and coordinate the work

- Holding too many meetings, which wastes valuable time

- Failing to control meetings, which makes them nonproductive

- Failing to exercise proper management control over field activities (more money and time are spent here than in any other part of the project)

Most of these tips start with the word *failing*—and that is what you are going to do if you do not avoid these pitfalls. Keep in mind that it takes about three or four good projects to wipe out the bad reputation acquired from one bad one!

CHAPTER 9

Construction-Project Execution

In prior chapters we have been discussing how to meet our project goals by *planning, organizing, and controlling* a construction project as individual operations. Much of the material we covered was background information on getting into the position of being the executing contractor. This chapter looks at the key operations one is likely to encounter while managing a construction project.

Using some imagination should allow readers of these discussions to place themselves in the role of either a member of the owner's team; the construction manager (CM), if a construction-management contract is in effect; or the general contractor (GC), if any one of a variety of third-party contracts is employed.

Construction-Project Execution

In this chapter we will walk through the major management of construction activities, from the notice to proceed to the final turnover of the facility to the owner. We will try to make the project activities broad enough to suit readers serving the major segments of the construction industry. However, the variations among the four major contracting strategies used in the execution of a construction project are great enough to require individual discussions in some instances.

The four basic contracting strategies we will cover are:

1. The owner awards a lump-sum construction contract based upon a competitive-bid situation where there is a completed design, plans, and specifications prepared by the owner's design consultants.

2. The owner contracts for a design-build project by engaging a single entity to design and build the project into one that meets the owner's demands and budget.

3. The owner engages the design consultant to provide a completed set of plans and specifications and then proceeds to subcontract the work managed by their own staff.

4. The owner hires a construction manager to act as the owner's agent to control the construction process, based either upon a fee as a percentage of final construction costs or upon a fee as a percentage of costs but with a guaranteed maximum price.

We set our starting point for this chapter at the receipt of a signed contract, letter of intent, or official notice to proceed with the work. It is important to note that upon the signing of the contract, the time clock starts and it becomes time to finalize the execution plan and build the project.

Basic Project Parameters

Many of the basic matters affecting the execution of the project should already be in place and, if not totally, at least partially resolved. The construction manager, project manager (PM), or owner's representative should have been selected and be available for duty on the project. The construction labor posture and the strategic end date should have been set. The overall budget or cost target should have been agreed to by the owner and contractors. Some sort of design basis, ranging from conceptual to final, should have been set, depending upon whether the project is design-build, design-bid-build, negotiated, or construction management.

The project should be suitably financed to service the planned cash flow. The quality goals should have been established by the design documents or the contract provisions, and some form of estimate prepared to act as a cost-control document.

Construction-Project Initiation

The initiation phase of the project is especially critical because resources vital to meeting the project goals can be squandered on false starts, poor planning, and organizational mistakes. Project initiation covers the brief period from when the contract is signed until the field office is up and running. During that period, the CM, PM, or owner's representative is on the run between the home office and the field and intensely involved with getting the project's administrative procedures into place. This can be a very chaotic period for the project managers if sound management and procedures are lacking.

Our first example will be a design-bid-build lump sum project with a completed set of plans and specifications. The project-initiation assets consist of

1. the contract

2. the cost estimate

3. a preliminary, baseline, or milestone schedule

4. the contracting plan and construction technology to be used

5. firm proposals from vendors and subcontractors

6. a preliminary project-execution plan

7. a preliminary project organization chart

8. a pool of home-office and field personnel dedicated to the project

9. a complete package of project-design documents

10. preliminary design of temporary field facilities and their location on-site

11. contacts with the owner and the design consultants—phone number, cell-phone number, e-mail address

12. the field inspection and labor survey

13. estimated field craft labor hours and staffing requirements

14. a list of required heavy construction equipment and small tools

15. corporate operating standards and manuals

16. other available corporate resources

17. project-proposal files

This list is quite comprehensive but still does not include all project-specific data already available to the team of the CM, PM, or owner's representative. There are a few other necessary project factors that must be in place before the project can be successful:

1. An effective owner's project team dedicated to the project

2. Input from the design team for approvals, design interpretations, quality reviews, etc., if not already spelled out in the specifications manual

3. An existing infrastructure of business and government services serving the project site

4. A competitive marketplace for necessary project goods and services

5. Adequate project-financing procedures

These are key factors that can make or break the project and are outside the contractor's control to a degree. It is important to note that while any of these five factors can seriously affect the project schedule, they are not force majeure. In other words, these are not "acts of God" but merely another set of factors, however difficult ones, that must be dealt with—and can be dealt with, by an aggressive, knowledgeable, and savvy CM, PM, or owner's representative (we continue to include an owner's representative because, as you recall, one of the four contracting methods involves the owner acting as its own general contractor).

The first thing the CM, PM, or owner's representative has to do is to make a project priority list for the key project-initiation activities, Such a list might look like this:

1. Get up to speed on existing project documentation.

2. See that contract numbers and accounting procedures are in place.

3. Order long-lead materials and equipment.

4. Place the one or two basic subcontractors required for site work, if this is a greenfield project. If an existing structure is to be demolished, obviously a demolition contractor will be first on the site after the necessary permits have been obtained.

5. Organize in-house and client project-kickoff meetings.

6. Finalize a detailed project-construction schedule.

7. Convert the project cost estimate to a control budget.

8. Finalize the organization chart and start bringing key people on board.

9. Finalize temporary site facilities and schedule their installation.

10. Start preparation of the field-procedure manual (FPM) and issue within three to four weeks.

11. Meet with personnel and labor-relations managers to assess field staff requirements and any site labor agreements.

12. Finalize the field overhead budget.

13. Initiate any heavy-equipment and small-tool sources and policies.

14. Ensure that the required field-insurance, bonding-coverage, and risk-management needs have been met.

15. Prominently display all required OSHA, federal nondiscrimination, and other policies and regulations.

16. Set up a contract-administration tickler file for contractual requirements, so as to be proactive in those matters.

The Project-Kickoff Meeting for Design-Bid-Build or Negotiated Projects

The initial project meeting will serve many purposes, particularly introducing all the team members, the owner's representatives, the design consultants, the CM or GC's field supervisory staff, and key subcontractors and vendors. In preparation for this meeting, the CM or PM should have thoroughly read and understood all provisions in the contract with the owner. If clarification of some provisions was not already obtained with the owner or design consultants, these issues can be discussed and finalized at this kickoff meeting. The main purposes of this meeting are to ensure that:

- Everyone understands their duties, obligations, and rights under their appropriate contract.

- Ambiguous design or contract issues are discussed and resolved—if not at this meeting, then at a subsequent one.

- Everyone understands when their requests for payments are required by the CM or PM and are to be submitted to the owner.

- Everyone understands when the CM or GC should expect payment from the owner in order to pay vendors and subcontractors.

- The schedule for future project meetings is distributed and required attendees are identified.

Although they should be spelled out in the project contract or specifications manual, it is always a good idea to make a list of the provisions important to contractor–subcontractor relations and ensure that all parties have been made aware of them once again and that they are documented in the project-meeting minutes. This will require another review of the owner's contract and the provisions in the specifications manual that apply across the board:

- Date when requisitions are to be submitted, and their format and content

- Restrictions on allowable overhead and profit on change orders to the general contractor and by second- and third-tier subcontractors

- Unit prices, if included in the contract, and how they are to be applied by the subcontractor when extra work is required

- Allowances, alternates, and methods of dealing with them

- Restrictions on the use of the contingency, if there is one included in any contract

- Provisions for liquidated damages, if included in the contract, and any bonus arrangements included in the contract

- Requirements for the general contractor to submit a list of personnel to be assigned to the project, and to obtain the owner's acceptance of same

- Requirements to submit the names of proposed subcontractors for the owner's review and comment prior to the award of subcontract agreements (some of the early required subcontractors may already have been approved and be in attendance at this meeting)

- Appointment of the owner's representative and the authority and responsibilities vested in that representative

- Requirement for noise abatement and restriction during working hours and any restrictions on working hours—in terms of late nights, weekends, or holidays

Record Drawings

Of vital importance to most construction projects is a record of the actual construction performance criteria, as opposed to the design performance—e.g., were the underground utility lines installed in the location as shown on the drawings and at the invert elevation as required? The purpose of the record or as-built drawings is twofold: to provide the design consultants with confirmation of design compliance and to provide the owner with an accurate record of exact locations in case repairs, replacement, or maintenance of enclosed items of work are required at some future date.

The importance of maintaining current and accurate records of such information must be constantly stressed throughout the project and checked daily or weekly, as required, by the field supervisory staff to ensure that the records are current and accurate. Generally, these record drawings contain the following:

- Record of all changes, whether due to field conditions or change orders

- Record of all changes due to acceptance of any alternates listed in the contract

- Dimensions, vertical and horizontal, that either confirm or correct the design dimensions of entire areas or components within designated areas

- Elevations relating to site and site-utility work—both line and grade, to include manhole-rim elevations and invert elevations of all buried site utilities and duct banks containing underground electrical conduits

- Floor-to-floor elevations and floor-to-ceiling elevations
- Locations of all concealed items; mechanical, electrical, and plumbing risers; branch piping; and branch wiring
- Locations of all plumbing valves (generally presented in chart form, listing the valve and its location)
- Location of all HVAC duct dampers and fire dampers
- Location of all pipe heat tracing, if installed

Inspections and Testing

A list of all inspections should be included in the minutes of the kickoff meeting, along with the record-drawing requirements. It is a good idea to repeat this list in the first three or four meeting minutes to impress attendees with their importance. These inspections and tests generally fall into the following categories but may vary from project to project:

- Earth-compaction inspections and tests.
- Concrete inspections, both visual and via preparation of test cylinders to confirm compression requirements. A visual inspection of ready mix coming down the chute of the truck may alert the superintendent to the fact that it contains too much water. The delivery ticket should be inspected to determine when the concrete was loaded onto the truck. Many specifications require rejection of the load if it has been in the truck more than two hours.
- Infiltration and exfiltration tests for underground storm-water lines.
- Mill reports from the manufacturer of structural steel beams, columns, etc.
- Welds and bolt-up connections. If tension-control bolts are used, a visual inspection should be made to ensure that the proper part has been sheared off during torquing.
- Shear-stud testing.
- Mortar-cube testing.
- HVAC and plumbing tests, by both the design engineers and the local authorities.
- Inspections of flashings—windows, roof, and exterior-wall penetrations.
- Inspections of acoustical or insulation batts, particularly those installed in concealed spaces.
- Inspections of various substrates prior to their enclosure.
- Inspections of fire-protection system—underground lines, risers, valves, and fire pumps, as well as pressure testing of filled lines.

The Project-Kickoff Meeting for a Design-Build Project

The design team takes the lead role in this situation, with construction playing an important supporting role. The project manager or construction manager on the construction side of the design-build team will have a minimal role as the architects and engineers begin to develop the concept formulated by the owner. Construction people will sit in on these initial meetings possibly to offer order-of-magnitude and square-foot pricing information as various types of structures are under consideration. As the design unfolds and requires more detail, the CM or PM will assume a greater role in the process. If specific designs emerge and are approved by the owner, the CM or PM may be in a position to begin assembling the estimate and even begin to obtain competitive pricing, if sufficient design information has been developed. For example, if the building's structure is to be precast concrete floors and decorative precast spandrel panels, the CM or PM can begin to solicit square-foot prices and decorative forming costs from the local precasters.

In this mode, attention has to be given to the design-documentation delivery dates. The plans and specifications need to proceed through the local building-official offices, unofficially, for review and comment even before they are 100 percent complete to ensure that some local code has been complied with. If cordial relationships have been previously established by either the design side or the builder side of the design-build team, building officials are often very helpful in providing an unofficial review of the drawings to pick up any changes that need to be made prior to the final review for issuance of a building permit. The fire marshal can also play an important role in this initial look-see to point out some details that ought to be added or changed.

The Project-Kickoff Meeting for a Construction-Management Project

Project-kickoff meetings in the construction-management mode are somewhat different because the CM is acting as the owner's representative, or agent, so it is as though the owner is sitting at the head of the table. Most of the same procedures that we have already discussed will occur.

There will be an owner's representative who will work closely with the CM during the subcontract-buyout process and the material and equipment purchases. If the CM was engaged during the design stage, he or she may have already assembled a team of subcontractors to work with, setting the scope of those bid packages and the schedule for receipt, review, and issuance of the subcontracts in the owner's name—after conferring with the owner and obtaining agreement as to the selection. The CM will have assigned field staff and made arrangements in his or her home office for the receipt, storage, and processing of all of the paperwork that will begin flowing in from the field. He or she will also have made arrangements in the field for all of the paperwork that will flow from the office.

The same procedures that we outlined for the kickoff meeting for a design-bid-build or negotiated-contract project will take place as the CM prepares for the first project meeting. Since the CM has been working with the owner and design team since early in design stage, he or she will have been able to prepare the bid packages more quickly and get the project off to a fast-tracked start. This is another advantage of the construction-management process.

Key Questions to Ask Subcontractors during Negotiations in Any Contract Mode

Whether it is the CM preparing subcontract bid packages or the builder having been awarded a contract after competitive bidding beginning the buyout of subcontracts, there are a number of considerations the CM or PM needs to take into account. Even though the subcontract bid package includes references to the appropriate plans and specification sections, along with division 1 (General Conditions) of the specifications manual and a statement that the subcontractor is bound to the contractor by the same terms and conditions by which the contractor is bound to the owner, there are some other, less obvious questions to consider:

- Since the time of the issuance of the bid or contract documents, has the architect issued any addenda to those documents? Were they issued to the appropriate subcontractor? Were there revised drawings issued during the bidding process? If so, has the subcontractor received the addenda, changes, and revised plans and specifications and included them in the price?

- If the project is exempt from state sales tax, has the subcontractor or vendor acknowledged this and excluded state tax from all materials? This seems like a simple thing, but you may be surprised if, in the rush to put the pricing together, the subcontractor failed to deduct the sales tax.

- Are the subcontractors aware of the baseline schedule, and do they accept that schedule, the amount of time dedicated to their various tasks, and their ability to meet their milestone dates? Have they included any float in their schedule for those glitches that always occur—and if so, how much?

- Have any exceptions been taken to the plans and specifications but left unsaid because the subcontractor is confident the substitutions that he or she is going to submit will be approved? If so, has the subcontractor based his or her price on that assumption (an assumption that usually does not pan out)?

Additional Items That May Need to Be Addressed with Mechanical, Electrical, and Plumbing Subcontractors

Some contracts use the word *contractor* in various sections of the specifications, and the question arises: "Does this refer to the general contractor or the subcontractor?" When the specifications do not include a definitions section where this distinction is addressed, the CM or GC needs to make the distinction very clear.

For example, in division 16, Electrical, a phrase may be included saying "the contractor shall perform all cutting and patching included in Division 16." Well, does this refer to the electrical contractor or the general contractor?

Other items to consider:

- Will the electrical contractor be required to provide temporary utilities for all trades during construction?
- Will the electrical contractor be required to provide temporary power to the general contractor's field office?
- Who will furnish starters for all mechanical equipment? Quite often this is not spelled out in either the electrical or mechanical specifications.
- Does each trade provide its own trenching and backfilling for underground work, or is this the responsibility of the general contractor?
- If concrete encasement of pipe or conduits is required, who furnishes the labor and materials?
- Is daily cleanup and trash removal clearly assigned to these trades or is it left vague?

Warranties and Guarantees for Electrical and Mechanical Work

Another often overlooked item relating to the electrical and mechanical scope of work deals with warranties and guarantees. Let us assume the building under construction is enclosed to the point where, in the summer, the air-conditioning units could be run to make it more comfortable for some workers and probably result in an increase in productivity; the same could be true in the winter when, by firing up the boilers, workers could be able to work in their shirtsleeves instead of all bundled up (which would make it easier to operate efficiently). We are not talking about 70°F in summer and winter, but enough heating or cooling to improve working conditions—particularly in the winter, when more costly propane heaters have to be fired up, moved around, and distributed where needed.

One-year warranties on these "contract" heating and cooling units are standard on most equipment, but when do these warranties or guarantees commence? They may start as soon as the equipment is up and running. If the completion of the project is two months away, will the general contractor have to pay to extend the warranties or guarantees for those two months, since the owner is expecting their equipment warranties to run from the date they are accepted by their mechanical and electrical engineers? This is just another concern to be addressed on the CM or PM's to-do list.

Implementing the Project Organization

This facet of the project initiation is again a function of the contracting mode, with design-build offering the least pressure and lump-sum design-bid-build offering the most, with construction management possibly right in the middle.

It is vital for the CM or PM to get the most-needed assistance on board first to help in setting up the project. People high on that list are the field superintendent, the scheduler, the cost-control people, and procurement and other field-office personnel. With these key people on board, the CM or PM can start delegating some of the high-priority project-initiation activities to them as shown in Table 9.1.

The number and type of early field-staff increase depends largely on the size and complexity of the project, the schedule, the site location, and the field indirect-cost budget. It is up to the CM or PM to exercise sound judgment in tailoring the early staff assignments to suit the contracting mode and to create a cost-effective management team during the vital project-initiation phase.

The field labor policy was agreed to with the owner during the project bidding and negotiations and is now ready for implementation. The owner may have requested a union job because the owner's own labor force has a collective-bargaining agreement, or the owner may have spelled out in the bidding documents that the project be a nonunion or merit shop.

The CM or PM will follow the owner's dictates during the buyout process and, whichever form of labor is employed—union, nonunion, or merit shop—will establish the harmonious environment necessary to provide a productive workplace. Labor relations will become a prime activity to be monitored by the field staff, the project superintendent, and the superintendent's assigned forepersons when the project is designated a merit-shop operation.

Key Person	Project-Initiation Responsibilities
Project manager or field superintendent (depending upon experience)	Design and execute temporary site facilities; set quality-control program; establish document-distribution system; set up files; monitor site-development work; comply with applicable government regulations
Field scheduler	Finalize construction schedule for approval and issuance; prepare field scheduling procedures for FPM
Field cost-control personnel	Prepare and issue approved project-control budget; prepare field cost-control procedures in FPM
Procurement personnel	Start materials-management program; ready first material and subcontract orders for approval and issuance
Field superintendent	Assist in project staffing; finalize site labor agreements; monitor government regulations, required postings, etc.
Field-office manager	Start to organize field-office staff, communication systems, and field-office procedures; publish approved FPM

TABLE 9.1 Project-Initiation Staff Assignments

If it is an all-union project, the field superintendent must become familiar with the various rules and restrictions in the various trade-union agreements to ensure that one trade is not encroaching on another's territory. A meeting with the various trades may be in order to determine what these restrictive practices are: Does the plumber or electrician provide his or her own labor to trench and backfill for underground pipes and conduits, or is this a task for the general contractor? Do all unions observe the same holidays or will some be working on a day that another union has negotiated as a holiday in its labor agreement?

What about starting times and finishing work times? Does the union allow some flexibility in starting time so that if project is located in a hot, humid geographic area, the field superintendent has the latitude of having work commence at 6:30 in the summer, thereby allowing workers to complete their requisite eight hours before temperatures rise later in the afternoon?

The Field-Procedure Manual

The FPM is the seminal document for organizing and controlling the field operations to meet the project goals. The contents were described in Ch. 8, so we will cover only timing and need here.

The main value of the FPM in the early stages is as an indoctrination tool for new people coming onto the project—so it should be issued early to be of value. Typically, an FPM is modeled on a previous version used on a similar project. The easiest way is to use the prior similar FPM and mark it up for the present project's conditions. That way the preliminary issuance can be made rather quickly and the FPM is then polished up as new project information becomes available. Holds can be placed on those areas that are not yet firm in the preliminary issuance. If the old model was produced on a computer, changes can be made easily and quickly.

On cost-reimbursable jobs, the FPM is issued for client approval to ensure that the owner agrees to the procedures set up to control the project performance and cost. That presents no problems, because cost plus a fixed fee normally involves an open-book format, with the owner having the right to audit the books at the end of the project to verify costs.

The main goals of the FPM are to set the project ground rules and to reduce about 90 percent of the field administrative activities to routine tasks. That allows for about a 10 percent demand of routine activities on the CM or PM's time—although as we all know, things do not always turn out as they are planned.

The FPM should be completed and issued in final form within the first 30 to 60 days after the project start. It is always subject to revision as the project progresses, and changes to the procedures may become necessary. Producing the document and issuing it in the form of a loose-leaf binder makes these updates and changes easy; and keeping the previous documents stapled behind the new versions creates an easily discernible trail of changes in procedures.

Finalizing the Project Budget and Schedule

These two essential control documents may have undergone some changes during the final contractor-selection process and contract negotiations. In other words, the management might have had to do some price negotiating and schedule revision to nail down the job. If the CM or PM was not present at those times, the changes may be startling, because they rarely become more favorable to the contractor.

If the contractor's negotiators did their job right, they may also have negotiated some corresponding scope revisions to suit the price and schedule changes. These revisions have to be incorporated into the project scope description, budget, and schedule to get everything in writing. The CM or PM needs to see that these matters have been taken care of and are reflected in the final project documents.

The details of these matters should be investigated by the newly appointed project cost and scheduling people. Any positive or negative effects of the changes should be investigated in detail. After the initial budget and schedule have been approved by management, they can officially be issued as project documents. These are the baseline documents for reaching the construction teams' project financial and time goals.

Risk Analysis

The construction manager or project manager must promptly analyze the project documents to ascertain those areas of risk that the firm has assumed in the contract. The chief and simplest of these risks are the ones covered by bonding and insurance; therefore, the bonding and insurance documents need another once-over to see that the proper coverage remains in effect.

Other, less well-defined areas also must be evaluated so that a defense mechanism can be activated as required. Among those areas are the scope-of-work definition, change-order procedures, budget and cost-control procedures, any schedule or budget penalties or incentives, quality-control requirements, and the like. None of these are fully coverable by insurance policies or performance bonds. These risks can be controlled only by the project manager's exercise of *total construction project management*, practiced under the watchful eye of their management staff to provide the leadership to infuse the company's management philosophy into all areas through the project execution.

Bonds and the Bonding Process

Bonds are a three-way arrangement between a surety (the bonding company), a principal (the contractors), and an obligee (the project owner). The most frequently employed bonds to limit risk are:

- Payment bond—also known as a labor-and-material bond, protects the owner against nonpayment of subcontractors and vendors by the general contractor

- Performance bond—protects the owner from nonperformance and financial exposure should the contractor default on the terms and conditions of the contract with the owner

The Miller Act, a federal law, requires the contractor on any federal project for a contract that exceeds $100,000 to post two bonds:

- A performance bond that the contracting officer considers adequate for the protection of the federal government.

- A payment bond for the protection of suppliers of labor and materials, the amount of which is to be equal to the total amount payable under the terms of the contract but not lower than the amount of the performance bond.

The surety company issuing these bonds must be listed on the Treasury Department's list of acceptable companies.

Most states have adopted similar laws for contractors working on state or city projects; these are known as Little Miller Acts.

The Bonding of Subcontractors

No worse scenario can be envisioned than having a key subcontractor default during the middle of a project. Bonding of subcontractors is not uncommon on large projects, but many general contractors and construction managers utilize another process known as Subguard, which is default insurance for subcontractors. This is generally applied on large projects employing substantial subcontractors; its sole purpose is to provide a financial remedy to an owner or general contractor in case of a subcontractor's non-performance. These funds are then used to help fund a successor subcontractor.

The cost of Subguard is generally less than that of a surety bond, but it also requires deductibles that can range into six figures. That is why it generally is only used on subcontractors with large contracts.

Subguard policies have these features:

- Indemnification for direct and indirect costs that result from the default in performance by the covered subcontractor or vendor
- Broad coverage for all enrolled subcontractors and vendors that qualify under the contractor's prequalification system (generally a part of the subcontractor-selection process)
- Full policy limits for each loss, even if those losses exceed the amount of the subcontract agreement

Owner- or Contractor-Controlled Insurance

Another risk-management tool is controlled insurance, known as a controlled insurance program or an owner-controlled insurance program. This is not a new type of insurance program; it has been around since the 1960s. However, it has gained popularity in the last 10 years because it presents a method to control the cost of insurance and still maintain adequate coverage.

When instituted by a general contractor, a controlled insurance program is a wraparound-type policy. It allows the general contractor to use his or her greater purchasing power to buy all of the required commercial general-liability insurance coverage for bodily injury and property damage arising out of operations (exclusive of automobile insurance) for the entire project, including subcontractors. Each subcontractor provides a credit for the cost of this insurance that he or she would have included in the bid if he or she had to provide it. The policy provided by the general contractor will generally cost less than the total of the individual policies provided by each subcontractor.

An owner-controlled insurance program works in much the same manner except that not only do subcontractors provide the credit for not providing the insurance, but the general contractor includes the premiums he or she would have paid as well. The owner then provides the required insurance coverage.

There are costs to administer either kind of controlled insurance program, and those costs are generally more than the savings on small projects. Unless the general contractor or the owner has a strong safety program in place, backed up by a full-time safety supervisor and an accident-prevention program that is tightly controlled, a series of accidents on the job can drive insurance premiums very high, effectively wiping out any cost savings and possibly significantly raising the insurance costs.

Change Orders

This is probably the appropriate place to discuss change orders, since they will become part of the risk-management process. Change orders are created for at least one of the following reasons:

- The owner desires to add to or deduct from the scope of work for the project.

- Actions by the owner or the owner's design consultants or other representatives cause a delay in the project which affects time, cost, or both.

- Unforeseen subsurface conditions become apparent.

- The general contractor, construction manager, or subcontractor discovers errors or omissions in the contract documents and requests additional money to correct them.

- A difference of opinion arises as to the interpretation of a responsibility or obligation that did not surface until well into the project.

The Owner Adds or Deducts Scope

Addition to or deduction from the scope of work by the owner is a rather straightforward approach to a change order. The CM or PM will make those adjustments or changes, carefully checking the additional cost, if any, and the impact on the schedule, if any. Deduction change orders often have a way of not including all the costs included in the change. Obviously if a drywall partition is deleted, the cost of taping and painting must be included. But complex deduction change-order requests need to be carefully scrutinized by the contractor to ensure that *all costs* associated with the deletion are included—particularly costs that may not be obvious unless the process of the deduction is traced step-by-step.

The Owner or Architect Delays in Responding to Requests for Information or Clarification

Delays caused by the owner or the owner's design consultants or other representatives enforce the need for proper and complete documentation. If a critical shop drawing has been delayed by the architect, that delay should be noted in either the weekly progress meetings or a letter or e-mail to the architect defining the delay and its impact on the schedule. This is not to be interpreted as an adversarial action, but it documents the request for review that, if ignored, may well turn into a dispute and a claim. Without prior notification of the delay and its impact, following through on a delay-of-project claim will be most difficult. So document, document, document.

Site-Work Unknowns Come Unburied

The site-work portion of the project is fraught with potential claims. We discussed in Ch. 5 the geotechnical report, its accompanying test borings, and the uncertainty they provide with respect to being representative of site conditions. Without X-ray vision, what the geotechnician wishes, in good faith, to convey to the contractor can vary significantly from what lies below the surface.

The site contractor may find that during mass excavation, the conditions encountered are very different from what a review of the geotechnical report predicted. There may be an abandoned trash dump not indicated on the drawings, or there may be a series of active streams also not indicated on the drawings. In effect, what is encountered differs *materially* from what could have been anticipated from the bid documents. There are certain established precedents about when something is *materially different*. Courts have established the 15 percent rule: If conditions vary more than 15 percent from what could have been anticipated, that may be cause to request a change order. Various courts have established different parameters, so it is wise to consult with a company attorney if such a situation arises. Once again, when something seems to go awry: Document the time, place, and activity. Take photographs and alert the appropriate parties—owner and architect—that you are encountering a potential problem.

Errors and Omissions in the Plans or Specifications Are Discovered

Some of the problems discovered in the contract plans are brought to light when the contractor is required to prepare a set of coordination drawings to assure the architect that everything will fit in the space allotted to it. That is the time when the duct fabricator says that the 18-by-26-in. duct will not fit in the 16-by-24-in. riser chase that must also carry three 3-in. electrical conduits. Also, the roof drainpiping needs more space in the ceiling under the roof, and the ceiling fixtures will not fit under the fire-protection branch risers. Some of these changes can be made with no impact on time and cost, but others may require changes that are not acceptable to the architect, such as lowering ceiling heights. The coordination-drawing preparation is a ripe field for potential extra work and should be approached carefully and cautiously. Most of the potential extra-cost issues arise as the subcontractors begin their shop-drawing preparation and really dig into the details of the project. That is the time to expect and either accept or deny their requests for extras.

Different Interpretations Arise of the Scope of Work Required of the Contractor or Subcontractor

The window schedule may contain some window sizes different from those shown on various floor plans, or the door schedule may reflect doors of one size and finish while the floor plan indicates a larger size and a different hardware set. The specifications may allow PVC pipe for all potable-water piping 1 in. or smaller in diameter, but the plans may indicate all copper pipe; one is obviously more expensive than the other. Which is required by the contract? And if the architect requires copper, should a change order be forthcoming from the contractor?

Many of these types of errors, omissions, and unclear details or specifications can be resolved amicably, possibly by trading off one higher-priced item that will work just as effectively as a lower-priced one—there the contractor can absorb the extra cost. But no matter how the issue is resolved, the problem must be documented all the way, in case tempers flare and a dispute and claim appear on the horizon.

Ensure That Change Orders Include All Costs

Depending upon the nature of the change and the potential costs, oftentimes not all costs associated with the change are included. A checklist created as a guide can be very useful when a change order of large scope and cost is being prepared.

There are three categories of costs associated with change orders: direct costs, indirect costs, and impact costs. The following partial checklist may prove helpful in assembling a change order:

1. Direct costs—those hard dollar costs required to complete the work

    *   Increase in bond premium, since the cost of the project will increase

    *   Subcontractor costs

    *   Equipment—whether idle or active, with or without an operator (remember, if equipment is rented, costs continue to accrue while it is idle)

- Costs to prepare the estimate
- Possible increase in insurance premiums, since the contract sum will increase
- Labor and labor burden (remember when figuring the hourly cost of labor that a worker only works 50 weeks a year but gets paid for vacation, so costs are based on a 52-week year)
- Materials, taxes, delivery costs, and costs to distribute
- Postage, express delivery, and costs of reproduction
- Safety equipment required
- Temporary heat and temporary protection (including both erection and dismantling)
- Utility costs

2. Indirect costs—those costs relating to field overhead and home-office overhead

Field overhead

- Project-management staff
- Project engineer
- Project superintendent
- Field-office secretarial help
- Field-office supplies

Home-office overhead

- Accounting and payroll costs
- Change-order preparation, research, and negotiations with subcontractors and vendors
- Office equipment and supplies

3. Impact costs—those costs associated with changes that can impact schedule or performance

- Loss of productivity due to trade stacking or disruption of operations and workflow
- Idle equipment and idle-equipment maintenance
- Underabsorbed corporate overhead, if a deduction change order of significant cost is planned
- Lack of availability of skilled trade workers
- Cost of extended warranties of equipment installed during the project, if the projected completion date will be extended due to the change

Handling the Paperwork

Proper application of management tools places the CM or PM in a position of monitoring the project reports for problem areas, combined with taking trips into the trenches to observe the actual work in progress. The CM or PM's routine should include regular attendance at weekly meetings. Either sitting in as an observer or participating, the CM or PM will take note of the items discussed and the inclusion of key events, questions, problems, and solutions that are (or should be) included in the meeting minutes. A checklist of the routine items the CM or PM should monitor to assure him- or herself that administrative matter are proceeding according to plan is as follows:

- Weekly schedule evaluation and report
- Weekly craft productivity and loading report
- Weekly staffing report
- Material status and expediting reports
- Change-order log
- Request-for-information log
- Quality-control issues and reports
- Costs versus budget
- Safety, accident, and security reports
- Purchase orders
- Invoice-payment report—submission of monthly requisition and tracking of payment
- Subcontractor status report
- Copies of incoming transmittals and correspondence
- Field-office cost report
- Daily log
- Accuracy and completeness of the meeting minutes

This list of key items, plus the day-to-day incoming and outgoing correspondence, creates a mountain of paperwork for the field supervisors and the CM or PM. They can get bogged down in the flow of paperwork if it is not dealt with promptly.

To counter the office-binding effects of monitoring routine activities on larger jobs, the CM or PM needs to set aside time each day to walk the site. This can be during the workday, which can provide an opportunity to interact with the various trades and make the CM or PM visible as a manager who knows what is going on on his or her project. The CM or PM should never walk from point A to point B without carefully and quickly observing the pace of the work, any work lacking the quality standards that are expected by contract, and any safety violations or unsafe conditions. He or she should have a pad and pencil in hand, noting any problems, potential delays, need for more workers and materials, or any other job concerns picked up during that site visit. Also, he or she should be quick to point out good work to a crew or comment on the fact that they are moving along at a good pace.

It is a good idea to walk the site with the project superintendent at least once a week, which will allow time to discuss the project with the superintendent and observe how he or she is relating to the various subcontractors and vendors. Future CMs and PMs can come up through the field-supervisory ranks, and these weekly walk-throughs will give you a better feeling as to their capability and give you an opportunity to possibly teach them a few things each week and prepare them for advancement, if they seem to be ready and anxious to move up the supervisory scale.

Discovering the Job's Critical Areas

Each type of construction project has its typical list of critical areas. It is up to the CM or PM to look at those likely problem areas before they go critical in accord with Murphy's law. Fortunately, the CM or PM has some management tools and many years of construction experience to draw on. Evaluating the project for critical areas forms the basis of building the CM or PM's ongoing project-priority list.

The first place to look is the CPM schedule, assuming that the project has one—not a Gantt- or bar-chart schedule. The list of items and milestones that are sorted with the lowest total float are the ones to watch. As we said in Ch. 4, the critical-path items will thread through the whole job. Naturally, the contracting plan also affects the critical path. If the project is being fast-tracked, for example, the design, procurement, and construction are intimately dependent upon each other. As the CM or PM is the manager of the last party in the project parade, that mode places the greatest stress on him or her. The CM or PM's successful performance is at the mercy of the delivery of the equipment and materials and of the performance of the subcontractors, but he or she does have some control of these events. And that is all in addition to the normal day-to-day construction problem areas.

For lump-sum bid-to-complete-design projects, the CM or PM faces the problem of early ordering of critical equipment, materials, and services to meet the schedule. There is usually a great deal of pressure on the construction schedule because the bidding procedures may have already eaten up a lot of float. A lump-sum contract places most of the contractual risks for cost and profit onto the construction contractor, with heavy emphasis on the cost-control side of the project. Close attention must be given to project scope, quality, change orders, and the like.

The construction-management contracting mode places emphasis on management practice because the contractual cost and scheduling risks are passed onto the owner. However, the CM must be mindful of the fact that if he or she cannot control either cost or schedule, he or she surely will have a tough time getting a good reference from the current client. The CM must be mindful of the qualifications and capabilities of the winning subcontractors to ensure a good overall project performance. The coordinating role also emphasizes the CM's strengths in human relations, problem solving, and leadership in administering the subcontracts. The CM's priority lists starts with the overall project goals, followed by the individual areas of the subcontractors.

The CM's intimate role in developing the construction schedule has the benefit of highlighting the project's high-priority problem areas. Running a CPM schedule as a project-execution model on a computer gives the most detailed analysis of impending problem areas. Doing some "what if" analysis on the CPM schedule before and during the project brings the problem areas into even sharper focus.

Discovering the job's problem areas is only half the battle. Effectively solving them is the other half. Problem solving and decision making are much too broad a subject for us to cover in these few paragraphs, but we will talk about them in Ch. 14. Solving problems calls for a single-minded approach while keeping the overall project goals in mind. The CM or PM must evaluate the risk-versus-reward ratio and a fallback position when determining the solution. Also, the CM or PM must draw heavily on his or her own experience and know-how and on his or her staff's in the decision-making process. After considering the time factor and all the data available, they must make their decisions.

Operating the Construction Quality-Control Systems

Quality-control and quality-assurance procedures are vital to the success of any construction project. Although they are delegated to the field supervisors, it is the CM or PM's responsibility to see that effective procedures are in place, are documented, and function correctly throughout the project.

Each type of construction project has its specific quality-control and quality-assurance hot buttons that must be recognized by the field supervisors and CM or PM. A good example is the construction of a plant that manufactures dissolvable surgical sutures. The humidity control is of paramount importance. Humidity that is not controlled precisely in accordance with the design engineer's tolerance could cause the sutures to begin to prematurely dissolve in their packages, and there is no need to describe the legal implications if that were to happen.

Mock-ups, models, and key assemblies may need to be inspected for quality compliance in the presence of the design consultants to ensure that the subcontractor responsible for the task is thoroughly aware of the quality standards required.

Documenting the Project

Why is it that sometimes the field supervisor, CM, or PM says, "Why didn't I write that down? Now I only have the owner's/design consultant's/subcontractor's verbal comments—and it is not what was said!"

Project documentation begins with the daily report, which should contain:

- The project name and date
- The subcontractors on-site, the number of workers, the locations where they are working, and a brief description of what operation they are performing
- If work is being self-performed, the names of the workers and activities they were performing
- The weather—not only whether it is sunny or raining, but temperatures, preferably at 7:00 A.M., noon, and 4:00 P.M., with unusually severe weather and its duration to be noted as well
- Visitors to the site—name, affiliation, time spent on-site, and reason for the visit
- Any unusual occurrences, labor disputes, or accidents (and an official accident report)
- Visits by building officials and any inspection that took place
- A short sentence or two on the status of the project—i.e., third-floor slab being poured, steel erection on eighth floor

In addition to all the operational benefits of paperwork, it is needed to form a project history or paper trail (or electronic trail, since most of these reports will probably be forms on the computer) for future needs. Documentation of important events is rather difficult to quantify. But look at it this way—is it possible that this event could turn into a dispute and then a claim? Would I be without documentation of the events that I think could be required if a dispute occurs? Have I recorded sufficient information to protect my company?

Some Danger Signs That Require Documentation

There are several danger signs that can portend trouble down the road. The field supervisor and CM or PM ought to be familiar with them and consider documenting those events, if and when they occur:

- A subcontractor who has previously staffed the project as required suddenly reduces personnel even though there is sufficient work for a larger crew. Maybe the subcontractor is having money problems and has to cut back on payroll.

- A subcontractor or vendor delays submitting shop drawings. In order to obtain shop drawings, an order must be placed. Perhaps that vendor or subcontractor is having some financial problems and is shopping around for lower prices even though the material or equipment is critical to the job.

- A subcontractor is unable to provide sufficient day-to-day materials. We are talking about two-by-fours, Sheetrock, gallons of paint. This may be another sign of financial problems.

- A subcontractor requests a joint check where no such request has been made before. This may be another indicator of a financial problem.

Besides documenting these events, the CM or PM should have a talk with the subcontractor, who may have a valid reason for a short-term shortage of funds—or the problem may be more serious. Subcontractor default in the midst of the project is a difficult and costly event to deal with. So be aware of the potential, and document.

Documentation Is All-Encompassing

Properly documenting a construction project involves virtually every aspect of the work. Most documentation is required by contract—i.e., logging requests for information and for clarification, logging change orders, keeping project-meeting minutes, and recording various testing procedures. Public-works projects have their own rules and requirements.

The project documentation is collected and organized in the various project files and memorialized in the CM or PM's project file. Both CMs and PMs should evaluate the existing systems as applied to the needs of their present project. Minor modifications or additions to the existing documentation systems may be required to ensure meeting the present project's goal but wholesale changes to the existing documentation system is to be avoided.

Why bother going to the expense of documenting a project? Could a project run without documentation? Perhaps, but even if it ran well, we still would need documentation for efficient day-to-day operations, not to mention the requirements as set forth in the contract with the owner and in the construction-specifications manuals.

Project operations feed on a continuous flow of accurate and timely information. If there is not a smooth flow of labor, materials, tools, and equipment to the field construction forces, productivity suffers badly and the project goals cannot be met. It would be virtually impossible to communicate the necessary information orally, because nobody could remember who said what to whom and who had the authority to say it.

In addition to all the operational benefits of paperwork, documentation is needed to form a project history or paper trail for past and future needs. When things go wrong, as they often do (schedules are not met, budgets overrun, quality is substandard), we need the paper trail to figure out the solution and proper corrective action.

It is the CM or PM's responsibility to see that the field activities are properly documented. People sometimes become lazy and forget to write minutes of meetings, confirm telephone calls, or respond to e-mails; each of these represents a potential information gap in the project documentation. Usually those are the documents that are key to winning or defending a claim or similar financial obligation. Meticulous and well-organized project documentation is just another of the hallmarks of a superior CM or PM.

Project documentation culminates in the monthly project-progress reports. The CM or PM is responsible for preparing or at least editing these documents. The quality and appearance of the monthly reports tend to set the tone for construction-project operations in general, so a good set of reports is essential. CMs and PMs must be sure that their reports will stand up to thorough scrutiny and will project a high-quality image.

A Picture Is Worth a Thousand Words

With the profusion of smartphones and small digital cameras today, taking a picture to document a delay or a questionable construction detail is a simple matter. Adding the time and date makes the documentation complete. Transmitting the photo to the architect for further advice or to inform him or her of a jobsite problem is a quick and easy method to create irrefutable documentation—and do it quickly. If possible, print these photos out and keep them in the file, in case they are needed for backup to a dispute or claim.

The Owner, Constructor, and Designer Interfaces

The interfaces among the owner, design consultants, and constructor are the most important communication channels on the project. The overarching need is to establish a relationship based upon trust, ethics, and professionalism. It is vital that these interfaces are handled effectively to further everyone's project goals.

The owner, design consultants, and constructor each have a vital stake in making the key interfaces work. However, it is the owner who has the most to gain in making them operate effectively to meet the project goals. Any discord among the key parties must be investigated by the responsible people and resolved before any permanent damage can occur.

A major factor contributing to the age-old antagonism between design and construction is often the absence of contractual relations between the two. An obvious exception to this is in the design-build mode, where both designer and constructor are the same firm.

Each party must come to the realization that the design and construction processes may, at times, be imperfect. In our 40 years or so of dealing with owners and design consultants, we have yet to see a perfect, error-free set of design documents. As designs become more sophisticated and complex, it is extremely difficult, perhaps impossible, to develop that perfect set of design documents. Couple that with the fact that not all obligations, responsibilities, and details can be incorporated into the design and written in the terms and conditions of the contract between owner and builder, and an element of good faith must be created to deal with those minor elements that have defied definition.

The owner, after all, does underwrite most of the project cost in the end. Constructors may question the designer's plans and specifications, and the design consultants may question the constructor's knowledge and experience; but this can be accomplished in a spirit of cooperation and willingness to get the job done.

Only when an environment of trust and a willingness to concede minor issues develop between the team members will the road to a successful project be traversed. From time to time, there may be a disagreement as to what the contract documents, plans, and specifications specifically state and what is up for interpretation. When those occasions arise, reasonable people should resolve those issues with some reasonable give-and-take.

As the leader of the field forces, the CM or PM is responsible for maintaining good client and project relations through the project's execution. He or she must maintain a proactive attitude from all members on all sides of the unified construction team. Personal disharmony can only hurt the success of any construction project and result in dashed expectations on all sides.

Obviously, none of the preconceived notions about owners, contractors, and designers are true, or the construction industry would not enjoy the success it does. However, some individuals do have difficulty getting beyond their preconceptions, which lies at the root of the problem. CMs and PMs must show leadership by submerging any negative impressions and encouraging the most cooperative feeling possible among their team members and the other players. Projects with good preconstruction planning input usually find that the design interface runs much more smoothly. The construction, design, and owner groups have effectively investigated and resolved any differences as to the various construction materials and constructability factors by the time the project gets into the field. That leaves the probability of design and construction errors as the remaining area of contention.

Design errors of omission and commission will always be with us, although recent improvements in computer software substantially reduce these problems, as we shall see in Ch. 10. Owners should be sure that their design team has an effective quality-control system in place before drawings are issued for construction. The most common areas are generally coordination problems, particularly when the architect subcontracts civil, structural, and mechanical, electrical, and plumbing design to other firms and is lax in keeping those designers up to date on architectural changes.

Even the contracting method can create an environment that fosters design deficiencies. The most common one occurs in the lump-sum bid-to-complete-design contracting mode. Many of the more obvious design inconsistencies are discovered and corrected during the bidding stage, if the project is a private concern.

In the interest of increasing competition in a private bid, the designers may be willing to accept variations on products or equipment. For example, a building design with large window-wall requirements may not be completely detailed or specified, to keep the application open for several window-wall suppliers. After the most competitive system has been selected, it is up to the general contractor or the systems supplier to complete the final installation details. That in turn may involve some additional cost that was not clear in the bidding documents. This is a fertile area for problems in the design–construct interface, even though a statement in the general conditions may mention the possibility in general terms.

Other problem areas are cost and specification matters arising during the approval process for shop drawings. When alternative products are used in lump-sum bids and are not specifically indicated as alternates to the specified products, the contractor will have a little explaining to do. This certainly starts off the project on the wrong foot, and will cause the contractor to have to overcome intense scrutiny by the design team when future shop drawings are furnished. The owner may demand that the contractor supply one of the specified products because the alternate, not clearly spelled out in the bid, is unacceptable.

Fast-Track Project Problems

Another case occurs when a fast-track scheduling approach is being used and the field is opened too soon to suit the design schedule—or the design schedule falls behind. In either case, the schedule and field-delay costs place additional pressure to get the design documents issued for construction, possibly *without proper checking and quality control*. That practice can lead to some horrendous problems when major field installations are installed per incorrect drawings. That situation will eventually evolve into an open, devastating breakdown in all facets of the interface among design, owner, and construction, resulting in serious unmet project goals and additional costs.

All three players have a stake in avoiding this sort of situation. The CM or PM is placed in the difficult situation of trying to rectify the resulting technical and financial problems in the field. The owner faces the bottom line of having to pay for the nonproductive extra costs involved. The design firm faces financial as well as reputational losses. There is very little chance of anyone coming out clean in this no-win situation, so avoid it at all costs.

However, as we discussed in Ch. 2, when a public-works project is being bid, the bidders are cautioned to prepare their bid based upon the specific requirements of the bid documents. If a bid is submitted taking exception to any of the design plans or specifications, or if the bidder offers a better way to do things, the bid may be challenged by other bidders and most likely will be disqualified. Only when an award is made will the successful bidder point out a deficiency which may result in a change order increasing or decreasing the contract sum.

Looking at the positive side of the field–design interface, it is important to have its operation well organized. A communication channel between the field technical representative and a designated person in the design office must be arranged to handle routine design-document interpretation and to deal with unforeseen problem areas. As we said earlier, there is a normal flow between these groups in the approval process for vendors' data, design-document revisions, and other technical matters. Problem areas not fitting into the routine flow must be handled separately as they arise, in the manner called for in the contract, and they must be well documented, including official approval or disapproval.

Change orders are a case that falls into the category of nonroutine matters. They also must be handled as called for in the contract and well documented, including official approval or disapproval. Do not let unresolved change orders accumulate, because they cost money and must be resolved one way or the other. CMs and PMs cannot ignore the change-order process and the need for keeping the change-order log current with the date the change order was submitted, the date its return was anticipated, and the date it was actually returned.

Project Safety and Security

The background, needs, and development of a sound safety-and-security program is discussed in Ch. 12, but here we wish to address the CM and PM's role in integrating those two important functions into the ongoing site operations.

Site safety is an important function regardless of project size. Certainly, larger and more complex construction sites do require more sophisticated safety programs. On small projects the CM or PM may even wear the safety supervisor's hat. On very large projects, the site-safety group is more sophisticated to meet the special needs of a project of that size.

Owners have a vested interest in safety and can play a major role in the program. As we previously discussed, an owner-controlled insurance program will require the owner to provide sufficient safety rules, regulations, and supervision to ensure that the premium costs he or she will be assuming do not increase. And from a public-relations point of view, no owner likes to have a construction site pictured on the front page of the newspaper or on local television describing a major accident that occurred.

Both federal and state OSHA organizations play an overseer role on construction sites, and the Environmental Protection Agency will also be looking over the CM or PM's shoulder for any potential violations of environmental laws and regulations.

CMs and PMs are ultimately responsible for site safety and accidents that occur on their jobs, even though a safety engineer or department handles the details of the site-safety program. As the saying goes, the buck stops there. As part of the CM or PM's ongoing management, he or she must constantly evaluate field operations and the efficacy of the site-safety program. The CM or PM must develop a habit of looking for safety issues no matter where he or she walks on the site; even as he or she walks through the building on the way to a meeting with the architect, the CM or PM must be constantly on the lookout for actual or potential safety violations.

Just as the firm's general safety programs are founded on the support of top management, the site-safety program is based on the CM or PM's full commitment. He or she must be proactive by word and deed to continuously promote safety on the site. The CM or PM should lead the way in setting up work-safety goals and in organizing an award program to reward successful achievement of those goals—an accident-free project.

The site-safety supervisors usually handle the details for the governmental safety inspections, involving both federal and state OSHA. The CM or PM must participate in resolving any shortcomings or dealing with citations raised by the inspectors. The cost of violations in both dollars and time, not to mention human misery, can be substantial.

A successful safety program relies heavily on a good on-site first-aid facility, backed up by an effective ambulance and hospital facility for more serious injuries. Dry-run testing of the first-aid and other emergency procedures is also effective in evaluating and maintaining their reliability.

Fire safety is another major part of overall site safety, particularly when flammables are present during construction. An extensive welding operation requires special training in the use of fire-prevention tools (extinguishers) and welding blankets, as well as a detailed program dealing with procedures to follow before, during, and after a welding operation.

Keeping close contract with the local fire department is essential in the case of fires exceeding the capacity of any on-site fire-protection capabilities. Remote sites and very large projects may even require a self-sufficient fire department capable of handling any construction-site fire emergency.

Site Security

Although security and safety are not directly related and have separate responsibilities, they seem to be grouped together in practice. Security of the site is necessary for a lot of reasons, such as

- general site safety
- control of access by unauthorized personnel
- protection of physical assets—tools, equipment, materials, and personal effects—against theft and vandalism
- site control during labor unrest
- site monitoring during nonworking hours.
- contacts with local security forces
- control of proprietary and classified information and equipment

Site security starts, where feasible, with a perimeter fence to control public access and limit entry and egress for all personnel working on the site. Perimeter lighting is a deterrent to illegal entry when the gate is closed and locked for the night and on weekends.

Technology has been involved in construction-site security for many years. Remotely placed cameras that provide real-time video recordings of portions of the site or building provide the evidence of a crime—theft or damage—and if the builder advertises that they have been installed in strategic spots around the site, can discourage would-be thieves.

Project-Completion Phase

The project-completion phase starts at the beginning of the project, when all subcontractors and vendors are alerted to the provisions in the specifications manual dealing with project closeout. The CM or PM must advise all attendees at the first project meeting that they need to review project closeout procedures to ensure that, as they progress through the job, they are addressing some of the items required at closeout.

Establishing the date of substantial completion affects a number of issues; at that point, all utility-operating costs are transferred from the contractor to the owner. That is also the date when, upon acceptance by the mechanical and electrical engineers, the contractor's warranties and guarantees are transferred to the owner. If this date is allowed to drag out, the contractor will continue to pay for utilities, which can be expensive, since the entire building may be going through the heating or cooling cycle depending upon the time of the year.

There is sometimes confusion over what is meant by *substantial completion*. This is defined as that period when the building has been completed to the point where it can be used for the purpose for which it was designed. This does not mean that the building is 100 percent complete. The building can still be used "for the purpose for which it was designed" without carpet in a few offices, even with a missing door on the mail room. The CM or PM should review the owner and architect's definition of substantial completion as it is occurring to ensure that there is a meeting of the minds as to what is required to meet that standard.

Project Completion When the Project Is a Process Facility

Winding up a process-facility construction project can be a hectic time if the closeout operations are not well organized. A great deal of coordination among the construction contractor, owner, design group, and plant-operating team is vital to the checkout, start-up, and turnover of the facility. Planning for these final events actually starts during the initial project-planning stage with the scheduling of the unit-completion sequence. The first units scheduled for completion are usually the utility systems, be they simple tie-ins or complex stand-alone units.

Utility tie-ins usually are simple activities that are done fairly early in the job so that permanent utilities can be used for construction purposes. This reduces the cost for temporary construction utilities and allows the systems to be debugged during construction. The same thinking applies to utility units, except that they come onstream later in the job. The other off-site and process units are scheduled in a follow-on sequence that suits the start-up plan.

The complexity of phasing out a process construction project is shown by the key construction procedures required to ensure project completion and a smooth start-up. These activities are all carried as punch-list items that must be closely coordinated with the start-up schedule. The CM or PM is responsible for timely completion of these key items, either directly or by subcontractor:

- Pressure testing and sign-off of piping systems
- Final Piping and Instrimentation Diagram (P&ID) check of all systems
- Checkout and energizing of electrical systems

- Calibration and checkout of instrumentation and control systems
- Flushing of piping systems and equipment
- Installation of oils and lubricants
- Rotational and mechanical testing of mechanical equipment and systems (often in conjunction with vendor's representatives)
- Installation of catalysts and chemicals
- Curing of field-installed refractory linings
- Turnover of the manufacturer's operating and maintenance manuals
- Turnover of units to operations for plant commissioning and start-up

It is readily apparent that the work listed here can be quite costly and labor intensive, involving personnel from the owner, contractor, subcontractors, and design staff. The CM or PM must pay close attention to the scope of work in this complete and detailed area.

Power-piping and boiler codes from the American National Standards Institute require that all piping systems, vessel, tanks, heat exchangers, and the like be hydrostatically tested to ensure leak-tight and safe operating systems.

Shop-built equipment is usually tested in the shop and does not require retesting in the field unless the item has been field-altered. By far the largest field-testing load is in the area of the field-installed piping systems, which can number in the thousands on a large process project.

Construction-Project General Phaseout Activities

The key to closing out the project is the receipt of the owner's final acceptance letter indicating that all construction and contractual matters have been properly completed. The contract typically describes in some detail how the acceptance will be handled. CMs and PMs should make sure the turnover procedure is followed in a proactive manner to ensure a smooth transition with a minimum of revision.

Consolidating the voluminous project files seems to be the most nagging of the project closeout procedures, especially when the applicable procedures are either vague or nonexistent. It is truly amazing how the volume of paper accumulates on a construction project, in spite of the number of computers, with their hard-drive documentation and CDs.

Some firms write a job history covering the positive and negative factors that were encountered in executing the project. Any special features dealing with the owner or specific industry should be discussed. What went wrong and why? What or how could have been done better? Where there any labor problems? If so, how were they solved?

After the project has been closed out, there is still the warranty period, generally one year in length (more in certain instances for various materials and equipment). Warranty items should be attended to promptly and efficiently; if you are in doubt whether the items is a warranty item or not, but it is insignificant to repair, doing so will certainly stand your company in good stead. The final impression you wish to leave with the client is one of competence, professionalism, and a willingness to help in any way possible.

Most construction-closeout lists include requirements relating to as-built drawings. If you recall, we mentioned that this activity begins at the start of the project, when each subcontractor responsible for as-built drawings is reminded at the very first project meeting that these drawings must be accurate and updated as work progresses.

Periodically, the project superintendent must review the status of each as-built, verifying for accuracy by actually checking one or two dimensions from the as-builts from a select few subcontractors. If inaccuracies are discovered, it would be wise to assemble all those who have this responsibility and inform them that less-than accurate will not be accepted and payment may be withheld until inaccurate as-builts are corrected.

Assembling and turning over to the owner the project documentation as defined in the contract is frequently a sore point in closing out the project. Problems occur if the responsible subcontractors let this area fall behind or if vendors and suppliers are late with their information. The owner will not release final payment until all the contractual requirements for documentation have been fulfilled.

There again, periodically during the job the field superintendent can check with the appropriate subcontractor: "Do you have sufficient attic stock or will you need to add more?" "Are those special tools for the sprinkler heads on-site now?" This will let all subcontractors and vendors know that the superintendent is paying close attention to all aspects of the project.

While winding down the technical aspects of the construction job, CMs and PMs cannot forget about closing out the administrative side of the work. The final goal of finishing the project *as specified, safely, on time, and within budget* can slip away during the last days of the project. Many field personnel are anxious to move on to a new job to which they have been assigned and may not pay full attention to closing out the current one. It is important at this stage to make sure that all outstanding subcontractor and supplier invoices and claims have been submitted.

Building Commissioning

Concurrent with the accumulation-of-paperwork aspect of closing out the project, the building commissioning is a critical element of the turnover process. Building commissioning is the process of ensuring that the owner's operational needs as designated in the contract documents have been met.

The goal of building commissioning is rather simple, although the process of achieving it can be complex:

- Commissioning is the process of delivering a facility that operates as it was designed and intended to operate.

- Commissioning ensure that the facility meets the needs of the building owner, the facility manager, and the occupants.

- Commissioning is the process whereby the facility operators are trained in the operation and maintenance of the building's systems.

This process generally takes place with the owner's facility manager, the owner's design engineer's representative, representatives from the suppliers of equipment and systems, and the construction manager or project manager.

During the 2005 National Conference on Building Commissioning, Charles E. Dorgan presented a synopsis of the ASHRAE/NIBS commissioning guidelines (new at the time). Selected portions of this presentation are excerpted here to provide a broad-based approach to the commissioning process that will prove helpful to CMs or PMs developing their own plans. ASHRAE guideline 0-2005 (Annex J) provides a general format for developing the owner's project requirements, and Guideline 1-200X-HVAC &R Technical Requirements for Commissioning.

Guideline 0-2005—An Abbreviated Version

The Commissioning Process is a quality-oriented process for achieving, verifying, and documenting that the performance of facilities, systems, and assemblies meets defined objectives and criteria.

The Commissioning Process assumes that owners, programmers, designers, contractors, and operations and maintenance entities are fully accountable for the quality of their work. The Commissioning Team uses methods and tools to verify that the project is achieving the Owner's Project Requirements throughout the delivery of the project. For example, the contractor is responsible for fully constructing, testing, and ensuring that its employees' work has provided the level of quality expected. The Commissioning Authority then randomly samples the contractor's work to verify that it is achieving the

Owner's Project Requirements. If systemic issues are identified, then the contractor is expected to recheck all of his/her work and correct any deficiencies. This quality-oriented process is different than when the Commissioning Authority does 100% checking or non-quality-based sampling. Guideline 0-2005 has been developed to present an approach based on these assumptions.

The Commissioning Process begins at project inception (during the Pre-Design Phase) and continues for the life of the facility (through the Occupancy and Operations Phase). Because this Guideline details a process, it can be applied to both new ands renovation projects. The Commissioning Process includes specific tasks to be conducted during each phase in order to verify that design, construction, and training meet the Owner's Project Requirements. This Guidelines describes the overall Commissioning Process in order to provide a uniform, integrated, and consistent approach for delivering and operating facilities that meet an owner's on-going requirements.

The Commissioning Process is a quality-based method that is adapted by an Owner to achieve successful construction projects.

The use of a common content organization and the focus upon specific information achieves a closely coordinated set of documents that can be used together or in any combination to accommodate varying owner requirements. This Commissioning Process guideline allows the technical commissioning guidelines to avoid repeating information on the commissioning process, making them more concise and focused relative to their technical requirements.

The fundamental objectives of the Commissioning Process are to:

(a) Clearly document Owner's Project Requirements;
(b) Provide documentation and tools to improve the quality of deliverables;
(c) Verify and document that systems and assemblies perform according to the Owner's Project Requirements;
(d) Verify that adequate and accurate system and assembly documentation is provided to the owner;
(e) Verify that operation and maintenance personnel and occupants are properly trained;
(f) Provide a uniform and effective process for delivery of construction projects;
(g) Deliver buildings and construction projects that meet the owner's needs, at the time of completion;
(h) Utilize quality-based sampling techniques to detect systemic problems, as such sampling provides high value, efficient verification, accurate results, and reduced project costs, and;
(i) Verify proper coordination among systems and assemblies, and among all contractors, sub-contractors, vendors, and manufacturers of furnished equipment and assemblies.

Emphasis is placed on documentation of the Owner's Project Requirements at the inception of the project and the proper transfer of this
information from one party to the next. Owners adopt the Commissioning Process to achieve their stated objectives and criteria—starting
with the inception of a project instead of after a facility is occupied.

While circumstances may require owners to adopt the Commissioning Process during the Design or Construction Phase of a project,
such later implementation must capture the information that would
have been developed had the Commissioning Process begun at project inception. Beginning the Commissioning Process at project inception will achieve the maximum benefits.

Annexes have been included to assist in further understanding
the Commissioning Process and to aid in the development of the
technical guidelines.

The Commissioning Process has been structured to coincide with
the phases of a generic project with Pre-Design, Design, Construction, and Occupancy and Operations phases.

This Guideline describes the Commissioning Process; the responsibilities of Commissioning Team participants; the role of the Commissioning Authority; and a model framework for developing a
Commissioning Plan, specifications, and reports. This Guideline also
describes the general requirements for a training program for continued successful system and assembly performance. Documentation
necessary to meet the Guidelines requirements is also described.

Accessing these ASHRAE guidelines should provide the construction manager and project manager with greater insight into the
commissioning process that is such a key element in leaving the
owner with a successful project.

A Word about Lien Waivers

Most subcontract agreements require the subcontractor to submit a
partial lien waiver indicating receipt of payment for the previous
application for payment. These lien waivers are required for every
payment except the first, since no prior payment has been received.
This lien-waiver requirement must be strictly enforced: *If the subcontractor does not submit a lien waiver with the second and all subsequent
applications for payment, the request will not be honored.*

Then there are the second- and third-tier subcontractors, who
must also furnish lien waivers to their prime subcontractor. Subcontractors should be directed to provide the project superintendent
with the names of all such lower-tier subcontractors, not only because
the superintendent needs to know not only who is working on the
project but also which lower-tier subcontractors must submit their
lien waivers to their prime at requisition time.

Final lien waivers from all subcontractors and vendors are generally submitted with their final application for payment or invoice. This lien waiver stipulates that it is activated once the subcontractor or vendor receives the amount of the payment included in that final lien waiver.

Summary

Getting the project off on the right foot requires the CM or PM and other supervisory personnel to become intimately familiar with the contract documents—rereading and highlighting the key areas that will play an important role in the smooth progression of putting work in place. The negotiation of subcontract agreements and vendor purchases will present several challenges: meeting the budget and selecting productive, efficient subcontractors and reliable suppliers that can meet a demanding schedule and produce a quality product. This can be accomplished by diligently pursuing the commitment to the owner as expressed in the contract and the design documents. A working relationship with the owner, the design consultants, the subcontractors, and the vendors should be based upon a foundation of trust, professionalism, the ability to view a problem from another's perspective, and a sense of reasonableness.

Hard work lies ahead, but the building of a team that is willing and able to work together, with possibly a little give-and-take, will make this a rewarding endeavor.

CHAPTER 10

Total Construction Project Management for the Twenty-First Century

New technologies which have come to the fore in the twenty-first century—and their impact—will afford construction managers and project managers more efficient and effective means to accomplish their total construction project management program. These means and methods will involve three basic trends:

- Interoperability
- Building Information Modeling (BIM)
- Sustainability

Interoperability

There are a few definitions of interoperability; one is simply the ability of software and hardware on different devices from different sources to talk to each other. Another definition of interoperability is the ability to exchange and manage electronic information seamlessly and to comprehend and integrate information across multiple software programs.

Today's projects are more complex than ever before and involve specialization beyond the traditional architectural, structural, mechanical, and electrical engineering disciplines. Other concerns have entered the picture—master planning, infrastructure interfaces, renewable energy, environmental issues, sustainability, health and safety of building occupants, information technology, and communications, to name a few. Now all of these concerns, needing to be addressed, also need to talk—communicate—with each other. The proliferation of various types of software, using different languages, has been likened to the Tower of Babel.

The techniques required by professionals to be most effective and efficient require the ability and technical means to collaborate. Different programs with their proprietary data structures, in many instances, did not formerly have the means to link their databases and share information—hence the need for interoperability. The National Institute of Standards and Technology has estimated that the cost of inadequate interoperability among U.S. government facilities was about $15.8 billion per year. The cost to the Architecture-Engineering-Construction (AEC) industry was unknown, but likely substantial; combining the two (design and construct) seems like it was one of the driving forces behind the creation of interoperable information-technology language.

This seamless flow of information, aided by interoperability, will allow architects, engineers, and builders to access and communicate data regarding products and systems during the conceptual-design stage in order to select products and equipment deemed appropriate for the project being developed. Quality, reliability, efficiency, interchangeability of equipment, and costs can be obtained from manufacturers and incorporated into the design. The old question of capital costs versus operating costs ought to be able to be explored as one product's characteristics are weighed against another. In fact, the initiative to resolve interoperability issues got strong support from proprietary software vendors; and as we shall see when discussing building information modeling, the ability for design consultants and equipment manufacturers and vendors to interface electronically became essential.

Interoperability is a critical ingredient in the process of building information modeling. The practicality of a structural engineer being able to design a 3-D steel framework and send it on to the architect and the electrical and mechanical engineers for advice and comment seems elementary. But what about a manufacturer of lighting fixtures and heating and cooling devices who, by entering into this BIM cycle of 3-D virtual building design, can participate in determining efficiency of operation and capital costs and pass those on to the architect and engineers for review and comment?

Unless all of these participants in the AEC community can correspond electronically using the same "language," the anticipated productivity gains and resultant savings cannot be attained.

National Institute of Standards and Technology

This government organization was at the forefront of recognizing the value of interoperability. In 2002 the Institute conducted a study to uncover some of the problems that would confront a construction industry, already suffering from inefficient information management, as building information modeling began to come to the forefront of the design profession and the cost of software began to decline. What they found only emphasized the need for the AEC industry to get its act together.

- Collaboration software was not integrated with a contractor's other systems but used as a stand-alone system, defeating the purpose of the software.

- Since each project is unique, the desire of a contractor to enter into a system that is compatible with other architects and engineers seemed not so critical.

- Life-cycle management processes were fragmented and not given the importance they deserved.

- A lack of data standards inhibited the transfer of data between different phases in the life cycle of the project and its associated systems and applications—again, that inability to talk to each other.

- In some firms, an estimated *40 percent of engineering time* was devoted to locating and validating information gathered from disparate sources.

- Many smaller construction firms did not employ, or made only limited use of, technology in managing their business processes and information.

International Organization for Standardization (ISO)

As the world began shrinking electronically and engineers in Europe or Asia started working with architects in the United States and vice versa, the need for a common language became apparent to a nongovernmental organization that would become the world's largest developer and publisher of international standards: the International Organization for Standardization (ISO). Although the organization's acronym would appear to be IOS, it actually refers to the Greek word *isos*, meaning "equal."

ISO is a developer of international standards. Its standards, as related to the construction industry, incorporate technical advice obtained from hundreds of experts worldwide. ISO recognizes that since the construction industry has such a great impact on natural resources worldwide, it should be closely associated with the sustainability movement. The organization recognizes the need for common terminology, rules of information exchanges, measurement techniques, and material descriptors as the globalization of construction and international trade expands.

ISO TC 59 is the assigned category for the standardization of terminology and the organization of information in the building and civil engineering processes. The scope of this standardization includes:

- General terminology for building and civil engineering
- The organization of information in the processes of design, manufacture, and construction
- The general geometric requirements for building, building elements, and components, including modular coordination and general rules for joints, tolerances, and fits
- The general rules for other performance requirements for buildings and building elements, including the coordination of these with performance requirements of building components to be used in building and civil engineering
- The geometric and performance requirements for components that are not in the scope of a separate ISO technical committee

Beginning with the planning stage, ISO has published standards for construction activities from the laying of foundations to building commissioning. Currently ISO is developing standards for the following building components:

- TC 21: Equipment for fire protection and firefighting
- TC 59: Building construction
- TC 71: Concrete, reinforced concrete, and prestressed concrete
- TC 74: Cement and lime
- TC 77: Products in fiber-reinforced concrete
- TC 89: Wood-based panels
- TC 92: Fire safety
- TC 98: Bases for design of structures
- TC 136: Furniture
- TC 160: Glass in buildings
- TC 162: Doors and windows
- TC 163: Thermal performance and energy use in the built environment
- TC 165: Timber structures
- TC 178: Lifts, escalators, and moving walks
- TC 182: Geotechnics
- TC 189: Ceramics

- TC 195: Building-construction machinery and equipment
- TC 205: Building-environment design
- TC 218: Timber

The full content of each of these building components can be explored at the ISO website (www.iso.org). But just looking at one building component—TC 71, concrete—will provide some insight into the depth of ISO's standards:

- TC 71/SC 1: Test methods for concrete
- TC 71/SC 3: Concrete production and execution of concrete structures
- TC 71/SC 4: Performance requirements for structural concrete
- TC 71/SC 5: Simplified design standard for concrete structures
- TC 71/SC 6: Nontraditional reinforcing materials for concrete structures
- TC 71/SC 7: Maintenance and repair of concrete structures
- TC 71/SC 8: Environmental management for concrete and concrete structures

ISO 10303, also known as STEP (Standard for the Exchange of Product) is the international standard to describe product data independently of any particular system throughout the life cycle of a product—from design to manufacturing, usage, and disposal.

CIMSteel Integration Standards, Version 2

The structural-steel industry has been a leader in interoperability. The CIMSteel Integration Standards were created to achieve a seamless flow between software systems for structural-steel designers, detailers, and fabricators. This software allows the structural engineer, detailer, and structural-steel subcontractor to significantly reduce the time between design, purchasing of the steel, and development of shop drawings, and it fits perfectly into the BIM scheme. As the steel is designed and sent to the fabricator, shop drawings are created electronically and sent back to the design engineer, who will approve or adjust them, thereby either eliminating or dramatically reducing the revise-and-resubmit review process. The fabricator can possibly suggest a few changes to save time and money; once again, these changes can be reviewed, accepted, or rejected very quickly. And fabrication can commence much faster, getting the steel to the jobsite sooner. The savings in shop drawing preparation, design review, product delivery, fabrication, and shipment to the jobsite can result in some real savings to all parties to the process.

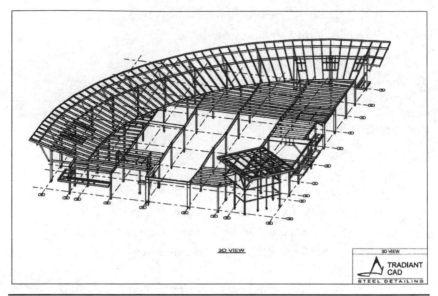

FIGURE 10.1   3-D representation of a structure. (*By permission of Tradiant CAD, Memphis, Tennessee.*)

Schedule compression can be achieved, for example, by the selection of a structural-steel contractor as the structure is being designed. Figure 10.1 3-D representation of that structure when transmitted to the fabricator can commence the production of shop drawings for review by the engineer, allowing the fabricator to order steel and begin fabrication while the mechanical, engineering, and plumbing designers begin their layouts and check for conflicts.

Building Information Modeling (BIM)

BIM, the creation of a three-dimensional digital structure—a *virtual building*—promotes two critical design functions:

- It creates a design in digital form that can be shared among all members of the design team and updated, changed, or modified easily to accommodate each member of the design team as they work through the process from project concept to construction drawings.

- It creates real-time analysis of the relationship of each design discipline in the design-build process, and coupled with a proven database of costs can allow tracking of design and budget in ways not available otherwise.

These 3-D models provide a project visualization; architects can show their clients how the new project will fit into the landscaped site, how the building will look as the owner drives around the side to the parking lot on the east side, and even how visitors will view the lobby and reception area.

In a 2012 survey conducted by FMI, a construction-management consulting firm located in Raleigh, North Carolina, respondents indicated that they used BIM on projects for the following reasons:

- detection of clashes in the design process
- shop-prefabrication management
- shop-drawing review
- marketing and sales development
- cost estimation
- safety analysis and management

Building information modeling is being used effectively in both the design and construction phases of projects. It also provides an owner with more latitude in the selection of a project delivery system. By engaging a construction manager during the design stage or negotiating a contract with a general contractor and allowing him or her to select a group of key subcontractors, the owner can control pricing to a greater degree and, as importantly, compress construction schedules. Since BIM allows quantity takeoffs daily as design progresses, costs can be monitored closely.

BIM Benefits During the Design Stage

When an architect is given a commission by an owner, he or she is given the design parameters and the budget. Unless the structure is being developed as a design-build project or the owner has elected to hire a construction manager who will work with the architect from concept to construction drawings, the architect has no real check on tracking design and cost. In a design-bid-build project, the architect proceeds with design, tapping into his or her own database of costs or using consultants to provide periodic estimates; both exercises add some degree of cost to the project. Where the project is a based on cost plus a fee with a guaranteed maximum price, the contractor may have been called in after an unsuccessful bid by the owner and wishes to negotiate those higher-than-budget project costs with the preferred contractor. Redesign—hopefully minor but possibly considerable—may be required, there again causing added costs in both design and time. The ability of BIM to generate quantity takeoffs, if it is used with a cost database available, can enable the architect to quickly and efficiently change a building component or system.

As the architect translates the owner's program into a building concept, he or she constantly faces the need to balance design with utility, scope, costs, and schedule. As the design progresses through its stages, traditional methods of 2-D presentation—with its inevitable changes, trade-offs, and value-engineering proposals and evaluations—all add costs to the project: not only physical costs to investigate and make the changes but also the time to incorporate the changes into the ongoing design work of the architect and of the civil, structural, electrical, mechanical, and systems engineers. The immediate dissemination of changes with BIM eliminates problems that often occur during 2-D design, such as when a change is not transmitted to all designers but ends up as a field problem.

Conflicts: 2-D Versus 3-D

Conflicts—for example, where a duct is blocked by a beam and has to be reconfigured, or a plumbing branch line needs to be moved so as not to hit an above-ceiling light fixture—are all real-life problems that sometimes occur in the 2-D design mode. Most are caught in time, and a few hours are spent rerouting the duct or pipe on paper. But some of these conflicts do not get resolved so easily, and appear instead during construction in the field. In a typical scenario, the subcontractor talks to the project superintendent, who relays the information to the architect in the form of a request for clarification or information. After a consultation with the mechanical, engineering, and plumbing engineer, a sketch is sent to the field and work progresses, hopefully at no additional cost other than time lost awaiting that revision. The same scenario utilizing BIM and a smartphone would probably be resolved after the afternoon coffee break.

Figure 10.2 is another example of how BIM can solve a problem, in this case a conflict between an 8-in. PVC pipe sleeve and a continuous footing. If the design had been created using the conventional 2-D process, this conflict may not have shown up until field operations commenced and the footing excavation and reinforcing bars were installed. Figure 10.3 shows how another potential conflict was discovered and cured via the installation of a sleeve through a grade beam; this will be indicated on the structural drawings, rather than possibly being discovered in the field. As in the previous scenarios, even with the ability of a project superintendent to convey these problems to the structural engineer quite rapidly by smartphone, time would have been wasted. Without BIM, an exchange of requests for information would be required to solve the problem. And time is money, as we all know; if there are delays, someone will have to pay for them. Along with the time to redesign and the break in the production flow, costing both

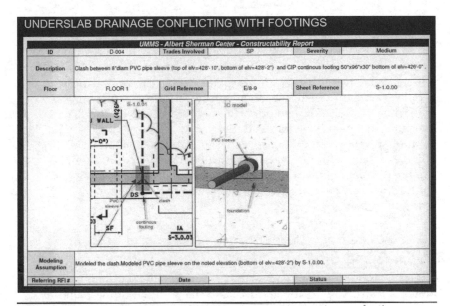

UNDERSLAB DRAINAGE CONFLICTING WITH FOOTINGS

		UMMS - Albert Sherman Center - Constructability Report			
ID	D-004	Trades Involved	SP	Severity	Medium
Description	Clash between 8"diam PVC pipe sleeve (top of elv=428'-10", bottom of elv=428'-2") and CIP continous footing 50"x96"x30" bottom of elv=426'-0" .				
Floor	FLOOR 1	Grid Reference	E/8-9	Sheet Reference	S-1.0.00
Modeling Assumption	Modeled the clash.Modeled PVC pipe sleeve on the noted elevation (bottom of elv=428'-2") by S-1.0.00.				
Referring RFI#	-	Date	-	Status	-

FIGURE 10.2   Conflict between an 8-in. pipe sleeve and a continuous footing.

SLEEVING THRU GRADE BEAMS

FIGURE 10.3   Conflict resolved by a sleeve through a grade beam.

architect and contractor money, the owner paying interest on the construction loan may become impatient with too many tweaks.

Building information modeling allows the project team to quickly make a change in, say, the structural design, which can also be reflected in all the other design disciplines with a click of the mouse. Because a series of "what if" changes can be performed rather quickly, the owner is afforded more opportunity to effect the most efficient structure—a bigger bang for the bucks.

BIM software actually enables design professionals to *virtually* build a project before it is ever physically constructed in the field. The big problem of coordination, often a very time-consuming and frustrating exercise, can either be totally eliminated or dramatically reduced.

What Are 4-D and 5-D?

4-D is the application of time. As the design progresses from foundation or substructure to superstructure to the addition of mechanical ducts, electrical conduits, and branch wiring to plumbing piping and, finally, partitions, fixture installations, and ceiling, flooring, and other finishes, the 4-D model captures each of these events—as they start, as they progress toward completion, and as they combine to form the completed structure. By creating, in effect, a snapshot as each building component starts and proceeds towards its finish, the BIM architect is producing a *virtual schedule*. After conferring with the construction manager or project manager as to the length of time for the project to complete a specific phase of construction, these snapshots can be used to create the actual schedule. When converted to a time-release mode with start and finish dates included, it can be used to compare *actual* progress with *planned* or *scheduled* progress.

An owner at a progress meeting can ask the architect, "Can you bring up where the structural-steel erection should be at this time?" The architect, having saved the progression of steel design and its passage through fabrication in his or her computer, could bring this up on the screen and say, "Well, according to the schedule we should be placing steel deck on the fourth floor." By looking out the field-office window, it would be rather easy to see if construction was up to the fourth floor, below it, or above it, which would indicate whether the project was on schedule or behind or ahead of schedule. That is the beauty of 4-D BIM.

The following figures are images of the construction of the Albert Sherman Center at the University of Massachusetts Medical School in Worcester, Massachusetts, taken from that hospital's Web site. In order to demonstrate the ability to the 4-D BIM approach to *virtual scheduling*, we have added hypothetical dates to each of these stages

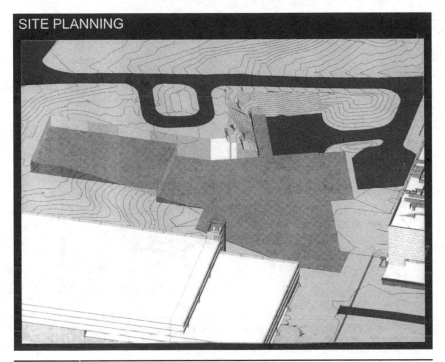

SITE PLANNING

Figure 10.4   April 15, 2012: The building platform is brought to grade and areas rough graded.

of construction as design progressed, solely to demonstrate our point. By adding these dates, we have created a virtual schedule. If changes are requested by the owner, they can, once approved, be added to the design. If owner and contractor agree, the necessary changes can then be made to the schedule.

Using these BIM images as a virtual schedule, we can deduce the following:

- Figure 10.4, April 15, 2012—the building platform has been brought to grade, areas have been designated as entrance driveways, and parking areas have been rough graded.

- Figure 10.5, September 5, 2012—the superstructure is complete except for roof framing.

- Figure 10.6, November 25, 2012—the building envelope is in progress on the structure in the foreground; on the rear structure, framing is in progress.

- Figure 10.7, January 18, 2013—the building envelope is complete on the foreground structure; exterior walls are in progress on the larger rear structure and columns to support the roof structure are in place.

CORE/SHELL STEEL STRUCTURE

Figure 10.5 September 5, 2012: The superstructure is complete except for roof framing.

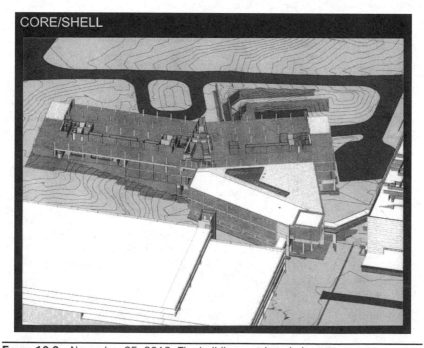

CORE/SHELL

Figure 10.6 November 25, 2012: The building envelope is in progress.

CORE/SHELL

FIGURE 10.7   January 18, 2013: The building envelope is complete on the foreground structure; the exterior wall is in progress on the rear structure.

This *virtual schedule* can be effectively used at a project meeting, when the owner asks, "How are we doing schedule-wise?" The architect or project manager can call up the image relating to the time frame under discussion and, by a short walk on the site or a look out of the field-office window, determine whether the project is on target or behind or ahead of schedule based upon the status as shown on the virtual schedule.

Progressing to 5-D adds the cost portion to BIM. As the building model progresses, the software can list the material quantities as the design develops, in effect creating a quantity takeoff commensurate with the stage of the design. Adding a unit cost of labor and materials and a production rate provides the owner and design consultants with a daily design-versus-cost matrix. Figure 10.8 shows this flow as quantities, costs, markups, and production rates not only create a moving budget but also develop the virtual schedule.

This entire process can progress very effectively if a construction manager is brought on board during the design stage and can contribute costs and production rates, or if the owner elects to negotiate

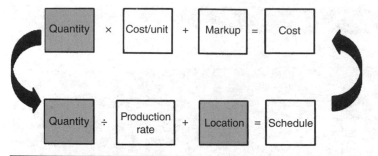

Figure 10.8 The pathway from 4-D scheduling to 5-D estimating when cost of materials and production rates are included allows participants to evaluate cost versus design almost instantaneously. (*By permission of Vico Software Inc., Miami, Florida*)

a contract with a general contractor and brings the contractor and his or her estimator onto the project in the early stages of design via BIM.

Progress to 6-D

The use of BIM for facilities management is referred to as 6-D. The longest and potentially most costly portion of a building's life is the period of operations and asset management. Facility managers need lots of information quickly after the commissioning of their building. By including data on life expectancy and replacement costs of building components and equipment in the BIM design process, owners can better understand how to track equipment costs and building systems to achieve better payback over the anticipated life of their new facility. Managing accurate as-built drawings and having the capacity to update them easily as the inevitable changes occur over the life of the building is of distinct advantage to the owner and becomes invaluable if the building is sold or leased to another owner. With the 6-D application of BIM, the facility manager can quickly retrieve the following documents:

- operating and maintenance manuals
- maintenance logs and associated records
- schedule of maintenance notifications
- balancing data achieved during commissioning and updates as they occur
- real-time building system-controls data
- security-system functions and updates

BIM is not just an advanced method of building design, it is the merging of technology, construction efficiency, and life-cycle building management.

The National Institute of Building Sciences (NIBS)

In the Housing and Community Development Act of 1974 (Public Law 93-383), the U.S. Congress established the National Institute of Building Sciences to provide a gateway between government and the private sector. NIBS is a nonprofit, nongovernmental organization whose mission is to support advances in building science and technology. Recognizing the problems and opportunities in the construction industry and the need to develop and encourage advanced technology in the industry, NIBS desires to act as a fulcrum around which design and construction can work together more efficiently and effectively.

The institute looks at BIM in three ways:

1. As an information-technology product with open standards so the technology can be shared

2. As a collaborative process whereby owners, design consultants, and contractors can work together to create the most efficient and effective facility

3. A facility life-cycle management mechanism

BIM becomes an *enterprise*, an industry-wide event encompassing real estate, facility owners, finance, all areas of architectural and engineering, manufacturing and fabrication, facility maintenance, operations and planning, regulatory compliance and environmental issues. With all of these parties working together in a collaborative fashion, ideas, concepts, and a virtual model can be figuratively and literally turned inside-out to produce the most effective, functional, and environmentally designed structure with optimum life-cycle operating costs.

A study of the construction industry in 2008 by NIBS indicated that total spending in the United States would be about $1.288 trillion, for that year, and a study by the Construction Industry Institute estimated that there was up to 57 percent non-value-added effort—waste—in current building business models. That's an incredible 734.16 billion in waste, pointing out an urgent need to maximize the value of the services provided and significantly reduce this waste.

When the current figures show that 44 percent of the world's raw materials are consumed by buildings, as is 40 percent of the world's energy and up to 65.2 percent of the total electricity consumed in the United States, and that buildings contribute substantially to carbon emissions and wasted materials, the construction industry—owners, architects, engineers, and builders—has a responsibility to act more responsibly.

The National Institute of Building Sciences recognized the value of BIM not only in creating new concepts and practices in design and construction but also in controlling facility life-cycle costs. The institute began to develop the National BIM Standard to address these areas. The original document stated the case for its existence very clearly: "A BIM is a digital representation of physical and functional characteristics of the facility. As such it serves as a shared knowledge resource for information about a facility forming a reliable basis for decisions during the life cycle from inception onward."

Three Slightly Different Definitions of BIM

The National Institute of Building Sciences defines BIM as the use of "cutting edge digital technology to establish a computable representation of the physical and functional characteristics of a facility."

The General Services Administration of the U.S. government defines BIM as an "intelligent and multi-faceted computer software data model to not only document a building design, but to simulate the construction and operation of a new capital facility or a recapitalized (modernized) facility."

The Associated General Contractors of America says that BIM allows for the "virtual construction of structures through the development and use of intelligent computer software that helps simulate construction."

Whichever definition you chose, BIM differs from the more conventional computer-assisted design in two ways: It uses new technology to create and modify design documents and it involves the use of a computer to store and analyze huge amounts of data in a collaborative, coordinated, detailed 3-D presentation involving designers, owners, general contractors, and subcontractors.

The National BIM Standard

In December 2007, the National Institute of Building Sciences announced the establishment of the National BIM Standard to set standard definitions for building-information exchanges and support key business contexts using standard semantics. NIBS indicated that BIM development, education, and implementation will become ingrained into the industry and fill a critical need to increase the efficiency of the construction process.

The entire charter for the National BIM Standard can be accessed at www.facilityinformationcouncil.org/bim/pdfs/NBIMS Charter. pdf. It describes BIM as a valuable tool that tends to be applied within vertically integrated business functions rather than horizontally across an entire facility life cycle. It calls for information gathered during the process to be gathered and applied throughout the entire project life cycle, which includes the operation and maintenance of the facility. It also calls for interoperability—seamless data exchange at the software level, where various software systems can talk to each other.

The Part 1 of the charter defines the minimum BIM standard for traditional vertical construction, i.e., office buildings. It also contains a proposal to measure the degree to which a mature BIM standard is developed. Subsequent sections include "Introduction to the National Building Information Modeling Standard," "Prologue to the National BIM Standard, Information Exchange Concepts," "Information Exchange Content," and "NBIM Standard Development Process."

The Real Value of BIM to Contractors and Owners

In the 1997 edition of document A201, General Conditions, from the American Institute of Architects, the contractor was required to review the drawings and notify the architect of any design errors or omissions; but the 2007 edition of A201 requires the contractor to conduct this review only in his or her capacity as a contractor, not as a licensed design professional.

Some architectural firms do not have all required design disciplines in house—i.e., civil, structural, mechanical, electrical, and plumbing— so they confer with these engineers as the design progresses from conceptual through construction drawings, but only on a 2-D basis. A 3-D design may reveal conflicts or coordination problems, saving valuable time and producing a set of drawings that are fully, or very close to fully, coordinated.

This need for coordination is evident in the standard specifications manual. A standard requirement generally found in division 1 of the specifications, taken from an actual set of specifications from a previous project, is as follows:

Coordination Drawings

General: Prepare coordination drawings for areas where close coordination is required for installation of products and material fabricated off-site by separate entities, and where limited space necessitates maximum utilization of space for efficient installation of different components.

1. Coordination drawings include, but are not limited to:
   a. Structure
   b. Partition/room layout
   c. Ceiling layout and heights
   d. Light fixtures
   e. Access panels
   f. Sheet metal, heating coils, boxes, grilles, diffusers, and similar items
   g. All heating piping and valves
   h. Smoke and fire dampers
   i. Soil, waste, and vent piping
   j. Major water
   k. Roof drainpiping
   l. Major electrical conduit runs, panelboards, feeder conduit, and racks of branch conduit
   m. Above-ceiling miscellaneous metal
   n. Sprinkler piping and heads
   o. All equipment, including items in contract as well as owner-furnished installed items
   p. Equipment located above finished ceiling requiring access for maintenance service (in locations where acoustical lay-in ceilings occur, indicate areas in which the required access area may be greater than the suspended grid system)
   q. Existing conditions, including but not limited to mechanical, plumbing, fire-protection, and electrical items
   r. Seismic restraints
2. Intent: Coordination drawings are for the construction manager's use during construction and are not to be construed as replacing shop drawings or record drawings. Architect's review of submitted coordination drawings shall not relieve the construction manager from his or her responsibility for the coordination of the work of the contract.

The need for coordination of all design disciplines is very clear. When 2-D drawings are produced, some conflicts generally occur, so the architect places the responsibility for uncovering any such problems squarely on the contractor by inserting this clause in the specifications.

BIM takes a major step in curing this problem that was so common when 2-D plans were produced and electronically transmitted to the engineer's design consultants for their work.

BIM Benefits During the Construction Phase

When a builder is involved in the BIM process, he or she brings along his or her experience in constructability issues and an up-to-date database of costs. If the building has a structural steel framework, then as the design is reviewed to ensure that beam sizes and depths are compatible with the material and equipment to be installed within the space allotted—both in floor-to-floor chases and between designated ceiling heights and the underside of the structural beams—a fast-track process can be pursued. The contractor can provide several structural-steel subcontractors with enough information to develop precise pricing or an estimate within an acceptable plus-or-minus cost percentage, thereby allowing the quick release of steel to the mill. All this can happen while the other drawings wend their way through the design phase. Foundation drawings can follow the same process, and a fast-track project is well under way.

Taking into account design costs, construction costs, and construction financing can really get the biggest bang for the buck.

The Disadvantages of BIM

As with all new technologies and all new products, lack of experience in the processes can create problems, both real and imagined. The melding of design among so many participants, including the general contractor and subcontractors, can blur the line between design and construction.

Lawyers are quick to enter the scene when design defects occur, and they will, even though BIM is touted as the cutting edge of technology. When a contractor actively collaborates in the design of a building and a problem arises with that design, does the contractor have or share any liability?

The *Spearin* Doctrine

There is a court case that many lawyers hang their hat on—*United States v. Spearin*, all the way back in 1918. Spearin, a contractor, bid on a U.S. Navy dry-dock project that included replacement of a 6-ft section of storm-sewer pipe, which he performed. This replacement sewer line proved inadequate to handle the volume and pressure of water runoff, and it broke due to internal pressure. The Navy said Spearin was responsible and directed him to replace the pipe, but Spearin said he would not because he had installed the pipe that was specified. The case went all the way to the U.S. Supreme Court—Mr. Spearin must have been a stubborn man! But the Supreme Court agreed with him and their decision became known as the *Spearin* doctrine. The court stated:

> If the contractor is bound to build according to plans and specifications prepared by the owner, the contractor will not be responsible for the consequences of defects in the plans and specifications. This responsibility of the owner is not overcome by the usual clauses

requiring bidders to visit the site, to check the plans, and to inform themselves of the requirements of the work.

Does any part of the *Spearin* doctrine apply to problems developed during BIM? In the 1988 case of *Haehn Management Co. v United States* (15 Cl. Ct. 50, 56), the Federal Court of Appeals stated that the *Spearin* warranty of specifications can be vitiated by the involvement of industry or the contractor's participation in the drafting and development of the specification absent superior knowledge on the part of the government. The contractor apparently changed the specifications, and therefore *Spearin* did not apply.

Who Owns the Design Documents?

Since the BIM design passes through lots of hands as it progresses from conceptual to construction documents, can one or more of the participants in the design process claim ownership of a portion along the way? This may well become an intellectual-property matter, and the BIM design may be capable of being copyrighted, so anyone using a portion of that design on another project may be in violation of someone's copyright. With possibly 20 or more contributors to the finished BIM product, though, who owns which portion of the design? Many building permits require the sealing of drawings; which entity will perform that, and does that entity then assume responsibility for the entire integrity and compliance of the final design?

Complying with local, federal, and international building codes in this international design-and-construct environment we live in is another concern. Design consultants may be based in different countries and must become familiar with the various building codes in unfamiliar geographic areas for which they are submitting their portion or portions of the design.

Ultimately these questions will be answered, but until they are, there may be numerous liability issues.

Insurance Questions

If BIM is a collaboration among several parties, there is a question of design liability and how one gets an insurance company to provide a policy that protects all parties. Can the contractor obtain insurance for professional services even though they are not licensed professional architects or engineers? The contractor's professional-liability endorsement protects against property damage and personal injury, but can it include an endorsement beyond that for, say, economic loss if a portion of the building fails due to a design error rather than substandard construction? As BIM becomes more prevalent in the marketplace, various liability issues will undoubtedly arise. Until they are settled by the courts, there will be several unknowns that test the process.

Sustainability

Sustainability involves the architecture, engineering, and construction industries. We will deal with the construction side now; the architect and engineer will have their chance to contribute to sustainability as we progress in the book.

Sustainability has several meanings, but as related to the construction industry it means development that meets the needs of the present without compromising the ability of future generations to meet their needs. This involves, as examples, use of fossil fuels to generate power, production of potable water to quench thirst, and preservation of forests for future generations.

When you look at the impact construction and the products it produces has on the environment, you can see that, if left unchecked, we will deplete these natural resources; the only question is when. Buildings are responsible for the following demands, according to data from the U.S. Environmental Protection Agency and the U.S. Energy Information Administration:

- 72 percent of all U.S. energy consumed
- 38 percent of all greenhouse gases produced
- 13.6 percent of all potable water consumed (that translates to 15 trillion gal per year)
- 40 percent of all raw materials used globally
- 136 million tons annually of building-related construction and demolition debris

The alarming rate of natural-resource depletion was the impetus for the creation of the green movement, which we will discuss in the chapter on Green Buildings and Sustainable Construction. Sustainable construction involves developing designs that seek to strike a balance between the project's short-term goals and its long-term goals of creating efficient operating systems that protect the environment and the country's natural resources. It also brings into focus the consideration of cost of capital equipment versus the costs to operate and maintain that equipment. Will a greater initial cost result in payback many times over due to much lower operating costs?

When we consider *sustainability*, for example, can we use fenestration to let in natural light and perhaps reduce the requirement for interior lighting? And if we use the latest window technology of insulated glass with inert gas between panes, coatings to reflect sunlight, and thermal-break aluminum window frames, will we have created another energy-efficient building-envelope component?

Sustainable products are those made of natural, renewable materials that can be recycled, such as:

- Exterior cladding made from
  - wood siding certified by the Forest Stewardship Council (FSC)
  - reclaimed wood siding
  - siding made from recycled paper, wood, or bamboo
  - oriented strand board
  - compressed hardboard, masonite, or medium-density fiberboard
  - fiber cement
  - aluminum or steel from recycled materials
- Roofing made from
  - recycled metal
  - slate or clay tiles
  - fiber-cement shingles
  - FSC-certified wood shingles and shakes
  - green roofs (actual rooftop gardens or grass)
- Insulation made from
  - recycled newspaper
  - recycled long-filament fibers from discarded clothing, such as blue jeans
  - recycled plastic
  - soy-based foam
- Flooring made from
  - bamboo
  - cork
  - FSC-certified wood planks
  - natural or recycled carpet
  - linoleum made from natural biodegradable materials
  - tiles made from 100 percent recycled glass
- Paint:
  - paints with low or no volatile organic compounds
  - natural paints
  - nontoxic stains
- Wall and ceiling drywall, caulking, and adhesives:
  - wallboard with 85 to 95 percent recycled-gypsum content
  - wallboard with 10 percent recycled paper
  - FSC-certified paneling
  - caulking and adhesives with low content of volatile organic compounds

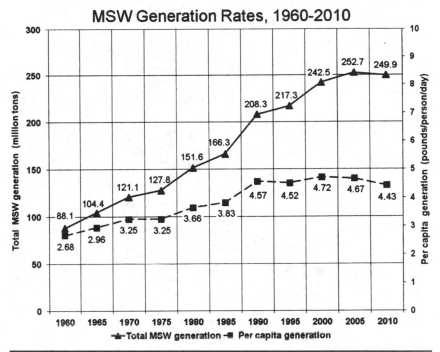

FIGURE 10.9 Increase in materials going into landfills, 1960–2010.

- soy-based glue
- soy-based caulking

Just looking at the municipal solid waste generated over the past four decades shows a staggering increase in materials going into landfills. Figure 10.9 reveals an increase from 88.1 million tons in 1960 to nearly 250 million tons in 2010. If we did not begin a strong recycling program, we would very quickly run out of places to dump our debris. One of the answers is recycling, a process that started slowly around 1960, as shown in Fig. 10.10, and accelerated rapidly by 2010, but it still has a way to go to gain the full participation of our population. Figure 10.11 contains the recycling rates of selected products and reinforces the need to increase the recycling of plastic bottles, especially since bottled water has become such a mainstay in many households.

The green movement has generated demand for environmentally friendly products and equipment. Contractors can join in the parade and become environmentally conscious by diverting construction waste or demolition debris at the jobsite:

- Up to 10 to 12 percent of a project's construction waste is cardboard. Contractors can direct their subcontractors and

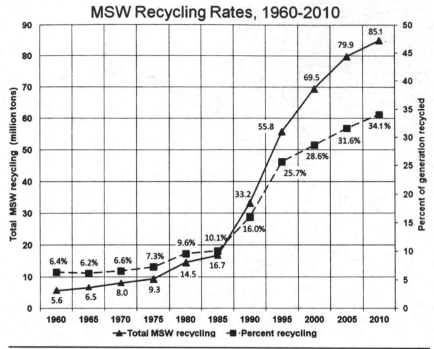

FIGURE 10.10   Recycling rates accelerating rapidly by 2010.

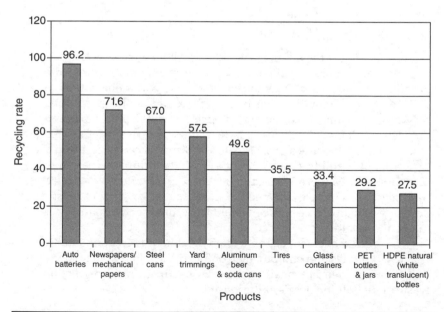

FIGURE 10.11   Recycling rates of selected products.

vendors to reduce extraneous packing. Avoid individual packaging if doing so will not cause damage to the product. Purchase in bulk.

- Use scrap in lieu of new materials. Do not let a carpenter cut 8 in. from an 8-ft two-by-four—make him or her look for scraps.
- Reuse nonreturnable containers like 5-gal taping cement pails.
- Establish a return or buyback policy for excess materials instead of taking them back to the yard, where they may spend the rest of their life.
- Work in smaller batches so that excess epoxy, as an example, will not be generated and dumped, or caulking buckets or tubes be thrown away while still containing product.
- Contact a local construction-and-demolition recycler who will accept commingled waste and might even pay a small amount to do so.

Environmentally conscious owners may begin to insert contract language requiring the contractor to present a plan to reduce construction and demolition waste as part of the bid documents. There are a number of ways that this can be inserted in the bid documents and a contractor who has thought ahead and begun to embrace some of these concepts may find him- or herself in a favorable position with private owners, selected perhaps even without being the low bidder.

An owner could

- require that specific percentages of demolition material be diverted from landfill with a provision in the demolition specifications stating that, say, 75 percent of concrete or asphalt demolition materials be recycled and diverted from the landfill.
- describe a desire to reduce materials going to the landfill and have each bidder submit a plan to do so, allowing the owner to pick the most effective and practical plan.
- have the contractor submit a construction-and-demolition waste-management plan and document actual waste diversion throughout the project's life (the owner may even offer a bonus if these quantities are reduced).

NOTE: *Table 10.1 contains listings of sources for 3-D modeling software, training, interoperability software, consulting services, and hardware.*

Software—3D Modeling (or 2D to 3D Conversion)

Product Name	Manufacturer	BIM Use	Manufacturer's Description	Supplier	Hardware Requirements	Approx. Cost
Revit Building	AutoDesk	Creating and reviewing 3D models	Autodesk® Revit® technology is Autodesk's platform for building information modeling. Built on the Revit platform, Autodesk® Revit® Building software is a complete, discipline-specific building design and documentation system supporting all phases of design and construction documentation. From conceptual studies through the most detailed construction drawings and schedules, Revit-based applications help provide immediate competitive advantage, deliver better coordination and quality, and can contribute to higher profitability for architects and the rest of the building team. At the heart of the Revit platform is the Revit parametric change engine, which automatically coordinates changes made any where—in model views or drawing sheets, schedules, sections, plans...you name it. Compatible with 2-D drawings	Various http://usa. autodesk.com/ adsk/servlet/inde x?siteID=123112 &id=2435651	System Requirements: Microsoft® Windows® XP (Professional, Home, Tablet PC Edition) or Microsoft® Windows® 2000 SP4 (or later) Intel® Pentium® 4 1.4 GHz or equivalent AMD® Athlon® processor 1 GB RAM with 1 GB free disk space 1024 x 768 monitor and display adapter capable of 24-bit color Internet connection for license registration Microsoft Internet Explorer 6.0 System Recommendations: Microsoft Windows XP Professional SP2 (or later) Processor: Intel Pentium 4 2.8 GHz or equivalent AMD Athlon processor RAM: 2 GB with two-button mouse with scroll wheel	~$4,000 to $6,000 per seat plus subscription
Virtual Construction	Graphisoft	3D Modeling/ Virtual Building	The Virtual Construction™ suite of technologies employs 3D modeling to virtually construct your project. This process identifies constructability issues during design and/ or preconstruction. The 3D model is further utilized to extract accurate estimating quantities and to analyze alternative construction sequences. Finally, design (3D), schedule (4D), and cost (5D) are all interlinked, so a change to any of the three automatically updates the other two.	Various based on region http://www. graphisoft. com/ products/ where_ to_buy/		~$8,000 per seat

			Graphisoft's construction tools include Change Manager. This solution automates revision of drawing changes and allows member's of a project team to easily identify, communicate and manage those changes so they have minimum impact on time, cost and schedule. 5D Construction Management			Contact Bentley sales for a customized quote: http://www.bentley.com/BentleyWebSite/Tools/sales_contact.aspx
Bentley Architecture	Bentley	Architecture Design	Bentley is committed to delivering complete BIM solutions that support the whole project delivery process for the entire lifecycle of facilities. Built on a single platform while supporting industry standards, these solutions focus on design rather than drafting, integrate design with engineering, facilitate multi-disciplinary collaboration, and allow distributed teams to "build as one" within a managed information environment. Bentley Architecture, an advanced Building Information Modeling (BIM) application, virtually creates 3D prototypes of buildings, thereby providing significant business-critical benefits for architectural design and AE/EA firms of all sizes. Able to deliver better buildings on time and on budget, they can substantially improve client services, reduce costs, and increase revenue. Prerequisites: MicroStation V8 (08.01.02.15 or higher) and MicroStation TriForma (08.01.01.30 or higher) Supports DGN File Formats	Bentley Architecture http://www.bentley.com/BentleyWebSite/Tools/sales_contact.aspx	Intel-compatible Pentium PCs running Microsoft Windows 2000, Microsoft Windows NT or Microsoft Windows XP Memory: Minimum 256 MB Hard disk: 200 MB minimum free Input device: mouse or digitizing tablet (tablet on Windows requires WINTAB driver or Bentley's Windows Digitizer Tablet Interface) Output Device: Majority of industry-standard devices supported; works with Windows-supported printers Video: Supported graphics cards (256 or more color card recommended for rendering); 16-bit color minimum for QuickVision GL); dual screen graphics supported for Windows NT4; multi-monitor configurations supported with Windows XP and Windows 2000	

Notes:

1. Products presented are not in any specific order.
2. Costs are typical and may vary based on configuration or other requirements.

TABLE 10.1 Source for BIM Software, Software Training, Software Interoperability, Software for Change Management, and Hardware Options.
(By permission from The Associated Contractors of America, Arlington, Virginia)

	Software—3D Modeling (or 2D to 3D Conversion) (Continued)			
Product Name	**Manufacturer**	**BIM Use**	**Manufacturer's Description**	**Supplier**	**Hardware Requirements**	**Approx. Cost**
Architect	VectorWorks	2D and 3D Production Management	For more than twenty years, VectorWorks Architect has provided building information modeling (BIM) capabilities that offer a true increase in productivity. With BIM, your 2D drawings, 3D drawings and project data are linked into a simple, yet powerful design environment. VectorWorks BIM technology allows you to design in 2D and 3D simultaneously. Changes made in one view are automatically updated in the other. And, because your drawings are linked, a change in one can automatically ripple through the entire drawing set, saving you time and reducing drawing errors. Plus, you can integrate information into your design to automatically generate schedules, as well as track materials and costs. And, you can export this information to use downstream in other costing and database programs.	Vector Works http://www.nemetschek.net/sales/index.php	VectorWorks ARCHITECT RAM: 256MB Hard drive space: 200MB VectorWorks ARCHITECT plus RenderWorks RAM: 256MB Hard drive space: 1GB Macintosh Operating System: Mac OS X 10.3.9 or later Other Software: QuickTime 6 or higher Processor: PowerPC G3 or newer Other Hardware: CD-ROM drive Screen Resolution: 1024 x 768 (XGA) Display color depth: 16 bit Windows Operating System: Windows 2000 SP4 or later; Windows XP Other Software: QuickTime 6.5.2 or 7.0.2 Processor: Pentium or newer Other Hardware: CD-ROM drive Screen Resolution: 1024 x 768 (XGA) Display color depth: 16 bit	~$1,400 per seat

Software—Training						
Revit	Autodesk	Software Fundamentals and Advanced Training	Tutorials A complete set of self-paced training exercises is available to help you become more productive using Autodesk® Revit® Building. Revit Classroom Training Autodesk offers Autodesk Revit Fundamentals and Autodesk Revit Advanced training courses from our offices in Waltham, Massachusetts and San Rafael, California. How-to Articles Through step-by-step lessons written by recognized experts in their fields, Autodesk how-to articles and tutorials can help you master the techniques you need to become more productive with Autodesk products and solutions. API Training and Consulting for Developers Visit the Autodesk Developer Center for the latest courses and schedules of hands-on API training. Also, get information about Autodesk API consulting services.	Autodesk	Tutorials, Classroom Training, and How-to Articles Prerequisites: Implementation of Revit	~$1,000

TABLE 10.1 Source for BIM software, Software Training, Software Interoperability, Software for Change Management, and Hardware Options. (*By permission from The Associated Contractors of America, Arlington, Virginia*) (*Continued*)

			Software—Training *(Continued)*			
Product Name	Manufacturer	BIM Use	Manufacturer's Description	Supplier	Hardware Requirements	Approx. Cost
Virtual Construction	Graphisoft	Transition Services	To both ease the transition to the new technology and to provide additional customer support during peak loads, Graphisoft Construction Services will utilize this product line under the direction of the Customer's project team. Services include: Construction Modeling Constructability Analysis Estimating Support 4D Sequencing Support Production and Procurement Planning Support Site Planning Support 5D Construction Simulation Presentation	Graphisoft	Services Include: Construction Modeling Constructability Analysis Estimating Support 4D Sequencing Support Production and Procurement Planning Support Site Planning Support 5D Construction Simulation Presentation Prerequisites: Implementation of Virtual Construction Suite	Varies based on need
Bentley Architecture	Bentley	Bentley Training Programs	Classroom Learning Led by experienced Bentley instructors, classroom learning is offered through scheduled courses at Bentley training facilities or as account-specific training in your office or nearby training facility convenient to your work location. Distance Learning Live, instructor-led distance learning is taught by experienced Bentley instructors via the Internet. Distance learning is available through scheduled courses or as account-specific courses tailored to your workflow. OnDemand eLearning OnDemand eLearning delivers professional training to every desk through recorded interactive courses and lectures. Hundreds of hours of OnDemand eLearning courses are available to Bentley LEARN and Enterprise Training Subscription users.	Bentley	Programs for every user, goal or budget Bentley Institute offers a variety of training programs that make it easy for any individual or organization to get professional training. Organizations can increase return on investment and train more people through the annual training subscription programs, Bentley LEARN and Enterprise Training Subscriptions. Prerequisites: Implementation of Bentley solution	Varies based on need

| Various | Kristine Fallon Associates, Inc. | BIM Training and Transition | KFA has developed training curricula and conducted training in multiple BIM products. KFA was principal author of Autodesk Revit Building 8 training curriculum and materials. KFA restructured the training curriculum for Autodesk Revit Building Essentials from a five-day to a three-day program and the Advanced curriculum from a three-day to a two-day program. As part of the documentation review, KFA verified all model sets used in each unit of the programs were complete and had the necessary building components and families for students to create models. The conventional training manuals were also reformatted and adapted for modified online versions of both training programs. | KFA | KFA has followed the evolution of intelligent building modeling technology for over a decade, developing academic programs using advanced modeling products, producing a Triforma white paper for Bentley Systems, evaluating the maturity and scalability of BIM systems for the Spallation Neutron Source project at Oak Ridge National Lab and assisting Revit Technology in market research and feature prioritization prior to their initial product release. | Varies based on need |
| Various | ERDC | BIM Training and Support | The ERDC Computer-Aided Design and Drafting/Geographic Information Systems Technology Center for Facilities, Infrastructure, and Environment (CADD/GIS Center) provides the expertise, standards, and onsite implementation support to execute BIM technology in the Federal design arena. The CADD/GIS Center is well-acquainted with the unique requirements of the Federal user, including long-term management, operation, and maintenance of facilities in the Federal environment and the impact of the President's current Management Agenda. | U.S. Army Engineer Research and Development Center | The Center offers training and implementation support for this software to the Federal user community.

Must be part of the Federal User community. | Varies based on need |

TABLE 10.1 Source for BIM Software, Software Training, Software Interoperability, Software for Change Management, and Hardware Options. (By permission from The Associated Contractors of America, Arlington, Virginia) (Continued)

			Software—Training (*Continued*)			
Product Name	**Manufacturer**	**BIM Use**	**Manufacturer's Description**	**Supplier**	**Hardware Requirements**	**Approx. Cost**
Various	Gehry Technologies	Digital Technology Integration	GT brings fifteen years' experience applying advanced digital technologies to complex building projects undertaken by Gehry Partners and other leading architecture and engineering companies. Our clients and partners are firms and building teams interested in moving beyond the limits of drafting and paper driven project management and into 21st century, digitally enabled design and construction practices.	Gehry Technologies	Gehry Technologies ("GT") is a building design and construction technology company that provides integrated, digitally driven construction practice tools and methodologies to companies and their projects.	Varies based on need
			Software—Interoperability			
JetStream v5	Navisworks	Combining and reviewing 3D models	Powerful and flexible yet affordable and easy to use, JetStream v5 saves time and reduces the risk of problems onsite, thus saving money. JetStream v5 gives access for all stakeholders to work effectively together on a project employing 3D design models on the desktop or over the Web, and regardless of size or file format. BIM compliant, a solution for Virtual Construction and an aid to LEAN construction techniques, JetStream v5 has become an integral part of many major companies' review processes.	Various http:// www. navisworks. com/resellers/ resellers.php	Minimum specification for NavisWorks would be: Pentium II processor (or equivalent) 64MB RAM Windows 95 or above	~$8,000 per seat

			Recommended specification for NavisWorks would be: Pentium III processor (or equivalent) Hardware-accelerated OpenGL graphics card 128MB RAM Windows NT 4 or above	Component of JetStream plugin architecture
		Our product is used in diverse markets including Engineering, Construction, Oil & Gas, Petrochemicals, Energy, Shipbuilding, Automotive, and Utilities, and offers proven benefits in saving time and money throughout the project lifecycle for both engineering and procurement contractors as well as owner operators, and for CAD and non-CAD user alike. Our fifth major release, JetStream v5, is scalable and easy to deploy and administer; fitting into existing processes and linking to third party software and databases.		
JetStream v5 Roamer	Navisworks	Combine Design into one model	Roamer can be used to navigate and explore the design free from the limitation of pre-programmed animation and without specialist skills or hardware. Fully compatible with the all major 3D design and laser scan formats, Roamer can quickly open and combine competing 3D files to create a single virtual model for review of geometry, object information and linked ODBC databases. Textures, materials and lights can also be viewed from .nwd files or, when used with the Presenter plug-in, other compatible formats. Stereo viewing support, collision detection, gravity and a third person view improve the reality of the Roamer experience. Allows collaboration using 3D design models regardless of format.	Various http://www. navisworks.com/ resellers/resellers. php	Minimum specification for NavisWorks would be: Pentium II processor (or equivalent) 64MB RAM Windows 95 or above Recommended specification for NavisWorks would be: Pentium III processor (or equivalent) Hardware-accelerated OpenGL graphics card 128MB RAM Windows NT 4 or above

TABLE 10.1 Source for BIM Software, Software Training, Software Interoperability, Software for Change Management, and Hardware Options. (*By permission from The Associated Contractors of America, Arlington, Virginia*) (*Continued*)

Software—Change Management

Product Name	Manufacturer	BIM Use	Description	Supplier	Hardware Requirements	Approx. Cost
JetStream v5 Clash Detective & TimeLiner	Navisworks	Report indifferences in 3D project models and track project status	**Key Attributes of JetStream v5 Clash Detective** • Check for intersecting, distance between & duplicate geometry. • Workflow planning based on time & space co-ordination. • SwitchBack to 3D design software for fault finding & fixing. • Full audit trail of detected clashes. • Extensive interference management & reporting capabilities. • Clash test XML import/export. • Point/Line Based Clashing. **Key Attributes of JetStream v5 TimeLiner** • Improve site planning & enables "what-if" scenarios for visual risk management. • Integrates with existing tools such as Primavera, MS Project & Asta Powerproject • Time-based work-flow planning when linked with Clash Detective	Various http://www.navisworks.com/resellers/resellers.php	JetStream v5 Roamer is required for use of these plugins	Both included with JetStream Roamer v5 Pro which is ~$8,000 per seat Sold separately: Clash Detective: ~$4,000 per seat TimeLiner: ~$1,500 **Indicated cost for budgetary purposes only.

Change Manager	Graphisoft	Identify, communicate and manage changes	Graphisoft Change Manager features include:
			• Easily define document sets & analyze changes.
			• Assign changed documents to project team members.
			• Review changes on a document-by-document basis:
			  * Adjustable contrast with color-coded identification of new, deleted, and changed entities.
			  * Slider: overlay of new and old documents with a slider control to shift the display.
			  * Shift: mouse control that shifts the position of the new document relative to the old to make it easy to understand the changes.
			• Cloud annotation with logging.
			• Project Log: Stores the list of changed documents, action assignments, change descriptions, and action completions for each document set.
			• Ignore Change: Stylistic changes such as fonts and repetitive changes such as revision numbers on title blocks can be identified once and ignored in all future documents.
			• AutoCAD® Release 14 DWG format and above.
				Graphisoft http://www.graphisoft.com/products/construction/products_and_services/change_manager/contact.html		~$895 + VAT Volume discounts available

			**For custom Interoperability options, please see the consulting section below.			

		Software—Consulting		
Revit	AutoDesk		Autodesk Consulting also provides consulting offerings for project assessments, process audits, and a range of implementation services. Custom consulting offerings are also available to meet your specific needs.	Various	Prerequisites: Implementation of Revit	Varies based on need
Various	Gehry Technologies	Digital Technology Integration	GT brings fifteen years' experience applying advanced digital technologies to complex building projects undertaken by Gehry Partners and other leading architecture and engineering companies. Our clients and partners are firms and building teams interested in moving beyond the limits of drafting and paper-driven project management and into 21st century, digitally enabled design and construction practices.	Gehry Technologies	Gehry Technologies ("GT") is a building design and construction technology company that provides integrated, digitally driven construction practice tools and methodologies to companies and their projects.	Varies based on need

Table 10.1 Source for BIM Software, Software Training, Software Interoperability, Software for Change Management, and Hardware Options. (*By permission from The Associated Contractors of America, Arlington, Virginia*) (*Continued*)

	Hardware Options				
Product Name	Manufacturer	Performance	Base Description	Product Information & Functionality	Approx. Cost
			Dell (www.dell.com)		
Dimension 1100	Dell	Low End	Processor: Intel® Celeron® D Processor up to 330 (2.66GHz, 533MHz FSB, 256KB L2 cache). Operating System: • Genuine Windows® XP Home Edition • Genuine Windows® XP Media Center Edition 2005 Memory: Up to 1GB (dual channel) Shared1 DDR SDRAM at 400MHz for superior performance (Note: 400MHz memory performs at 333 MHz with 533 FSB processors). Storage: Ultra ATA Hard drives up to 160 GB2 to meet your storage needs.	Basic Essential Productivity at an Affordable Price	Starting base at ~$350 Fully loaded at ~$1,000 (Price may change as product is customized)
Optiplex GX620	Dell	Mid Range	Processor: Intel® Pentium® D Dual Core with up to 2MB L2 cache, Intel Pentium® 4 HT & XD Security up to 2MB L2 cache, or Intel Celeron® D Processor Operating System: • Genuine Windows® XP Professional • Genuine Windows® XP Home Edition • Windows VistaTM capable Memory: Up to 4GB1 of 533MHz Dual Channel Shared2 DDR2 SDRAM	Maximum performance and scalability that meets your most demanding needs today and provides investment protection for the future. The GX620 also shares a common image and BIOS with the GX520, providing unsurpassed choice while minimizing ownership costs.	Starting base at ~$1,100 Fully loaded at ~$2,250 (Price may change as product is customized)

Precision 690	Dell	High End	Processor: 64-bit Dual-Core Intel® Xeon® Processors (Up to 3.73 GHz, 2x2MB L2 Cache, 1066MHz FSB) or Up to 3.0 GHz, 4MB L2 Cache, 1333MHz FSB) Operating System: • Genuine Windows® XP Professional • Genuine Windows® XP Professional x 64 Edition • Windows VistaTM -capable1 • Red Hat® Enterprise Linux WS v.4 (EM64T) Memory: Up to 64GB3 quad-channel4 architecture DDR2 Fully-Buffered DIMM 533 and 667 MHz ECC memory Storage: Up to 2TB8 of SATA integrated storage or up to 1.2TB8 of integrated SAS storage, both with integrated support for RAID 0 and 1; up to 1.7TB8 of storage with 500GB8 SATA boot drive plus 4 SAS drives, Optional PERC 5/i PCI-e RAID card supports RAID 0, 1, 5	The Dell Precision 690 is an ultra-high-performance workstation that maximizes performance and scalability in an innovative, new, customer-driven chassis design. The 690 offers up to two 64-bit Dual-Core Intel® Xeon® 5000 and 5100 series processors, up to 64GB1 of memory in four fully-buffered DIMM channels (with an optional memory riser card) and a wide range of high-performance OpenGL graphics cards. The Dell Precision 690 is designed for the most business-critical, computer-intensive and graphically demanding workstation environments.	Starting base at ~$2,358 Fully loaded at ~$8,500 (Price may change as product is customized)
	Gateway (www.gateway.com)				
E-1500D SB	Gateway	Low End	Intel® Celeron® D Processor 346 (3.06GHz, 533MHz FSB, 256K L2 cache) Genuine Microsoft® Windows® XP Professional Edition (SP2) ATI® Xpress 200 Chipset 512MB DDR2 PC4200 533MHz SDRAM (2-DIMM) 80GB Ultra ATA100 7200rpm hard drive w/2MB cache	When it comes to affordable computing solutions, why compromise on performance? That's why the Gateway® E-1500C SB Desktop is an invaluable business asset for organizations keeping an eye on the bottom line.	Starting base at ~$645 Fully loaded at ~$1,165 (Price may change as product is customized)

TABLE 10.1 Source for BIM Software, Software Training, Software Interoperability, Software for Change Management, and Hardware Options. (*By permission from The Associated Contractors of America, Arlington, Virginia*) (*Continued*)

Hardware Options

Product Name	Manufacturer	Performance	Base Description	Product Information & Functionality	Approx. Cost
			Gateway (www.gateway.com) (Continued)		
E-6610 Series	Gateway	High End	Intel® Core™2 Duo E6300 (1.86GHz 1066MHz FSB 2MB cache, non-HT) Genuine Microsoft® Windows® XP Professional Edition (SP2) Intel® 975X Chipset with DDR2 and Intel® Core™ Duo support 1024MB PC5300 Dual-Channel DDR2 667MHz SDRAM (2-512MB modules) 160GB Serial ATA II/300 7200RPM w/Raid 0 (2-80GB Hard Drives - Striping)	The Gateway® E-6610 Desktop is designed for optimal performance with workstation-class configurability. Benefit from the top-notch level of performance and features, including the latest Intel® 975X chipset, OpenGL graphics, ultrahigh-speed hard drives, ECC or non-ECC memory and RAID options. Complementing these high-end features is BTX Technology with ultra-quiet dual-fan cooling for improved reliability.	Starting base at ~$1,350 Fully loaded at ~$3,530 (Price may change as product is customized)
			Hewlett-Packard (www.hp.com)		
Dx2200	Hewlett-Packard	Low End	Operating systems installed Genuine Windows® XP Home SP2 Processor type Intel® Celeron® D 326 Processor features 2.53GHz, 256-KB L2 Advanced Transfer Cache, 533MHz Front Side Bus Memory 256MB PC2-5300 (DDR-667) Hard drive, internal 80GB SATA Hard Drive (7200 rpm) 3.0 GBps	The HP Compaq dx2200 combines essential business features and proven business technology for a PC that is ideal for mainstream business applications and environments.	Starting base at ~$349 Fully loaded at ~$850 (Price may change as product is customized)

| Dc7600 | Hewlett-Packard | High End | Operating systems installed
Genuine Windows® XP Professional

Processor type
Intel® Pentium® 4 Processor 640

Processor features
3.2 GHz, 2-MB L2 Cache, 800-MHz Front Side Bus

Memory
512 MB (533 MHz)DDR2 (2x256)

Hard drive, internal
80GB 3.0Gb/s Hard Drive (7200 rpm) | The innovative design of the HP Compaq dc7600 Convertible Minitower provides increased expandability, flexibility, convenience, and savings. With easy conversion from a minitower to a desktop, the CMT offers maximum expandability and performance options. Security, stability, and manageability features add IT peace of mind. | Starting base at ~$849
Fully loaded at ~$1,500
(Price may change as product is customized) |

TABLE 10.1 Source for BIM Software, Software Training, Software Interoperability, Software for Change Management, and Hardware Options. (By permission from The Associated Contractors of America, Arlington, Virginia)

Summary

The information-technology era has been embraced by the AEC community. Construction managers and project managers will come face-to-face with more projects using building information modeling to create *virtual buildings* that will increase productivity and lower costs.

The interoperability of previously disparate software systems can now permit the AEC community to communicate effectively with each other and with the providers of materials and equipment for the structure. As owners, design consultants, and builders have come to recognize the need to preserve our diminishing natural resources and provide an environmentally friendly community, the sustainable-construction movement has also gained considerable traction.

These environmental concerns form the basis of the green building movement. Construction managers and project managers must gain more familiarity with these advances in design and construction as part of their total construction project management education.

CHAPTER 11

Green Buildings and Sustainable Construction

The Washington, DC–based United States Green Building Council (USGBC) is a nonprofit organization, founded by David Gottfried and Rick Fedrizzi in 1993, whose mission is to create a sustainable future through cost-efficient and energy-saving buildings. USGBC has set out to transform the way in which buildings are designed, constructed, operated, and maintained in an environmentally and socially responsible way.

Although we touched on sustainability in the previous chapter– Total Construction Project Management for the Twenty-First Century, we will review it later in this chapter as an integral part of the green building movement.

When one looks at the impact of construction worldwide, it is rather easy to see the need, and the urgency, to create a more environmentally friendly way to design and construct our buildings. In the United States, building accounts for:

- 36 percent of total energy use
- 65 percent of electricity use
- 30 percent of greenhouse-gas emissions
- 30 percent of raw-materials use
- 30 percent of waste output
- 12 percent of potable-water use

In Europe, the environmental concern is fivefold: energy supply, security, climate-change mitigation, health and well-being of citizens, and economic growth and availability of financial resources. Society looks to the Organization for Economic Cooperation and

Development (OECD), a 50-year old organization originally created to tackle economic and social challenges in Europe, for assistance in addressing these goals. The OECD lists the following concerns and challenges:

- 40 percent of Europe's total energy consumption comes from its 160 million buildings.

- 3.3 million barrels of oil could be saved each day if buildings were more energy efficient.

- 460 million tons of carbon dioxide emissions could be saved each year through more effective energy measures.

- OECD countries are responsible for 30 to 40 percent of solid-waste generation.

- OECD countries account for 30 percent of raw-materials usage.

What Is LEED?

The process for developing environmentally friendly design and construction involved in creating a rating program that awards points for accomplishing certain specific goals. Developed by the USGBC in 2000 via a consensus-based process, the program is titled Leadership in Energy and Environmental Design, known more familiarly as LEED. What started out as a small movement in 2000 has evolved into program with 35,000 projects currently participating, representing 4.5 billion ft^2 of construction in all 50 states and 91 countries. The purpose of the LEED process is to develop environmental and sustainable standards for various types of building structures.

Construction managers (CMs) or project managers (PMs) should familiarize themselves with the green-building concept and incorporate this information in their total construction project management curriculum, because at some future date they most likely will become involved in a green building project. The U.S. Environmental Protection Agency defines green building as follows:

Green building is the practice of creating structures and using processes that are environmentally responsible and resource efficient throughout a building's life-cycle from siting to design, construction, operation, maintenance, renovation and deconstruction. This practice expands and complements the classical building design concerns of economy, utility, durability and comfort. Green building is also known as a sustainable high performance building.

How Does LEED Work?

LEED is a point-based system where points are awarded for satisfying various green building criteria within five categories:

1. Sustainable sites
2. Water efficiency
3. Energy and atmosphere
4. Materials and resources
5. Indoor environmental quality

An additional category, innovation in design, relates to sustainable building expertise and design innovations not covered under the five environmental categories. Six additional bonus points can be obtained by implementing these strategies.

LEED certification is available for all building types, including:

- New construction and major renovation
- Existing construction
- Commercial interiors
- Core and shell
- Schools
- Retail
- Health care
- Homes
- Neighborhood development

Points are awarded for compliance and when finally compiled are referred to as a LEED certification. The lowest level of certification is Certified and the highest is Platinum. As an example, for new construction, the following point totals are required to achieve these various LEED certifications:

- Certified: 40–49 points
- Silver: 50–59 points
- Gold: 60–79 points
- Platinum: 80–110 points

Typical LEED checklists are shown in Figs. 11.1 and 11.2. The LEED process involves the building site, envelope, interior environment, and energy-efficient lighting, heating, and cooling systems.

LEED for New Construction v 2.2
Registered Project Checklist

Project Name: _____

Project Address: _____

Yes	?	No		
			Project Totals (Pre-Certification Estimates)	**69 Points**
			Certified: 26-32 points **Silver**: 33-38 points **Gold**: 39-51 points **Platinum**: 52-69 points	

Yes	?	No			
			Sustainable Sites		**14 Points**
Yes			Prereq 1	**Construction Activity Pollution Prevention**	Required
☑	☑	☑	Credit 1	**Site Selection**	1
☑	☑	☑	Credit 2	**Development Density & Community Connectivity**	1
☑	☑	☑	Credit 3	**Brownfield Redevelopment**	1
☑	☑	☑	Credit 4.1	**Alternative Transportation**, Public Transportation	1
☑	☑	☑	Credit 4.2	**Alternative Transportation**, Bicycle Storage & Changing Rooms	1
☑	☑	☑	Credit 4.3	**Alternative Transportation**, Low-Emitting & Fuel Efficient Vehicles	1
☑	☑	☑	Credit 4.4	**Alternative Transportation**, Parking Capacity	1
☑	☑	☑	Credit 5.1	**Site Development**, Protect or Restore Habitat	1
☑	☑	☑	Credit 5.2	**Site Development**, Maximize Open Space	1
☑	☑	☑	Credit 6.1	**Stormwater Design**, Quantity Control	1
☑	☑	☑	Credit 6.2	**Stormwater Design**, Quality Control	1
☑	☑	☑	Credit 7.1	**Heat Island Effect**, Non-Roof	1
☑	☑	☑	Credit 7.2	**Heat Island Effect**, Roof	1
☑	☑	☑	Credit 8	**Light Pollution Reduction**	1

Yes	?	No			
			Water Efficiency		**5 Points**
☑	☑	☑	Credit 1.1	**Water Efficient Landscaping**, Reduce by 50%	1
☑	☑	☑	Credit 1.2	**Water Efficient Landscaping**, No Potable Use or No Irrigation	1
☑	☑	☑	Credit 2	**Innovative Wastewater Technologies**	1
☑	☑	☑	Credit 3.1	**Water Use Reduction**, 20% Reduction	1
☑	☑	☑	Credit 3.2	**Water Use Reduction**, 30% Reduction	1

Powered by
Adobe· LiveCycle·

FIGURE 11.1   LEED for new construction.

LEED for New Construction v 2.2
Registered Project Checklist

Yes	?	No			
			Energy & Atmosphere		**17 Points**
Yes			Prereq 1	**Fundamental Commissioning of the Building Energy Systems**	Required
Yes			Prereq 1	**Minimum Energy Performance**	Required
Yes			Prereq 1	**Fundamental Refrigerant Management**	Required

*Note for EAc1: All LEED for New Construction projects registered after June 26, 2007 are required to achieve at least two (2) points.

Yes	?	No			
			Credit 1	**Optimize Energy Performance**	1 to 10
			Credit 1.1	10.5% New Buildings / 3.5% Existing Building Renovations	1
			Credit 1.2	14% New Buildings / 7% Existing Building Renovations	2
			Credit 1.3	17.5% New Buildings / 10.5% Existing Building Renovations	3
			Credit 1.4	21% New Buildings / 14% Existing Building Renovations	4
			Credit 1.5	24.5% New Buildings / 17.5% Existing Building Renovations	5
			Credit 1.6	28% New Buildings / 21% Existing Building Renovations	6
			Credit 1.7	31.5% New Buildings / 24.5% Existing Building Renovations	7
			Credit 1.8	35% New Buildings / 28% Existing Building Renovations	8
			Credit 1.9	38.5% New Buildings / 31.5% Existing Building Renovations	9
			Credit 1.10	42% New Buildings / 35% Existing Building Renovations	10
			Credit 2	**On-Site Renewable Energy**	1 to 3
			Credit 2.1	2.5% Renewable Energy	1
			Credit 2.2	7.5% Renewable Energy	2
			Credit 2.3	12.5% Renewable Energy	3
			Credit 3	**Enhanced Commissioning**	1
			Credit 4	**Enhanced Refrigerant Management**	1
			Credit 5	**Measurement & Verification**	1
			Credit 6	**Green Power**	1

FIGURE 11.1   (*Continued*)

LEED for New Construction v 2.2
Registered Project Checklist

Yes	?	No	Materials & Resources	13 Points
Yes			Prereq 1 **Storage & Collection of Recyclables**	Required
			Credit 1.1 **Building Reuse,** Maintain 75% of Existing Walls, Floors & Roof	1
			Credit 1.2 **Building Reuse,** Maintain 95% of Existing Walls, Floors & Roof	1
			Credit 1.3 **Building Reuse,** Maintain 50% of Interior Non-Structural Elements	1
			Credit 2.1 **Construction Waste Management,** Divert 50% from Disposal	1
			Credit 2.2 **Construction Waste Management,** Divert 75% from Disposal	1
			Credit 3.1 **Materials Reuse,** 5%	1
			Credit 3.2 **Materials Reuse,** 10%	1
			Credit 4.1 **Recycled Content,** 10% (post-consumer + 1/2 pre-consumer)	1
			Credit 4.2 **Recycled Content,** 20% (post-consumer + 1/2 pre-consumer)	1
			Credit 5.1 **Regional Materials,** 10% Extracted, Processed & Manufactured	1
			Credit 5.2 **Regional Materials,** 20% Extracted, Processed & Manufactured	1
			Credit 6 **Rapidly Renewable Materials**	1
			Credit 7 **Certified Wood**	1

Yes	?	No	Indoor Environmental Quality	15 Points
Yes			Prereq 1 **Minimum IAQ Performance**	Required
Yes			Prereq 2 **Environmental Tobacco Smoke (ETS) Control**	Required
			Credit 1 **Outdoor Air Delivery Monitoring**	1
			Credit 2 **Increased Ventilation**	1
			Credit 3.1 **Construction IAQ Management Plan,** During Construction	1
			Credit 3.2 **Construction IAQ Management Plan,** Before Occupancy	1
			Credit 4.1 **Low-Emitting Materials,** Adhesives & Sealants	1
			Credit 4.2 **Low-Emitting Materials,** Paints & Coatings	1
			Credit 4.3 **Low-Emitting Materials,** Carpet Systems	1
			Credit 4.4 **Low-Emitting Materials,** Composite Wood & Agrifiber Products	1
			Credit 5 **Indoor Chemical & Pollutant Source Control**	1
			Credit 6.1 **Controllability of Systems,** Lighting	1
			Credit 6.2 **Controllability of Systems,** Thermal Comfort	1
			Credit 7.1 **Thermal Comfort,** Design	1
			Credit 7.2 **Thermal Comfort,** Verification	1
			Credit 8.1 **Daylight & Views,** Daylight 75% of Spaces	1
			Credit 8.2 **Daylight & Views,** Views for 90% of Spaces	1

Powered by
Adobe® **LiveCycle**™

Figure 11.1 (Continued)

LEED for New Construction v 2.2
Registered Project Checklist

Yes	?	No			
			Innovation & Design Process		**5 Points**
			Credit 1.1	**Innovation in Design:**	1
			Credit 1.2	**Innovation in Design:**	1
			Credit 1.3	**Innovation in Design:**	1
			Credit 1.4	**Innovation in Design:**	1
			Credit 2	**LEED® Accredited Professional**	1

FIGURE 11.1   *(Continued)*

LEED for Existing Buildings: Operations & Maintenance
Registered Project Checklist

Project Name: _____

Project Address: _____

Yes	?	No		
			Project Totals (Pre-Certification Estimates)	**92 Points**
			Certified: 34-42 points Silver: 43-50 points Gold: 51-67 points Platinum: 68-92 points	

Yes	?	No			
			Sustainable Sites		**12 Points**
▼	▼	▼	Credit 1	**LEED Certified Design and Construction**	1
▼	▼	▼	Credit 2	**Building Exterior and Hardscape Management Plan**	1
▼	▼	▼	Credit 3	**Integrated Pest Mgmt, Erosion Control, and Landscape Mgmt Plan**	1
▼	▼	▼	Credit 4	**Alternative Commuting Transportation**	1 to 4
			Credit 4.1	10% Reduction	1
			Credit 4.2	25% Reduction	2
			Credit 4.3	50% Reduction	3
			Credit 4.4	75% Reduction or greater	4
▼	▼	▼	Credit 5	**Reduced Site Disturbance**, Protect or Restore Open Space	1
▼	▼	▼	Credit 6	**Stormwater Management**	1
▼	▼	▼	Credit 7.1	**Heat Island Reduction**, Non-Roof	1
▼	▼	▼	Credit 7.2	**Heat Island Reduction**, Roof	1
▼	▼	▼	Credit 8	**Light Pollution Reduction**	1

FIGURE 11.2   LEED for existing buildings.

LEED for Existing Buildings: Operations & Maintenance
Registered Project Checklist

Yes	?	No			
			Water Efficiency		**10 Points**
Yes			Prereq 1	**Minimum Indoor Plumbing Fixture & Fitting Efficiency**	Required
☑	☑	☑	Credit 1.1	**Water Performance Measurement**, Whole Building Metering	1
☑	☑	☑	Credit 1.2	**Water Performance Measurement**, Submetering	1
☑	☑	☑	Credit 2	**Additional Indoor Plumbing Fixture and Fitting Efficiency**	1 to 3
			Credit 2.1	10% Reduction	1
			Credit 2.2	20% Reduction	2
			Credit 2.3	30% Reduction	3
☑	☑	☑	Credit 3	**Water Efficient Landscaping**	1 to 3
			Credit 3.1	50% Reduction	1
			Credit 3.2	75% Reduction	2
			Credit 3.3	100% Reduction	3
☑	☑	☑	Credit 4.1	**Cooling Tower Water Mgmt**, Chemical Management	1
☑	☑	☑	Credit 4.2	**Cooling Tower Water Mgmt**, Non-Potable Water Source Use	1

FIGURE 11.2   (*Continued*)

LEED for Existing Buildings: Operations & Maintenance
Registered Project Checklist

Yes	?	No			
			Energy & Atmosphere		**30 Points**

Yes			Prereq 1	**Energy Efficiency Best Management Practices**	Required
Yes			Prereq 1	**Minimum Energy Efficiency Performance**	Required
Yes			Prereq 1	**Refrigerant Management,** Ozone Protection	Required

*NOTE for EAc1: All LEED for Existing Building projects registered after June 26th, 2007 are required to achieve at least two (2) points under EAc1.

▼	▼	▼	Credit 1	**Optimize Energy Efficiency Performance**	1 to 15
				ENERGY STAR Rating: 65 / Alternative Score: 15% Above Nat'l Average	Required
			Credit 1.1	ENERGY STAR 67 / Alternative Score: 17% Above Average	1
			Credit 1.2	ENERGY STAR 69 / Alternative Score: 19% Above Average	2
			Credit 1.3	ENERGY STAR 71 / Alternative Score: 21% Above Average	3
			Credit 1.4	ENERGY STAR 73 / Alternative Score: 23% Above Average	4
			Credit 1.5	ENERGY STAR 75 / Alternative Score: 25% Above Average	5
			Credit 1.6	ENERGY STAR 77 / Alternative Score: 27% Above Average	6
			Credit 1.7	ENERGY STAR 79 / Alternative Score: 29% Above Average	7
			Credit 1.8	ENERGY STAR 81 / Alternative Score: 31% Above Average	8
			Credit 1.9	ENERGY STAR 83 / Alternative Score: 33% Above Average	9
			Credit 1.10	ENERGY STAR 85 / Alternative Score: 35% Above Average	10
			Credit 1.11	ENERGY STAR 87 / Alternative Score: 37% Above Average	11
			Credit 1.12	ENERGY STAR 89 / Alternative Score: 39% Above Average	12
			Credit 1.13	ENERGY STAR 91 / Alternative Score: 41% Above Average	13
			Credit 1.14	ENERGY STAR 93 / Alternative Score: 43% Above Average	14
			Credit 1.15	ENERGY STAR 95+ / Alternative Score: 45%+ Above Average	15

FIGURE 11.2 *(Continued)*

LEED for Existing Buildings: Operations & Maintenance
Registered Project Checklist

			Energy & Atmosphere, continued		
			Existing Building Commissioning		
			Credit 2.1	**Investigation and Analysis**	2
			Credit 2.2	**Implementation**	2
			Credit 2.3	**Ongoing Commissioning**	2
			Performance Measurement		
			Credit 3.1	**Building Automation System**	1
			Credit 3.2-3.3	**System Level Metering**	1 to 2
				Credit 3.2 40% Metered	1
				Credit 3.3 80% Metered	2
			Other		
			Credit 4	**Renewable Energy**	1 to 4
				Credit 4.1 On-site 3% / Off-site 25%	1
				Credit 4.2 On-site 6% / Off-site 50%	2
				Credit 4.3 On-site 9% / Off-site 75%	3
				Credit 4.4 On-site 12% / Off-site 100%	4
			Credit 5	**Refrigerant Management**	1
			Credit 6	**Emissions Reduction Reporting**	1

FIGURE 11.2   (*Continued*)

LEED for Existing Buildings: Operations & Maintenance
Registered Project Checklist

Yes	?	No			
			Materials & Resources		**14 Points**
Yes			Prereq 1	**Sustainable Purchasing Policy**	Required
Yes			Prereq 2	**Solid Waste Management Policy**	Required
			Sustainable Purchasing		
			Credit 1	**Ongoing Consumables**	1 to 3
				Credit 1.1 40% of Purchases	1
				Credit 1.2 60% of Purchases	2
				Credit 1.3 80% of Purchases	3
			Credit 2.1	**Durable Goods,** Electric	1
			Credit 2.2	**Durable Goods,** Furniture	1
			Credit 3	**Facility Alterations and Additions**	1
			Credit 4	**Reduced Mercury in Lamps**	1 to 2
				Credit 4.1 90 pg/lum-hr	1
				Credit 4.2 70 pg/lum-hr	2
			Credit 5	**Food**	1
			Solid Waste Management		
			Credit 6	**Waste Stream Audit**	1
			Credit 7	**Ongoing Consumables**	1 to 2
				Credit 7.1 50% Waste Diversion	1
				Credit 7.2 70% Waste Diversion	2
			Credit 8	**Durable Goods**	1
			Credit 9	**Facility Alterations and Additions**	1

FIGURE 11.2   (Continued)

LEED for Existing Buildings: Operations & Maintenance
Registered Project Checklist

Yes	?	No			
			Indoor Environmental Quality		**19 Points**
Yes			Prereq 1	**Outdoor Air Introduction and Exhaust Systems**	Required
Yes			Prereq 2	**Environmental Tobacco Smoke (ETS) Control**	Required
Yes			Prereq 3	**Green Cleaning Policy**	Required
			IAQ Best Management Practices		
			Credit 1.1	**IAQ Management Program**	1
			Credit 1.2	**Outdoor Air Delivery Monitoring**	1
			Credit 1.3	**Increased Ventilation**	1
			Credit 1.4	**Reduce Particulates in Air Distribution**	1
			Credit 1.5	**Facility Alterations and Additions**	1
			Occupant Comfort		
			Credit 2.1	**Occupant Survey**	1
			Credit 2.2	**Occupant Controlled Lighting**	1
			Credit 2.3	**Thermal Comfort Monitoring**	1
			Credit 2.4-2.5	**Daylight and Views**	1 to 2
				Credit 2.4 50% Daylight / 45% Views	1
				Credit 2.5 75% Daylight / 90% Views	2
			Green Cleaning		
			Credit 3.1	**High Performance Cleaning Program**	1
			Credit 3.2-3.3	**Custodial Effectiveness Assessment**	1 to 2
				Credit 3.2 Score of ≤ 3	1
				Credit 3.3 Score of ≤ 2	2
			Credit 3.4 -3.6	**Sustainable Cleaning Products and Materials**	1 to 3
				Credit 3.4 30% of Purchases	1
				Credit 3.5 60% of Purchases	2
				Credit 3.6 90% of Purchases	3
			Credit 3.7	**Sustainable Cleaning Equipment**	1
			Credit 3.8	**Entryway Systems**	1
			Credit 3.9	**Indoor Integrated Pest Management**	1

FIGURE 11.2   (Continued)

LEED for Existing Buildings: Operations & Maintenance
Registered Project Checklist

Yes	?	No			
			Innovation in Operations		**7 Points**
			Credit 1.1	**Innovation in Operations:**	1
			Credit 1.2	**Innovation in Operations:**	1
			Credit 1.3	**Innovation in Operations:**	1
			Credit 1.4	**Innovation in Operations:**	1
			Credit 2	**LEED® Accredited Professional**	1
			Credit 3	**Documenting Sustainable Building Cost Impacts**	2

FIGURE 11.2 *(Continued)*

The building envelope, being the largest single building component, requires attention in the design stage. Even the color of the exterior will have an impact on the building's efficiency, affecting both solar gain and solar loss. A lighter color will lower solar absorption; according to one study, a 30 percent reduction in solar absorption can result in a 12.6 percent savings in annual cooling costs.

Exterior-wall insulation, as we all know, contributes to energy savings and is an integral part of green building design. Glazing systems

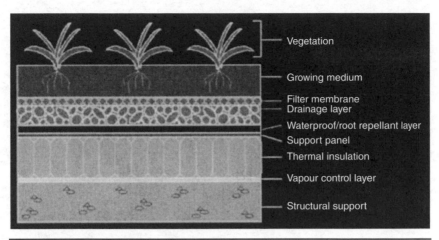

FIGURE 11.3   Section through a green roof.

and high-performance windows and doors are also a major area of an energy-efficient building.

Roof insulation is another critical element of a green building, so why not have a green roof? A green roof is exactly what it sounds like. Figure 11.3 shows a typical section through a green roof, from roof structure to the top layer of vegetation. Figure 11.4 shows the top U.S. metropolitan areas where green roofs have been installed as of 2011. The advantages of these types of roofs are rather straightforward:

- Water is stored in the substrate and returned to the atmosphere via evaporation.

- In summer, green roofs retain 90 percent of the precipitation; in winter, 25 to 40 percent.

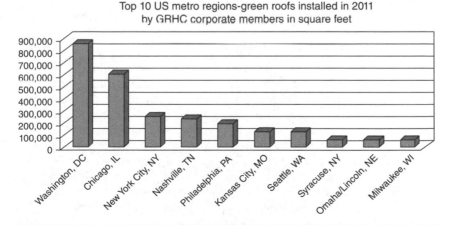

FIGURE 11.4   Top U.S. metropolitan areas with green roofs.

- They retain rainwater and act as natural filters for any water runoff.

- Green roofs reduce the amount of storm-water runoff, delaying the time during which runoff occurs and thereby resulting in decreased stress on sewer systems when peak loads occur.

- Due to the daily dew and evaporation benefits, plants on the green roof are able to produce some cooling effect and reduce the "urban heat island" effect that conventional roofing systems produce.

As of 2010, about 4,577,935 ft^2 of green roofs had been installed, up from 4,341,394 ft^2 in 2009.

Another Term You Will Hear When Dealing with Green Buildings: FSC

The Forest Stewardship Council (FSC) is a nongovernment agency established in 1993 to deal with the problem of global deforestation. This also created the Forest Certification Resource Center, whose mission is to assure consumers that the forest products they buy have come from operations that manage forests to a determined set of environmental, social, and economic standards. The FSC is modeled on many of the principles established by the International Organization for Standardization. With forest certification, this organization develops standards of good forest management and verifies that wood and paper products come from responsibly managed forests. A third-party certification is required before the stamp is put on FSC wood or paper products, so as a part of the green building movement and associated LEED certification program, construction managers and project managers need to add another term to their professional vocabulary: *FSC certification*.

Recycling of Industrial Materials

For many years, contractors have recycled asphalt from milling operations or from demolishing driveways and parking lots when working on upgrading or renovating an existing structure. This waste asphalt has been recycled into a component of the new product. Concrete from old walkways, building slabs, and foundation walls has also been sent to private centers, where it is crushed to provide structural fill under new concrete slabs, driveways, and assorted miscellaneous site structures. Wood products such as chips have been used for year on oriented strand board and in medium-density fiberboard. Recycling industrial materials preserves our natural resources by reducing demand on the fresh materials.

Recycling also conserves energy and in the process reduces greenhouse-gas emissions by decreasing the demand for those products made through an energy-intensive manufacturing process.

Money is saved by letting recyclers pick up the materials, either at no cost or for a small sum from them in return for reusing or reselling the materials. This process saves the general contractor (GC) money by removing the need to truck these waste materials to a municipal or private waste depot.

Coal fly ash, slag, and spent foundry sand are other waste materials that can end up in ready-mix concrete or as flowable fill or embankment stabilization. Coal ash is used in the manufacture of cement and ceiling tiles, and flue-gas-desulfurization gypsum foundry sand and pulp and paper by-products can end up in the manufacture of soil and agricultural amendments.

Recycled lumber often is sold at a premium price, especially when it is milled as a wood flooring product. Some recovered long-leaf pine has been sold for $20 per square foot or more, and the presence of nail holes or other blemishes often raises the price even higher because of its perceived aesthetic value. A recent trend in recovering logs from the Great Lakes and other large bodies of water, submerged for decades after timber-transporting ships lost some or all of their cargos during violent storms. Once recovered and milled for flooring or other decorative products, these reclaimed logs command premium prices.

Residential Housing

The National Association of Home Builders (NAHB) lists 17 things for a green residential builder to consider:

1. Size the building appropriately. The average size of new homes in the United States more than doubled over the second half of the twentieth century, from 1,100 ft^2 in 1950 to 2,414 ft^2 in 2004.

2. Orient the building to allow the sun to light the house during the day and warm it in the winter. If the roof's eaves hang over, the sun's rays in the summer will be blocked, which will help keep the house cooler.

3. Buy locally, thereby saving on fuel and delivery costs and related emissions.

4. Recycle construction waste and all recyclable packaging wastes.

5. Use dual-flush toilets. Forty percent of water used in the home goes right down the drain. Dual-flush toilets can save 6,000 gal per year for a family of four.

6. Locate residences near public transportation or within walking distance to work or stores. This can improve the quality of life. The average U.S. American spends over 100 hours annually commuting, and a little daily walking exercise cannot hurt.

7. Save money with effective insulation: R-50 in the attic, R-30 in walls. Inefficient windows and doors cost U.S. Americans $40 billion annually in higher utility costs.

8. Use energy-efficient lights. Compact fluorescent bulbs use 65 percent less energy and last 5000 hours. Several manufacturers are developing additional types of high-efficiency bulbs.

9. Choose appliances with the Energy Star symbol; they use 10 to 50 percent less energy than ones without that designation.

10. Consider the old-fashioned root cellar. It is gaining acceptance again, lowering the cost of refrigeration.

11. Use solar panels wherever practical.

12. Install geothermal heating and cooling where practical. It uses about 50 percent less energy to heat and 25 percent less to cool.

13. Use primers, paint, and caulking materials without volatile organic compounds.

14. Avoid PVC, which produces toxic dioxins.

15. Buy FSC-certified lumber, which means the wood was not taken from fragile ecosystems.

16. Buy green power (power generated from renewable sources such as wind, solar, etc.) if it is offered in your state.

17. Buy used and salvaged materials whenever possible.

What Are the Costs and Benefits of Green Building Construction?

The most definable cost of green building is the registration fee with the USGBC, which ranges from $450 to $600 (membership fees vary for different organizations).

The city of Bloomington, Indiana, published a paper in 2008 on green-building costs which contained a July 2007 report by the construction-consulting company Davis Langdon. Davis Langdon compared the cost of LEED-seeking buildings to conventionally designed buildings, looking at 83 green and 138 conventional buildings. The report concluded that there was no "one size fits all" answer; some buildings achieved their goals without any additional funding, while others required added funding for specific sustainable products like photovoltaic systems. The city of Seattle reported an average incremental cost for meeting LEED Silver standards across all projects of 1.7 percent.

Gregory H. Kats of the Massachusetts Technology Collaborative (in Westborough, Massachusetts), in his white paper *Green Building Costs and Financial Benefits*, indicates that the average premium for green buildings is less than 2 percent, or $3 to $5 per square foot. Kats attributes the majority of this excess cost to the increased architectural and engineering design time, modeling costs, and the time required to integrate sustainable building products into projects.

Information received from USGBC-Capital Analysis report and included in Kats's report indicates that on average the green cost premium for LEED certification at various levels for offices and schools are as follows:

- Certified (eight buildings included)—the average cost premium was 0.66 percent.
- Silver (18 buildings included)—the average cost premium was 2.11 percent.
- Gold (six buildings included)—the average cost premium was 1.82 percent.
- Platinum (one building included)—the average cost was 6.50 percent.

The study also revealed the advantages of a LEED-certified building over a conventional one:

- An average of 25 to 30 percent better energy efficiency
- Even lower peak electricity consumption
- A greater likelihood of generating renewable energy on-site
- A greater likelihood of purchasing grid power generated from renewable energy sources

The higher the level of LEED certification, the more energy-efficient the building is when compared to conventional buildings:

- Certified—18 percent more efficient
- Silver—30 percent more efficient
- Gold—37 percent more efficient
- Platinum—28 percent more efficient

Effect on Employee Productivity

According to a study prepared by F & L Building Maintenance—a major company involved in building-maintenance services, with offices located in Bellevue and Tacoma, Washington—more than

**Financial benefits of green buildings
summary of findings (per ft²)**

Category	20-year net present value
Energy savings	$5.80
Emissions savings	$1.20
Water savings	$0.50
Operations and maintenance savings	$8.50
Productivity and health value	$36.90 to $55.30
Subtotal	**$52.90 to $71.30**
Average extra cost of building green	(−3.00 to −$5.00)
Total 20-year net benefit	**$50 to $65**

Source: Capital E analysis

FIGURE 11.5   Financial benefits of green roofs.

30 percent of U.S. workers may suffer from health problems associated with what is known as "sick-building syndrome." Green buildings promote a healthier work environment.

Figure 11.5 lists the financial benefits of green buildings, including productivity and health. This list of benefits was compiled by Capital E, a company headquartered in Washington, DC. Capital E is involved in, among other ventures, developing an open platform in collaboration with sponsors including USGBC and GBData.org, an organization for the collection of green-building performance data.

Although the energy, emissions, and water savings are significant, along with operations and maintenance saving, it is the impact on employee productivity that is overwhelming and could be the determining factor for some clients in going green.

Risks Associated with Green Building and Risk Mitigation

Risks associated with the construction of green buildings differ from those associated with conventional construction practices. These risks range from site selection to materials used in construction to insurance coverage. Construction managers and project managers must become aware of the subtle but stark differences that face them when they become involved with green buildings.

Let us start with site selection, which often involves seeking out land in close proximity to public transportation and pathways that are conducive to bicycle riders living within a reasonable distance from the proposed site; both add points to the green-building site checklist. Most likely this means an area with population density in or near an urban center. The sustainable-site checklist also awards the owner one point for brownfield development, which means that the site may have been an abandoned industrial or commercial site where environmental contamination could exist in full bloom or lurk around the corner. Buried materials containing various amounts of PCB, mercury, lead, and chlorinated hydrocarbons that the owner assumed had been mitigated may not have been entirely removed.

Although any remediation costs for hazardous materials not specifically noted will more than likely fall upon the owner to absorb, and most construction contracts include a provision absolving the contractor from dealing with hazardous materials, the CM or PM must take into account the impact any such discoveries will have on the project schedule. Proper and prompt documentation if and when hazardous materials are discovered by the CM or GC can start the time clock ticking on schedule-delay notification.

Southern Builders v. Shaw Development

This lawsuit, filed in February 2007 in the state of Maryland, has been deemed the first legal action taken with respect to a green building. Shaw Development hired Southern Builders to build Captain's Galley, a $7.5 million, 23-unit condominium building in Crisfield on Maryland's eastern shore. It was to achieve LEED Silver certification. At the time, Maryland was offering credits of up to 8 percent of the total project cost for LEED Silver buildings that exceeded 20,000 ft^2 in size. If the certificate of occupancy was obtained after the expiration date of the final credit certificate, the credits would go back into the program's pool.

The contract between Shaw and Southern Builders was a standard AIA A101 stipulated-sum agreement, but it did not specifically address the tax credits relating to the deal Shaw had with the state. The contract did, however, require Southern Builders to deliver the project in 336 days, and it did incorporate green building requirements in the project manual.

The project was delivered beyond the contract completion date, and Captain's Galley failed to obtain LEED Silver certification. Shaw sued Southern for late delivery and failure to attain LEED Silver certification. Southern Builders filed a $54,000 mechanics lien against the property, which was later reduced to $12,000, and Shaw countered with a $1.3 million lawsuit, claiming damages that included $635,000 in lost tax credits. Shaw sold only six of the 23 condos, the Captain's Galley Restaurant was forced to close, and Shaw ultimately filed for Chapter 11 bankruptcy.

This case illustrates a couple dangers inherent in green building that contract language might have forestalled, but failed to address:

1. The contract did not specify which party was responsible for obtaining the LEED Silver certification, nor did it hold the GC responsible in case that certification was not achieved.

2. Shaw's financing relied on obtaining Maryland tax credits of 8 percent of the total project cost, which in turn was also based on completion of the project within the strict credit-expiration deadline.

The situation with Shaw versus Southern is an example of developing a risk-averse contract when a new form of construction, tied to tax credits or other federal, state, or local incentives, is part of the deal. Unstated in the legal machinations was the builder's experience in constructing buildings designed to meet certain LEED certification levels. The lesson learned from this lawsuit is that caution and experience, on the part of the owner, designer, and builder, are necessary to properly evaluate the risks in venturing into green building projects and to manage those risks with the appropriate contract language.

Long-Term Performance in Green Buildings

Because of the relatively short period of time that green buildings have been in operation, it is difficult to quantify the long-term savings projected in the initial design documents. The longevity, performance, and warranties of some of these new products make it difficult to determine whether the payback anticipated will be achieved. And if it does not, are there any liability issues that may or may not be brought to bear? By whom and to what effect? Are there statutes of limitation that will come into play and prevent recovery of such losses?

Owners can obtain insurance products to address environmental liabilities, and perhaps this is a question to be asked by the CM or PM while preparing the bid. Although a builder's risk policy may be required by the owner, such a policy may not cover all of the aspects of a LEED project, so it is best to seek the advice of the company's insurance broker before putting a price in the builder's risk, if required in the contract.

Material reuse poses another potential environmental concern. The materials-and-resources section of LEED 2009 for new construction lists various points to be awarded for reuse of existing walls, floors, roofing, interior nonstructural elements, and recycled content. Will any of the existing surfaces to be retained contain hazardous materials, such as paint containing mercury or lead? What about those steel beams that can remain in place? If they contain lead paint—and depending upon the age of the building, they probably do—do the bid documents call for removal and repainting of the beams or is encapsulation an option? Or is there no mention at all of the existence or remediation of lead paint? Plywood or other products containing formaldehyde may also exist in the building, and disposal is probably the only answer there to meet Environmental Protection Agency (EPA) regulations. If there is any doubt, the contract can stipulate that until a chemical analysis determines otherwise, the bidder assumes that these materials contain hazardous materials and the costs for removal and disposal are excluded from the proposal.

The section of LEED 2009 on indoor environmental quality provides credits for an indoor air-quality management plan during construction, awarding 3.1 credits for proper management. CMs and PMs must be familiar with the requirements to meet this goal.

Will there be limitations on the amount of air exchange that is necessary for a healthy worker environment during painting, applications of floor or wall mastic for tile, and covering walls with vinyl?

Steps a CM or PM Can Take to Reduce Liability Issues

This green building movement, which only began to gather momentum in the first decade of the twenty-first century, is still relatively unfamiliar to many construction managers and project managers who have either no or limited experience working on a green project.

Owners, also reading about the advantages of going green, may be expecting more than is possible for the dollars they are likely to spend. As we stated previously, the long-term impact of savings accrued by LEED certification may simply not have had enough time to develop a strong database.

A design team without several green buildings under their belt may not know what it is they do not know. So we in the construction business need to tread carefully. There are some steps that the CM or GC can take to reduce some of these uncertainties:

- Request a prebid LEED meeting where owners, architects, engineers, and builders can discuss the proposed project and have the designers elaborate on the unique aspects of green construction in general and those design features in the project at hand.

- Advise the owner that these types of projects may require additional time due to the innovative design, and possibly a higher fee to compensate for the added time spent with the design consultants or searching for the proper materials or equipment.

- Advise the owner that achieving a certain LEED certification level may not *guarantee* the savings anticipated.

- Negotiate a fair but carefully prepared contract that addresses guarantees, timelines, and indemnities. Do not make promises unless you are sure you can deliver them, such as performance. The law will probably only hold you to the standard-of-care provision.

- Prepare the schedule with careful thought about contingencies or float and—as we pointed out with regard to the *Southern Builders v. Shaw Development* case—any tie-ins to federal, state, or local subsidies, tax credits, and the like.

- Fully understand what you are building. If certain products are not familiar, contact the manufacturer and request a meeting with their technical people to learn as much as you can about the material or equipment.

- When subcontracting the work, make certain that the subcontractor has ample experience in the component of work in question. Get a list of the subcontractor's projects and try to get an appointment to meet with the owner and maintenance people to discuss performance.

- Take the commissioning process seriously as a critical phase of green building construction. The owner must be educated on how the building works and what responsibilities he or she has in the process.

Sustainability and Sustainable Design Go Hand in Hand with Going Green

On July 29, 2005, the U.S. government passed the Energy Policy Act, requiring federal agencies to achieve mandated sustainability goals to protect the environment. The act requires a minimum of 30 percent improvements in energy-cost savings from a baseline established with the American Society of Heating, Refrigeration, and Air Conditioning (ASHRAE) 90.1. It also contained a memorandum of understanding committing 19 federal agencies to leadership in the design, construction, and operation of high-performance and sustainable facilities. This surely will have an impact on the private sector as well.

The memorandum of understanding contains specific, measurable requirements for design and construction of all new construction and major renovations of buildings owned or leased by the government, grouped into five main areas:

1. Employ integrated design principles
   - Integrated design
   - Commissioning
2. Optimize energy performance
   - Energy efficiency
   - Measurement and verifications

3. Protect and conserve water
   - Indoor water
   - Outdoor water
4. Enhance indoor environmental quality
   - Ventilation and thermal comfort
   - Moisture control
   - Day lighting
   - Indoor air quality during construction
5. Reduce environmental impact of materials
   - Recycled content
   - Low-emitting materials
   - Biobased content
   - Construction waste
   - Ozone-depleting compounds and substances

Sustainable Design Principles

The goal of sustainable design is to eliminate the negative impact that construction has through skillful and sensitive design. There are common principles that apply to sustainable design:

- Use low-impact materials, which are nontoxic and sustainably produced (materials that can reproduce themselves, i.e., wood, paper) and require little energy to produce.
- Use manufacturing processes that produce products that require less energy.
- Provide longer-lasting and better-functioning products that require replacement less frequently.
- Design to incorporate the potential for reuse or recycling.
- Design to include the total carbon footprint.
- Purchase renewable materials from local sources to reduce emissions created by long-haul trucking.
- Purchase renewable materials from sustainably managed renewable sources.
- Emphasize "healthy buildings" that provide occupants with high indoor air quality.

Sustainable design standards and project design criteria are increasingly available from public and private sources and should be investigated during the conceptual design phase. As an example, the

Construction Industry Compliance Assistance Center (www.cicacenter
.org) provides a state-by-state tool to locate regulatory and compliance
information for construction-and-demolition disposal.

Expo 2000—Hanover, Germany

A bill of rights also known as the Hannover Principles was developed
by William McDonough Architects (Charlottesville, Virginia) that is a
model of design principles for sustainability:

1. Insist on the right of humanity and nature to co-exist in a healthy,
   supportive, diverse, and sustainable condition.
2. Recognize interdependence. The elements of human design interact with and depend upon the natural world. . . .
3. Respect relationships between spirit and matter. . . .
4. Accept responsibility for the consequences of design decisions
   upon human well-being, the viability of natural systems and their
   right to co-exist.
5. Create safe objects of long-term value. Do not burden future generations with requirements for maintenance or vigilant administration of potential danger due to the careless creations of products,
   processes, or standards.
6. Eliminate the concept of waste. Evaluate and optimize the full life-
   cycle of products and processes, to approach the state of natural
   systems, in which there is no waste.
7. Rely on natural energy flows. Human designs should, like the
   living world, derive their creative forces from perpetual solar
   income. Incorporate this energy efficiently and safely for responsible use.
8. Understand the limitations of design. No human creation lasts
   forever and design does not solve all problems. Those who create
   and plan should practice humility in the face of nature. Treat
   nature as a model and mentor, not as an inconvenience to be
   evaded or controlled.
9. Seek constant improvement by the sharing of knowledge. Encourage direct and open communication between colleagues, patrons,
   manufactures, and users to link long-term sustainable considerations with ethical responsibility, and re-establish the integral relationship between natural processes and human activity.

Sources for Regulations, Guidance, and Policy

The Federal Facilities Environmental Stewardship and Compliance
Assistance Center in Washington, DC (www.fedcenter.gov) is the
source for many federal regulations and green-building materials
relating to high-performance buildings and contains a plethora of
all types of information to assist the builder, architect, and engineer.

Listed here are some of the materials linked to at the center's Web site:

- Regulations, guidance, and policy
  - Department of the Army memorandum: Sustainable Design and Development Policy Update
  - Executive Order 13423: Strengthening Federal Environmental, Energy, and Transportation Management
  - Executive Order 13514: Federal Leadership in Environmental, Energy, and Economic Performance
  - EPA's Strategic Agenda to Protect Water and Build More Livable Communities Through Green Infrastructure
  - High Performance and Sustainable Buildings Guidance
- Supporting information and tools
  - Building Energy Software Tools
  - Building Life-Cycle Cost Program
  - Construction Waste Reduction Potential
  - Energy Benchmarking Tool for Buildings
  - FEMP Energy and Cost Savings Calculators for Energy-Efficient Products
  - Green Homes
  - GSA Sustainable Facilities Tool
  - High Performance Buildings Website and Database
  - Room Air Conditioner (RAC) Cost Estimators
  - Roof Savings Calculator (RSC)
- Directories, catalogs, and newsletters
  - *Building E2 News*—quarterly newsletter from the Department of Energy Building Technologies Program
  - *Environmental Building News*—subscription-based monthly newsletter from BuildingGreen.com
  - *Forest Stewardship Council (FSC)-US Newsletter*
  - *High Performance Buildings*—quarterly magazine
  - Innovation and Design Credit Catalog—listing of proven green building strategies
  - Sustainable Building Resource Directory—Web site-based resource directory for the mid-Atlantic region
  - Green Buildings (EPA)—EPA Web site with links to tools and resources regarding energy efficiency; renewable, green building materials; and indoor-environment issues

- Sustainable Building Sourcebook—online searchable directory of green-building professionals
- Sustainable Federal Buildings Database—compilation of federal-agency policies and guidelines on energy efficiency and sustainable government policies
- Organizations and programs
  - Building Materials Reuse Association
  - Energy Efficient Building Association
  - Energy Star—New Building Design
  - Green Roofs for Healthy Cities
  - National Institute of Building Sciences
  - Smart Communities Network (DOE Energy Efficiency and Renewable Energy)
  - Sustainable Building Industry Council
  - U.S. Green Building Council
- Lessons learned
  - Lifecycle Building Challenge Winners
  - Beneficial Landscaping Resources in the Northwest and Alaska
  - Landscaping with Native Plants
  - Green Landscaping
- Case studies
  - Better Buildings Case Studies
  - Commercial Building Energy Goals for Lodging
  - DOE's High Performance Building Research Case Studies
  - GreenSpec High Performance Building Case Studies
  - High Performance Buildings for Labs, Data Centers, and Cleanrooms
  - Low Impact Development for Sustainable Installations: Stormwater design and Planning Guidance for Development within Army Training Areas
  - Managing Wet Weather with Green Infrastructure Handbook Series
- Construction design
  - Advanced Buildings and Technologies
  - Combined Heat and Power: Effective Energy Solutions for a Sustainable Future

- Guiding Principles of Sustainable Design (NPS)
- Water-Saving, High Efficiency Toilets
- Construction guidelines and criteria
  - Advanced Energy Design Guide for Large Hospitals
  - Advanced Energy Design Guide: Medium to Small Buildings
  - Consolidated Sustainable Facility Checklist
  - Existing Building Evaluation and Prioritization Matrix
  - GSA LEED Cost Study
  - High Performance Buildings Guidelines
  - Water Efficient Products
  - Whole Building Design Guide
- Indoor air quality
  - Building Air Quality Action Plan
  - EPA's Indoor Air Quality (IAQ) Home Page
  - Healthy Indoor Environmental Protocols for Home Energy Upgrades
  - Indoor Air Quality (AIQ) in Large Buildings
  - Indoor Air Quality Building Education and Assessment Model
- Planning
  - Federal Green Construction Guide for Specifiers
  - Funding Programs for Green Buildings
  - Green Infrastructure
  - Handbook for Planning and Conducting Charrettes for High Performance Projects
  - Low Impact Development
- Training, presentations, and briefings
  - Art and Science of Stormwater Retrofitting
  - Benchmarking Energy Use and Generating a Federal High Performance Sustainable Building Checklist in ENERGY STAR Portfolio Manager
  - Building Commissioning: Ensuring High Performance Green Buildings
  - High Performance School Buildings Training
  - HVAC Management for High Performance Existing Buildings

- Navigating the Whole Building Design Guide
- Re-Tuning Commercial Buildings
- Training Offered by Sustainable Building Industry Council
- USGBC LEED Training Workshop Calendar
- Water Efficiency for High Performance Commercial and Institutional Buildings

Summary

The increased attention to promoting a healthy environment has focused on the design-and-construction industry, and the increased interest in green building has placed new responsibilities on construction managers and general contractors. Their project managers must gain familiarity with the LEED certification process and the new demands that will be placed upon them when they contract to build in this variation on conventional construction.

Learn the new terminology, components of construction, and risks that accompany this new era of building. Sustainability is one of the terms construction managers and project managers will have already heard about or will surely become familiar with. This movement incorporates the philosophy of designing and building structures utilizing materials that create energy reduction and do not deplete the fragile resources of the world in which we live.

CHAPTER 12

Construction Safety and Health

The importance of construction safety has many facets:

- It is the law per the Occupational Safety and Health Act.
- Workers should be shielded from accidents and injuries.
- Preventing accidents that could disrupt the efficiency and productivity of the crew maintains a harmonious and efficient workforce.
- A good safety record reduces the contractor's on-site insurance costs and corporate overhead.

So jobsite health and safety is clearly an integral part of total construction project management. Keeping all of these factors in mind, one can see why it is so important to structure a sound jobsite-safety plan and monitor it diligently. We might also add that an owner is not particularly keen on hiring a contractor with a poor safety record—so this becomes a sales-development issue as well.

Construction's "Fatal Four"

According to statistics available from the Occupational Safety and Health Administration (OSHA) as of 2010, out of 4,070 worker fatalities in private industry during 2010, 18.5 percent—nearly one fifth—were in the construction industry. The "fatal four" contributors were:

- Falls—260 of 751 total deaths in construction (35%)
- Electrocutions—76 (10%)
- Being struck by an object—63 (8%)
- Being caught in between—32 (4%)

Eliminating these four causes of death would have saved 431 workers' lives in the United States. This can only be accomplished by instituting an effective safety plan and monitoring the safety practices of workers on a daily basis.

The 10 Most Frequently Cited OSHA Standards Violated in 2011

1. Scaffolding, general requirements construction
2. Fall protection, construction
3. Hazard communication standard
4. Respiratory protection
5. Control of hazardous energy (lockout/tagout)
6. Electrical, wiring methods, components, and equipment
7. Powered industrial trucks
8. Ladders
9. Electrical systems—design
10. Machines

Contractors have gotten the message and OSHA in the past four decades has made a dramatic difference in workplace safety. Since 1970, workplace fatalities have been reduced by more than 65 percent and occupational injuries and illnesses have declined by 67 percent. During this period, employment doubled, so the percentage reductions have even more meaning. Worker deaths are down from about 38 workers per day in 1970 to just about 12 a day in 2010. Worker injuries and illnesses have also decreased, from 10.9 incidents per 100 workers in 1972 to less than 4 per 100 in 2010. OSHA has increased citations for violations over the years, and a review of the penalties for violations shows the administration means business.

OSHA Violations

The area director from the appropriate office will determine what, if any, citations will be issued after an inspection of a jobsite. These are the type of violations that may be cited and the penalties that may be proposed:

- Other-than-serious violation—a violation that has a direct relationship to job safety and health, but probably would not cause death or serious injury. A proposed penalty of up to $7,000 for each violation is discretionary.

- Serious violation—a violation where there is substantial probability that death or serious physical harm could result and the employer knows or should have known of the hazard. A mandatory penalty of up to $7,000 per violation is proposed.

- Willful violation—a violation that the employer commits knowingly or with indifference to the law. The employer either knows that what he or she is doing constitutes a violation or is aware that a hazardous condition exists and makes no reasonable effort to eliminate it. Penalties are proposed of up to $70,000 for each violation.

- Failure to abate prior violation—may bring a penalty of $7,000 for each day it continues.

It is obvious that OSHA meant business when they enacted the contractor program referred to as 29 CFR 1926 Subpart C Standards.

Why Are Construction-Safety Programs Important?

There are many factors supporting the need for effective construction-safety programs today. Generally they fall into humanitarian and economic factors. The hard-nosed economic factors of safety have forced even the most indifferent managers into taking a more humanitarian stance on construction safety. We say this because of the construction industry's widespread tough image, which tends to play down minor injuries so that employees see no need to advise management of their occurrence.

Present-day construction programs must also stress *accountability* for safety throughout the organization. Owners and contractor managers must initiate the concern for safety, and the resulting systems must be clear as to who is accountable for carrying out the program. Because either the construction manager (CM) or project manager (PM) will have sole responsibility for carrying out the project goals, he or she is held accountable for the success or failure of the site's safety performance on the project.

Economic Factors in Safety

There is no need to discuss the humanitarian aspect of safety because that is quite clear. The need to eliminate the suffering resulting from an accident or the trauma to a relative advised of the death of one of their family members on the job needs no further explanation.

The economic factors in safety are often overlooked, but they pose a real threat to a company's competitiveness and bottom line. The direct costs of a high accident rate are fairly obvious and easy to evaluate; some of these direct costs attributable to high accident rates are:

1. Higher workers'-compensation insurance rates

2. Higher liability-insurance rates

3. Losses not covered by insurance policies

4. Government agency fines

5. Depressed craft-labor productivity rates

6. Costs of investigating and filing accident reports

7. Stoppage of all work in the area of the accident

Indirect costs are a bit harder to evaluate but typically add up to more than direct costs. Some of the more obvious indirect costs are:

1. Increased employee turnover

2. Lost time of injured workers

3. Training costs of replacement workers

4. Time lost on the schedule (possible incurring liquidated damages)

5. Lowered worker and supervisor morale

6. Loss of worker efficiency due to worker demobilization and later re-mobilization

7. Damage to the owner's property

8. Damage to tools and equipment

9. Litigation support costs not covered by insurance

10. Loss of new business and damage to corporate image.

Worker's Compensation Costs

The more dangerous the industry (or even the craft), the higher the workers'-compensation rate per hour. For example, for the classification "Construction Misc.," the hourly contribution rate for workers' compensation in 2011 was $1.8476; in 2012 it increased by 4 percent to $1.921.

Workers'-compensation insurance premiums are calculated by the following equation:

$$WCIP = EMR \times \text{Manual rate} \times \text{Payroll units}$$

EMR is the experience modification risk, the multiplier determined by the previous work experience of the contractor. The manual rate is the rate structure assigned to each type of work performed. More dangerous crafts have higher hourly costs, going up to $4.7511 per hour for masons and all the way up to $6.7263 for roofers. A payroll unit is a number determined by dividing the contractor's annual direct labor costs by 100.

These workers'-compensation insurance rates will increase if the employer has a number of accidents on the job, and it takes three years of good accident experience to wipe out a bad year. So you can see how a poor safety record adds on to the cost of labor and can affect a contractor's competitiveness.

Other Costs Related to Construction Accidents

Higher liability-insurance premiums results when the insured's claim history reveals a high rate of accidents. These liability claims can involve people and physical property such as tools, construction equipment, and the physical plant. Again, these rates start out pretty elevated due to the nature of the construction industry, so a percentage rating increase can be significant. It is another add-on to the company's overhead.

Uninsured losses represent another direct job cost, covering deductible charges and uncovered losses. These losses can be substantial, especially on projects where the project risk analysis has not been well thought out.

The effect of high accident rates on field productivity is probably the most significant direct cost listed. Some list this item as an indirect cost, but it can be estimated that assigning an overall percentage rate such as 1 to 5 percent to the overall field labor hours would be appropriate. Also, remember to include the field supervisory hours in this cost, because their productivity is significantly lowered by poor safety performance.

The time spent on researching the cause and effect of site accidents can be substantial and is an extra nonreimbursable cost to the project budget. Typically this is high-priced, nonproductive time spent by supervisors who can ill afford it. The time requirement also percolates through the entire reporting process into the home office and government agencies, setting off an ongoing, time-consuming investigatory process. In addition to the project charges, accident reporting increases the firm's overhead, which eats away at its competitive position in the industry.

Indirect Costs

Because indirect costs are not listed separately in the project accounts, they are difficult to assess accurately. Increased employee turnover can result in projects with inferior safety records. High accident or near-miss rates go right to the core of poor morale on any construction site. That in turn makes the better, more concerned employees move to another job, leaving the less competent, accident-prone workers behind. Such a turn of events can exacerbate an already dangerous and unsafe working environment. When that happens, indirect site costs can increase geometrically, with disastrous results in project productivity and profitability. If immediate corrective action on the safety program is not taken, the CM or PM often faces a personally untenable position and can expect to be replaced.

The lost time of injured workers can upset the routine of work crews while the injured workers are away from their jobs due to the injury, in addition to any direct salary or wages that may be paid. If the accident is fatal or the injured employee does not return, the cost of the training and learning curve of the employee's replacement is an indirect cost to the job. Both of these items also adversely affect the field labor productivity.

Delays created by injuries and lost time have an adverse effect on the project schedule, jeopardizing one of the major project goals. If the contract schedule clause specifies liquidated damages, the adverse effect translates directly into poor profitability. Most contracts specify the contractor's safety performance, so it is difficult to blame the owner or outside interests for poor safety performance as a way of evading liquidated damages, which can run as high as $10,000 per day.

We have already mentioned the insidious effects of poor safety on the morale, efficiency, and productivity of the craftspeople, but the same is true of the field superintendent's supervisory staff.

Although we have already discussed the direct costs resulting from damage to construction equipment and the owner's property, there are indirect costs as well. Damage to tools and construction equipment results in lost time while units are repaired or replaced. If the works themselves are damaged from an accident, delays while the causes and effects are assessed—along with the time spent on actual repair—are significant. Because accidents and repairs are not classed as force majeure, the contract grants no relief for resulting schedule delays.

When litigation results from an accident, some of the direct costs for defending the suit are covered by insurance. However, there are expenses contributed by the defendant firm in the form of defense-preparation costs. The people involved must contribute specialist support in the form of depositions, technical advice, and the like, which become part of the firm's overhead. The cost to litigate, in most cases, exceeds the initial anticipated costs.

Perhaps the least-recognized indirect cost is the damage to the company's corporate image in the eye of the marketplace. Many owners have become intensely safety conscious in recent years and now make special efforts to select only *safe contractors*. A poor safety record is often ground for a contractor's not making the bidders list despite having an otherwise excellent track record.

Naturally, the "no free lunch" rule also applies to construction safety. We are not going to reduce all of these direct and indirect costs without spending some money. It costs about 2.5 percent of direct field labor costs to establish and operate a basic safety program, according to some sources. That figure includes those costs mandated by government regulations that must be spent on safety as a minimum and are often considered as direct job costs.

The cost of maintaining staff safety engineers, developing and maintaining safety standards and procedures, operating safety-training programs, carrying out site-safety inspection trips, purchasing safety equipment, etc., is directly proportional to the size and depth of the safety program. Most owners and contractors have gone through the initial setup expense and are now in the maintaining mode. Those that have few or no safety procedures in place will have to take the initial financial plunge. However, safety programs can be built up gradually to spread the cost over time.

In the final analysis, all safety-program expenses eventually are borne by the owner in both lump-sum and cost-plus-a-fee contract settings. Because lump-sum contractors absorb safety costs in their overheads, safety practice is sometimes diminished by the competitive nature of that sector of the construction business. Contractors should pay attention to the owner's safety requirements in the contract when they are bidding the job, regardless of the contracting basis.

Upper managers for the owner and contractor are responsible for setting and approving the home-office safety department's operating budget. This is their opportunity to put money where their mouths are in supporting construction safety. Some managements are willing to talk effective site safety but not to back it up.

Who Are the Main Players In Construction Safety?

The contributors to a successful construction-safety program fall into two groups, divided into line and staff functions.

Line Functions	Staff Functions
Owner management	Field safety engineers
Contractor management	Home-office safety groups
Contractor field supervisors	Trade unions
Construction or project manager	Government/OSHA/Mine Safety and Health Administration (MSHA)/Environmental Protection Agency
Field forepersons	Trade associations
Craft labor	Academia

The line functions are responsible for implementing and applying the safety regulations that are developed by the staff organizations. The line people are directly responsible for the execution of any results from the safety program at the job site.

The corporate staff functions are responsible for formulating, maintaining, and supporting the safety programs at the site. The corporate safety groups act as intermediaries between upper management and the field organization as well as outside safety-oriented organizations. The outside groups are the government, trade unions, trade associations, and academia.

The Owner's Contribution to Site Safety

A strong construction-safety program has to start with the owner's top management. They must issue a vigorous charter to the corporate safety department setting the goals and tone of the program. The charter must be followed with *continuing and meaningful reinforcement* every step along the way. The corporate safety group can then develop and implement an effective safety program down through the contractor's management team and into the field. At that, point the construction manager or project manager empowers the safety program down to the craft level by ensuring that the field safety group is delivering an effective program. If there is a single weak link in the safety chain, the humanitarian and cost-reduction goals for the safety program will not be met.

In accepting management's charter, the owner's corporate safety department makes itself responsible for implementing a humanitarian and cost-effective corporate safety program. Generally, this group is responsible for in-house as well as construction safety. Therefore, the construction-safety branch is typically a specialized group specifically oriented to the field construction environment. This arrangement works out well on those expansion and revamp projects that are done while the plant is in operation. Construction is an already hazardous operations environment that poses the utmost problems for execution of the safety-program plan.

The Construction Manager or Project Manager's Contribution to Site Safety

The contractor's management must at least accept the safe-operating challenge originated by the owner and infuse it into their own organization. We say, "at least accept" because many construction firms are even more proactive on safety than some owners. Accountability for safety throughout the organization is a vital part of any corporate safety program. The safety group is obligated to commit the corporate resources, support, and leadership necessary to meet the goals.

Effective training techniques for all levels of the field organization lie at the heart of any successful field safety program. The CM or PM and field safety supervisor usually analyze the proposed field-operation methods and the hazards to be encountered on that specific construction site. They will then work with the construction-safety group to formulate and implement the safety program on the site. It is important to have the field safety program carefully planned, with specific duties and responsibilities assigned to the participants to meet the proposed project-safety goals.

The CM and PM are responsible for the implementation and periodic monitoring of the safety program in the field to ensure that it is meeting the pre-established goals.

The Construction Supervisors' Contribution to Site Safety

This supervisory level is the cutting edge of the site-safety program. If these people do not make the safety program work, then all the management backing, sophisticated safety procedures, and good intentions will have been wasted. These people must be properly trained in delivering the correct safety message to the craftspeople working under their direction. They also need to continuously monitor their areas to ensure that proper safety procedures are practiced during construction.

When they are planning the execution of their work, they have to keep safety in mind to ensure that trained workers using safe equipment are available to safely perform the tasks planned. Field-labor productivity can be seriously hampered when the construction process must be stopped for safety reasons after work has begun. The field safety engineer, the superintendent, and the CM or PM have the authority to stop work if unsafe construction practices are observed.

Much information is passed on to the field via weekly toolbox talks that have been in practice for years. Each week, another tool or process is discussed and the safe practice related to that tool or process is reviewed in a short but concise manner. The field supervisor presenting the toolbox talk must have all the information at his or her fingertips in case a worker asks a question related to the talk but not included in it.

The Craft Labor's Contribution to Site Safety

We have now reached the level with the greatest exposure to danger. At this level, all workers must be just as aware of the safe way to perform their task as of the task itself. Although most craft labor is self-interested enough to be safety conscious, there will always be a few who want to take a shortcut and bypass a safety rule. It then falls to their fellow workers to straighten them out. Workers who repeatedly disregard a safety rule need to be reported to the field supervisor. There is no telling when a lax worker's actions will backfire and injure other members of the crew.

The safety program we will present further on in this chapter will include a graduated system to report safety violations. All workers must be convinced it is in their best interest to report safety violators.

The Government's Contribution to Site Safety

With the enactment of the Occupational Safety and Health (OSH) Act of 1970, the federal government put itself squarely in the middle of industrial safety programs. Also known as the Williams–Steiger Act, the OSH Act established rules and regulations governing safe working conditions in a variety of industries, including the construction industry.

The manual of rules and regulations, known as 29 CFR 1926, is divided into Part 1903, Inspections and Citation; Part 1904, Records and Reporting of Occupational Injuries and Illnesses; Part 1910, General Construction; and Part 1926, Safety and Health Regulations for Construction. Part 1926 is divided into Subparts A through X, each one dealing with a specific segment or operation in construction.

The U.S. Department of Labor's Bureau of Labor Statistics publishes a series of statistical data pertaining to industrial accidents and construction-related accidents. The most recent set of statistics spans the years 1992–2010 and should be carefully reviewed by CMs and PMs, because it provides industry trends and insight into the prevalent types of accidents and the age and demographics of workers who sustain occupational injuries and fatalities.

Table 12.1 shows fatal occupational injuries, including those in the construction industry.

Table 12.2 reveals the age and ethnicity of workers who experienced fatal injuries.

Figure 12.1 shows the four most frequent work-related fatal injuries for the years 1992–2010.

Figure 12.2 contains the number and rate of fatal occupational injuries by industry sector for the year 2010.

Developing the Company Safety Program

The structure of a company safety program just needs to include these rather simple components:

- A statement of the company's policy toward safety
- The objective and the program for preventing accidents and injuries
- The appointment, responsibilities, and duties of a safety director or safety coordinator
- The responsibilities of the field supervisors in administering the safety plan and their relationship with the safety director or safety coordinator
- The procedures for reporting job-related injuries and illnesses
- The working rules and regulations incorporated in the safety program
- A hazard-communication program as required by OSHA
- The procedures and penalties to deal with safety violations and violators

Industry (1)	Fatal Injuries		Selected Event or Exposure (2) (Percent of Total for Industry)			
	Number	Percent	Highway (3)	Homicides	Falls	Struck by Object
Total	4,547	100	21	11	14	9
Private industry	4,070	90	21	10	15	9
Goods producing	1,839	40	13	2	19	13
Natural resources and mining	768	17	14	1	5	17
Agriculture, forestry, fishing and hunting	596	13	12	1	5	18
Crop production	312	7	12	1	6	16
Animal production	151	3	10	–	7	11
Forestry and logging	70	2	17	–	–	54
Mining (4)	172	4	23	–	6	15
Mining, except oil and gas	61	1	–	–	–	13
Support activities for mining	99	2	35	–	7	16
Construction	751	17	11	1	35	8
Construction of buildings	157	3	7	2	50	9
Heavy and civil engineering construction	143	3	16	–	9	10
Specialty trade contractors	430	9	11	1	37	8
Manufacturing	320	7	16	4	13	15
Food manufacturing	53	1	21	9	21	–
Fabricated metal product manufacturing	47	1	11	–	21	26
Service providing	2,231	49	26	18	11	6
Trade, transportation, and utilities	1,141	25	36	17	8	6
Wholesale trade	185	4	32	12	8	13

Source: Bureau of Labor Statistics.

TABLE 12.1   Fatal Occupational Injuries, Including Those in the Construction Industry

Characteristic	Fatal Injuries		Selected Event or Exposure (1) (Percent of Total for Characteristic Category)			
	Number	Percent	Highway (2)	Homicides	Falls	Struck by Object
Total	4,547	100	21	11	14	9
Employee status						
Wage and salary (3)	3,548	78	24	10	14	8
self-employed (4)	999	22	11	14	14	13
Sex						
Men	4,192	92	21	10	14	9
Women	355	8	27	26	13	2
Age (5)						
Under 16 years	16	(6)	19	–	–	–
16–17 years	19	(6)	21	–	–	–
18–19 years	53	1	15	11	6	8
20–24 years	240	5	19	12	9	10
25–34 years	756	17	24	12	9	8
35–44 years	849	19	19	13	11	9
45–54 years	1,124	25	22	11	16	7
55–64 years	921	20	22	11	18	9
65 years and older	565	12	20	7	18	12
Race or ethnic origin (7)						
White	3,279	72	22	8	14	10
Black or African-American	384	8	27	26	9	5
Hispanic or Latino	682	15	15	13	18	9
American Indian or Alaska Native	31	1	23	–	16	–
Asian	136	3	15	43	8	2
Native Hawaiian or Pacific Islander	5	(6)	–	–	–	–
Multiple races	7	(6)	–	43	43	–
Other or not reported	23	1	17	13	13	–

Source: Bureau of Labor Statistics.

TABLE 12.2 The Age and Ethnicity of Fatal Injuries by selected worker characteristics. (*Source:* Bureau of Labor Statistics)

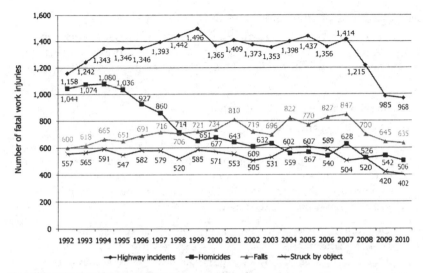

FIGURE 12.1   Four most frequent work-related fatal injuries for the years 1992–2010.

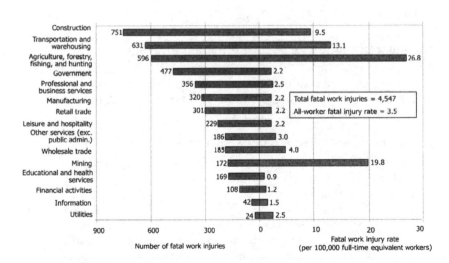

FIGURE 12.2   Number and rate of fatal occupational injuries by industry sector for the year 2010.

The Statement of the Company's Safety Policy

This can merely be a statement that outlines the reason why the company has instituted a safety policy and program and its firm commitment behind the project. This statement can be as simple as this:

Statement of the Company's Safety Policy

The Concord Construction Company's commitment to an accident-prevention program is unwavering. We see it as a task of organization and education that can and will be administered to avoid pain and suffering to all of our employees and to reduce lost time and operating costs incurred by our company. Therefore I state and pledge my full support and commitment to the following:

- Our company intends to fully comply with *all* safety laws, regulations, and ordinances.
- The safety of our employees and the public is paramount in our corporate philosophy.
- Safety shall take precedence over expediency and shortcuts.
- Every attempt will be made to reduce the potential for accident and injury occurrence.

Sincerely,

John Adams
President

The Objective of the Accident-Prevention Program

The objective of the accident-prevention program is similar to the company's statement of policy, but it goes into much greater detail and will follow this template (more definitive statements can of course only strengthen the objective):

The Objective of the Accident-Prevention Program

The Concord Construction Company recognizes that accident prevention is a problem that requires organization and education in order for it to be effective. And it needs strong leadership to guide it. For it to be successful, our policy must be administered intelligently, faithfully, and vigorously to reduce lost time, avoid pain and suffering to our employees, and, last but not least, to reduce the cost of insurance and operating costs and continue to make our company competitive. We rely primarily upon our supervisory personnel in both the field and office to furnish the sincere and constant cooperation required to administer this program.

Effectiveness of the accident-prevention program will depend upon the participation and cooperation of management, the supervision of employees, and a coordinated effort to carry out the following basic procedures:

1. Plan all work to minimize losses due to personnel injury and property damage.
2. Maintain a system that allows for the prompt detection and correction of any unsafe practices and conditions.
3. Make available, and enforce the use of, personal protective equipment and physical and mechanical equipment guards.
4. Maintain an effective system of tool and equipment inspection and maintenance, and "red tag" any defective tools or equipment.
5. Establish an educational program that instructs all participants in the basics of accident control and prevention by instituting:
   a. New-employee orientation training
   b. Periodic safety meetings
   c. Use and distribution of safety bulletins and related materials
   d. Introduction in the proper and prompt reporting of all accidents
   e. A system that institutes an immediate investigation into reported events to determine the cause of the accident and take steps to prevent any recurrence.
   f. Use of personal protective equipment that meets or exceeds minimum OSHA regulations.

Inspection of Tools and Equipment

All equipment and tools must meet or exceed minimum OSHA regulations. Equipment and tool inspection programs must also meet or exceed minimum OSHA requirements, including the maintenance of required records and other documentation.

General Safety Reference

The reference material for this safety program is contained in the U.S. Department of Labor's Occupational Safety and Health Administration (OSHA) Manual 29 CFR 1926.

The Safety Director or Safety Coordinator

The duties of the safety director or coordinator include the following:

1. Coordinate and monitor the accident-prevention program.
   a. Oversee accident investigations. The safety director is to personally investigate all serious injuries that could have resulted from a serous accident.
   b. Oversee the proper use of safety equipment.
   c. Perform frequent and unannounced jobsite safety inspections.
   d. Attend and participate in regular safety meetings.
2. Continually review job-safety reports and the preparation and dissemination of monthly summaries of safety violations, field inspections, and general program-administration items.

3. Immediately document critical conditions and the steps to be taken (and by whom) to correct these conditions.

4. Maintain communication with insurance carriers regarding accident-prevention programs and suggestions.

5. Review and take action, when required, when safety violations and violators are reported.

The Responsibilities of Field Supervisors and Their Relationship with the Safety Director

The field supervisor (superintendent) who is present daily at the jobsite is designated as the first line of defense in any accident-prevention program. These supervisors are required to obtain formal training in accident prevention, and they need to develop the means to communicate the safety program to everyone on the site—both the company's own employees and those of subcontractors and vendors.

The safety manual must include provisions for weekly toolbox talks, the content of which can be furnished by most insurance carriers. These simple and concise meetings generally focus on only one or two safety-related items each week. When conducted in small groups, they can be very effective in getting their message across.

Procedures in the Field Relating to Injury and Illness Reporting

This part of the safety program where injuries and illnesses in the field are handled must be very clear and concise about the way in which these events are reported. It should be stressed that the field superintendents are not to use their judgment as to which accidents or illnesses are to be reported; *all* accidents and illnesses are to be reported—and promptly.

Two types of accident- and illness-reporting forms are required:

1. The report required by the company's safety director

2. The report that complies with OSHA reporting requirements

The field supervisors should have both types of reports in their field office and be familiar with the way in which they are filled out and where they are to be forwarded. Figure 12.3 is a typical contractor accident/incident report. This one was obtained from the University of Cincinnati, but it is a simple thing to change references to the university to your company name.

Working Rules and Regulations of the Safety Program

The rules and regulations of the safety program are its nuts and bolts; they outline specific items of personal protective equipment designed for general and for specialized use. As an example, goggles or face shields must be used for all metal-cutting operations, and ear protection must be used when operating certain pieces of equipment. When powder-actuated tools are employed, a separate training program outlining the proper use of those tools is required.

UNIVERSITY OF **Cincinnati**

Division of Administration and Finance
Planning + Design + Construction
PO Box 210186
Cincinnati, Ohio 45221-0186

CONTRACTOR ACCIDENT / INCIDENT REPORT

Please Select Type:

☐ Employee Injury   ☐ Accident   ☐ Property Damage/Stolen Property
☐ Subcontractor Injury   ☐ Incident

INJURED PARTY/CLAIMANT:

Name: _____   SS#: _____

Address: _____   Home Phone #: _____

_____

City, State, Zip: _____   Date of Birth: _____

Employer: _____

Occupation When Injured: _____

TIME AND PLACE OF ACCIDENT / INCIDENT

Did Accident Occur on University Premises? ☐ Yes ☐ No

Accident Location (Job Name): _____   Job No.: _____

Address: _____   State/Zip: _____

Date: _____ Time: _____   Lost Time: ☐ Yes ☐ No

Name of Foreman/Supervisor: _____

Last Day Worked: _____   Return to Work: _____

Reported to Employer: _____

To Whom was Accident Reported: _____

Were University Personnel On Site When the Accident/Incident Occurred? ☐ Yes ☐ No

DESCRIPTION OF ACCIDENT / INCIDENT (completed by employee)   See Attached   Employee Description

WITNESS CONTACT INFORMATION   See Attached   Witness Contact Info

MEDICAL ATTENTION:
Was Medical Attention Provided: ☐ Yes ☐ No   When: _____

Name of Doctor/Hospital: _____   Phone No.: _____

Address of Doctor/Hospital: _____

Did this accident/incident meet the criteria for a post-accident/incident drug and alcohol test as defined by the COATS Substance Abuse Program? ☐ Yes ☐ No

If yes, were applicable drug and alcohol tests performed and submitted to the COATS database administrator? ☐ Yes ☐ No

If no, explain why: _____

SIGNING THIS REPORT DOES NOT CONSTITUTE CERTIFICATION OF AN INDUSTRIAL CLAIM (signatures)

Employee Signature   Date   University Project Administrator Signature   Date

Employee (typed or printed)   University Project Administrator (typed or printed)   Phone

NOTE: THIS REPORT MUST BE TRANSMITTED TO THE UNIVERSITY WITHIN 24 HOURS OF THE ACCIDENT/INCIDENT

cc: University Environmental Health & Safety

FIGURE 12.3 Typical contractor accident/incident report.

UNIVERSITY OF

Division of Administration and Finance
Planning + Design + Construction
PO Box 210186
Cincinnati, Ohio 45221-0186

DESCRIPTION OF ACCIDENT / INCIDENT (completed by employee)

Date

Accident Location (Job Name): 0 _____   Job No.: 0 _____

Describe in Detail what Occurred:

Exact Nature and Part of Body Affected (e.g., fracture of right hand, cut finger, etc.) (If applicable):

Property Damage (if applicable):

Have you ever had any Other Medical Treatment or Injury to Part(s) of Body Listed Above, Either Before or After this Injury? If so, Explain in Detail and Give the Name of Treating Physician.

Are you Reporting this Accident as an Industrial (work related) Injury?   ☐ Yes   ☐ No

FIGURE 12.3   (*Continued*)

UNIVERSITY OF Cincinnati

Division of Administration and Finance
Planning + Design + Construction
PO Box 210186
Cincinnati, Ohio 45221-0186

CONTRACTOR
ACCIDENT / INCIDENT REPORT

ACCIDENT / INCIDENT IN QUESTION:

_____   _____
University Project Administrator (typed or printed)   Date

Is the Injured Party a Contractor's Employee? ☐ Yes ☐ No   Subcontractor? ☐ Yes ☐ No

Please Provide a Description (In Detail) of Occurrence:

Did Anything Contribute to the Accident / Incident? (i.e., Environmental or external factors, another contractor, carelessness, lack of sleep, etc.)

Had this Contributing Factor Been Discussed in Pre-Construction or Gang Box Meetings: ☐ Yes ☐ No

CORRECTIVE ACTION INVOLVED:

Initial Response/Action to Incident (Please Describe):

Long Term Corrective Measures (Please Describe):

How was Corrective Actions Communicated to the Workers?

PERSONAL PROTECTION EQUIPMENT USED AT TIME OF INCIDENT (BY INJURED PARTY) – PLEASE CHECK ALL THAT APPLY

☐ Hard Hat   ☐ Full Body Harness and Lanyard
☐ Safety Glasses / Goggles   ☐ Hearing Protection
☐ Face Shield   ☐ Respiratory Protection
☐ Work Boots   ☐ Other: _____
☐ Gloves

SAFETY DEPARTMENT COMMENTS / FOLLOW UP:

┌───┐
│   │
│   │
│   │
│   │
└───┘

NOTE: THIS REPORT MUST BE SUBMITTED TO THE UNIVERSITY WITHIN 24 HOURS OF THE ACCIDENT/INCIDENT

CC: UC Environmental Health & Safety, File 0031A

FIGURE 12.3   *(Continued)*

Division of Administration and Finance
Planning + Design + Construction
PO Box 210186
Cincinnati, Ohio 45221-0186

**CONTRACTOR ACCIDENT / INCIDENT REPORT
WITNESS CONTACT INFORMATION**

Date

Accident Location (Job Name):   0 _____   Job No.: 0 _____

Name: _____

Address: _____

_____

City, Zip: _____

Phone #: _____

Name: _____

Address: _____

_____

City, Zip: _____

Phone #: _____

Name: _____

Address: _____

_____

City, Zip: _____

Phone #: _____

Name: _____

Address: _____

_____

City, Zip: _____

Phone #: _____

FIGURE 12.3 *(Continued)*

Division of Administration and Finance
Planning + Design + Construction
PO Box 210186
Cincinnati, Ohio 45221-0186

WITNESS STATEMENT

Name: _____   Title: _____

Social Security Number: _____   Date: _____ Time: _____

Employer: _____

Address: _____   Phone No.: _____

_____

Location at Time of Accident / Incident:

Describe to the best of your knowledge what happened before, during, and after the accident:

_____
Signature

Attach to Accident / Incident Report

cc: UC Environmental Health & Sa
      File 0031A

FIGURE 12.3   *(Continued)*

Details related to "red tagging" (taking defective equipment out of use) should be included in this section of the safety plan, and it is wise to include detailed procedures for the inspection and conditions that warrant "red tagging."

The Hazard-Communication (HazCom) Program

The hazard-communication program is another requirement that is often not given the attention it deserves, which is why it ranks among the top 25 Violations cited by OSHA.

The program was initiated when OSHA became aware of products used on construction sites, or incorporated into projects, which contain hazardous material. The improper handling, storage, and use of these hazardous products caused a significant number of injuries and job-related illnesses—hence the creation of the hazard-communication program.

Subpart H of OSHA Regulations 29 CFR deals with hazardous materials:

> This occupational safety and health standard is intended to address comprehensively the issue of evaluating the potential hazards of chemicals and communicating information concerning hazards and appropriate protective measures to employees, and to preempt any legal requirements of a state or political subdivision of a state pertaining to this subject. Evaluating the potential hazards of chemicals and communicating information concerning hazards and appropriate protective measures may include, for example, but is not limited to, provisions for developing and maintaining a written hazard communication program for the workplace.

The Material Safety Data Sheet (MSDS)

Manufacturers of materials deemed to contain hazardous substances are required, by law, to prepare a material safety data sheet (MSDS) for the product, describing its hazardous nature, the proper handling and storage of the product or material, and, in case of contact or ingestion by a worker, the necessary first-aid and medical procedures to follow.

Prior to the shipment to the site of any hazardous materials or products, the supplier of the materials or products is required to transmit an MSDS to the construction office for dissemination to the field. When the product arrives at the site, the provisions of the corresponding MSDS are to be followed as far as handling and storage are concerned. Workers using the material or product addressed on the MSDS should be given a copy of the sheet to read before it is filed away in a handy place where it can be retrieved immediately if the worker comes in contact with the material or ingests it.

It is the responsibility of the field superintendent or a designated alternate to keep these data sheets in an orderly fashion and ensure that all products covered by each applicable MSDS are stored and handled according to the provisions of the data sheet.

Figure 12.4 is a typical hazard-communication program that can be implemented by your company.

Figure 12.5 (see page 379) is a sample of a hazardous-substance inventory, which should be updated as required when hazardous substances are shipped to the field after receipt of the MSDS.

SAMPLE

HAZARD COMMUNICATION PROGRAM

Name of Agency/Institution/University Campus_____

Date Prepared _____

I.   PURPOSE

The purpose of this Written Hazard Communication program is to ensure that:

1. Hazardous substances present in the work place are properly identified and labeled.

2. Employees have access to information on the hazards of these substances.

3. Employees are provided with information on how to prevent injuries or illnesses due to exposure to these substances.

4. Identify by job title who has the responsibility for maintaining the program, the MSDS sheets, conduct training, etc.

Note: This program will be available to all employees for review and a copy will be located in the following area(s):

Location:
1.
2.
3.

II.   AUTHORITY AND REFERENCE

Occupational Safety and Health Administration (OSHA) 29 CFR 1910.1200

Dept. of Commerce (Chapter 32) (COMM) 32.15

III.   HAZARD DETERMINATION

A. A "hazardous substance" is a physical or health hazard that is listed as such in either:

1. 29 CFR Part 1910, Subpart Z, *Toxic and Hazardous Substances*, Occupational Safety and Health Administration.

revised 5/12/02   1

FIGURE 12.4   Typical hazard-communication program.

2. *Threshold Limit Values for Chemical Substances and Physical Agents in the Work Environment* (latest edition), American Conference of Governmental Industrial Hygienists (ACGIH).

B. A "hazardous substance" is regarded as a carcinogen or potential carcinogen if it is identified as such by:

   1. National Toxicology Program (NTP), *Annual Report on Carcinogens* (latest edition).

   2. International Agency for Research on Cancer (IARC) *Monographs* (latest edition).

   3. 29 CFR Part 1910, Subpart Z, *Toxic and Hazardous Substances*, Occupational Safety and Health Administration.

C. Manufacturers, importers and distributors will be relied upon to perform the appropriate hazard determination for the substances they produce or sell.

D. The following materials are not covered by the Hazard Communication Standard:

   1. Any hazardous waste as defined by the Solid Waste Disposal Act, as amended by the Resource Conservation and Recovery Act of 1976, as amended (42 USC 6901 et seq.) when subject to regulations issued under that act by the Environmental Protection Agency.

   2. Tobacco or tobacco products.

   3. Wood or wood products. **NOTE:** Wood dust is not exempt since the hazards of wood dust are not "self-evident" as are the hazards of wood or wood products.

   4. Consumer products (including pens, pencils, adhesive tape) used in the work place under typical consumer usage.

   5. Articles (i.e. plastic chairs).

   6. Foods, drugs, or cosmetics intended for personal consumption by employees while in the work place.

   7. Foods, drugs, cosmetics, or alcoholic beverages in retail stores packaged for retail sale.

   8. Any drug in solid form used for direct administration to the patient (i.e. tablets or pills).

IV.   APPLICATION

revised 5/12/02   2

FIGURE 12.4   *(Continued)*

This program applies to the use of any hazardous substance which is known to be present in the workplace in such a manner that employees may be exposed under normal conditions of use <u>or</u> in a foreseeable emergency.

V.   **RESPONSIBILITY FOR COMPLIANCE**

A. The administration of this program will be the responsibility of (person/position designated). The administrative responsibilities of this individual/position will include:

1. Identification of the employees to be included in the Hazard Communication Program.

2. Development and maintenance of a hazardous substance master inventory.

3. Coordination and supervision of employee training.

4. Coordination and supervision of the facility's container labeling program.

5. Coordination of any necessary exposure monitoring.

6. Coordination and supervision of required recordkeeping.

7. Periodic evaluation of the overall program.

B. Employees are responsible for following all safe work practices and using proper precautions required by the guidelines in this program.

VI.   **HAZARDOUS SUBSTANCE INVENTORY**

A. (person/position designated) is responsible for compiling, maintaining, and updating, when necessary, a master list of hazardous substances used or produced in the facility. The inventory list will include the common identity or trade name of the product and the name and address of the manufacturer. Hazardous substances will be listed alphabetically by manufacturer. Substances which are not in containers will also be included on the inventory list, e.g., welding fumes, carbon monoxide from a fork lift, etc. (*See* Form #1)

VII.   **LABELING**

A. (person/position designated) is responsible for evaluating labels on incoming containers. Each label must contain the following information:

1. Identity of the substance.

2. Appropriate hazard warning.

revised 5/12/02   3

FIGURE 12.4 (*Continued*)

3. Name and address of the manufacturer.

B. If the label is not appropriate, (person/position designated) will notify the manufacturer (or supplier) that the label is not adequate. (*See* Form #2)

(person/position designated) will send a second request to the manufacturer if the correct label is not received within 30 days. (*See* Form #3)

(person/position designated) is responsible for preparing an appropriate label if one is not supplied by the manufacturer within the second 30 days.

A container will not be released for use until an appropriate label is affixed to the container.

C. Labels will be removed if they are incorrect. When the container is empty it may be used for other materials provided it is properly cleaned and relabled.

D. Each department supervisor is responsible for ensuring that all containers used in his/her department are labeled properly and remain legible. Defacing labels or using them improperly is prohibited.

E. Unlabled portable containers, such as pails and buckets, should be used by one employee and emptied at the end of each shift. If the secondary containers are used by more than one employee and/or its contents are not emptied at the end of the shift, the department supervisor is responsible for labeling the container with either a copy of the original label or with a generic label which has a space available for appropriate hazard warnings.

*F. Piping systems shall be painted at access points and every 10 feet where the piping is 8 feet or closer to employee contact.

   1. Piping shall be painted as follows:

      a. (substance)(color)
      b. (e.g., oxygen) (e.g., green)

VIII. **MATERIAL SAFETY DATA SHEETS**

A. MSDS's will be available to the employees on all hazardous substances to which there is potential or actual exposure. (person/position designated) is responsible for ensuring that MSDS are available on all incoming products. A product will not be released for use until a completed MSDS is on file. (*See* Form #4)

If the MSDS is not available, (person/position designated) will notify the manufacturer that MSDS is needed. (*See* Form #5).

revised 5/12/02   4

FIGURE 12.4   *(Continued)*

(person/position designated) will send a second request to the manufacturer if the MSDS is not received within 30 days. (*See* Form #6)

B. (person/position designated) is responsible for the review of all incoming MSDS's. If the MSDS is not complete, it will be returned to the manufacturer with a request for the missing **information.** (*See* Form #7)

(person/position designated) will send a second request for the missing information if a complete MSDS is not received within 30 days. (*See* Form #8)

*C. (person/position designated) will request an MSDS on the purchase orders of all new products. (*See* Form #9)

D. (person/position designated) is responsible for compiling and updating the master MSDS file. This file will be kept at (Name of location).

Copies of MSDS's will be kept in the following areas:

Department   **Location**

E. Employees will have access to these MSDS's during all work shifts. Copies will be made available upon request to (person/position designated). (*See* Form #10)

F. (person/position designated) is responsible for updating the data sheets to include new information as it is received. A notice will be posted to inform employees that revised information has been received. (*See* Form #11)

IX. EMPLOYEE TRAINING

A. Prior to starting work with hazardous substances, each employee will attend a Hazard Communication Training Session where they will receive information on the following topics:

1. Policies and procedures related to the Hazard Communication Standard.

2. Location of the written Hazard Communication Program.

3. How to read and interpret an MSDS.

4. Location of MSDS's.

5. Physical and health hazards of hazardous substances in their work area.

6. Methods and observation techniques to determine the presence or release of hazardous chemicals.

revised 5/12/02   5

FIGURE 12.4   (*Continued*)

7. Work practices that may result in exposure.

8. How to prevent or reduce exposure to hazardous substances.

9. Personal protective equipment.

10. Procedures to follow if exposure occurs.

11. Emergency response procedures for hazardous chemical spills.

B. Upon completion of the training program, each employee will sign a form documenting that he/she has received the training. (*See* Form 12)

C. Whenever a <u>new</u> employee is transferred or hired, he/she will be provided training regarding the Hazard Communication Standard. The training session will be conducted by _____ before the start of his/her employment if possible.

D. (person/position designated) is responsible for identifying and listing any non-routine hazardous task performed at this facility. (person/position designated) will conduct training on the specific hazards of the job and the appropriate personal protective equipment and safety precautions and procedures. (*See* Form 13)

E. When a new substance is added to the inventory list, (person/position designated) is responsible for reviewing the MSDS for potential health effects. If the product presents a new health hazard (causes health effects unlike those covered in the training session), the ((person/position designated)) is responsible for notifying all affected employees about the new health effects which result from exposure to the new substance.

*A copy of the new Material Safety Data Sheet (MSDS) will be posted by (person/position designated) for 30 days. Both the new Material Safety Data Sheet and the Employees New Substance Signature Form will be placed above or near the MSDS information binder. Each affected employee must read the MSDS and sign the signature form. (*See* Form #11)

X. **INFORMATION TO CONTRACTORS**

A. (person/position designated) is responsible for providing outside contractors with the following information:

1. Hazardous chemicals to which they may be exposed as a result of working in this facility.

2. Suggestions for appropriate protective measures.

FIGURE 12.4   (*Continued*)

B. Contractors that are potentially exposed to hazardous chemicals present at the facility will <u>not</u> be allowed to begin work until they have been provided information concerning these hazards and have signed a form to document this exchange. (*See* Form #14)

C. (person/position designated) is responsible for obtaining information from contractors on all hazardous substances to which State employees may be exposed as a result of the contractor's work at the facility. (*See* Form #15). (person/position designated) will notify affected employees about the health affects that may result from exposure to each substance.

XI. PERSONNEL POLICIES

When an employee is not following safety and health rules regarding working with a hazardous substance, disciplinary action will be taken.

XII. RECORD KEEPING

A. All MSDS's will be kept for a period of _____ years after the use of the substance has been discontinued. **EXCEPTION:** If an employee exposure to a particular hazardous chemical occurs, the MSDS for that product will become part of the employee's medical records.
Medical records must be kept for **30** years.

Note: "Exposure" or "exposed" means that an employee is subjected to a toxic substance or harmful physical agent in the course of employment through any route of entry (inhalation, ingestion, skin contact or absorbtion, etc.), and includes past exposure, but does not include situations where the employer can demonstrate that the toxic substance or harmful physical agent is not used, handled, stored, generated, or present in the workplace in any manner different from typical non-occupational situations.

*B. The master inventory list will also be kept for _____ years.

XIII. COMMUNITY HAZARD COMMUNICATION

(person/position designated) is responsible for responding to requests from members of the community on hazardous substances used in the facility.

* XIV. EMERGENCY RESPONSE PROCEDURES FOR HAZARDOUS CHEMICAL SPILLS

A. When a hazardous chemical spill occurs, follow these procedures:

1. Move all employees away from spill to a safe environment.

FIGURE 12.4 (*Continued*)

2. Call 911 <u>or</u> the designated emergency response number in your area to notify the necessary response team for the hazardous chemical spill.

3. Retrieve the Hazard Communication Information Binder, if possible.

    a. Locate the MSDS for the hazardous chemical which spilled.

    b. If requested, provide the MSDS to the Emergency Response Team.

> **Note:** Do <u>not</u> try to contain the spill. The Emergency or Hazardous Material Response Team is trained to deal with hazardous chemical spills.

XV.   PROGRAM EVALUATION

<u>(person/position designated)</u> will conduct an evaluation of the Hazard Communication program annually. The individual responsible for the items identified for improvement will be notified in writing. It is expected that action will be taken to correct the item within five working days. (*See* Form #16)

*   At least annually, ___(indicate number) employees will be interviewed to determine the effectiveness of the Hazard Communication Program. Each interview will access the employee's retention of information given during the training session, use of MSDS's and response to chemical spills (if applicable). The results of each interview will be recorded on the Employee Interview Form. (*See* Form #17) The Employee Interview Form will be retained on file for 12 months.

This written program has been developed by the Bureau of State Risk Management, Department of Administration and is available on computer disk. (File name *a:\hazcom.doc*). It may be adapted to fit the particular needs of your facility. The program was adapted from a written program originally developed by the Occupational Safety and Health Administration (OSHA).

Note: When there is an asterisk (*) placed in front of a guideline, then this policy is not required by the Hazard Communication Standard or the Employees Right-To-Know Law.

FIGURE 12.4   *(Continued)*

Dealing with Safety Violators

One of the safety director's more important responsibilities and challenges is monitoring safety violations and dealing with violators. The safety director must drive home the fact to the project managers and project superintendents that *all* job-related injuries and illnesses are to be reported promptly. If workers understand that the company has a firm commitment to safety but will not report violations and deal with violators quickly and fairly, the program will become a toothless tiger.

Both project manager and project superintendent must also be aware that in this litigious society in which we live and work, violations that lead to dismissal can result in a lawsuit unless the facts and the process have been followed to a T.

The safety program should be reviewed from time to time with long-term employees to refresh their memory, and it should become

Form #1

HAZARDOUS SUBSTANCE INVENTORY

Organization : _____   Location: _____

Manufacturer	Product Name	Quantity	MSDS Yes/No	Work Area

Completed By: _____   Date: _____

a:\hcforms.doc

revised 5/12/02

10

FIGURE 12.5 Sample hazardous-substance inventory.

an integral part of any new-employee orientation program. After an employee, whether new or long-term, has reviewed the safety program, he or she should sign and date a copy of it so that it can be placed in his or her employee file. This will prevent any complaints from a worker who has committed a serious safety violation but says, "Gee, I didn't know that was a safety violation."

Drugs and Controlled Substances

The rules regarding the use of unauthorized drugs or controlled substances must be approached carefully, with concern for wrongful accusation. There have been stories of an employee being fired on the spot because he or she was accused of being "spaced out" on drugs. Unfortunately, the worker was merely having a reaction to a drug prescribed by a doctor, and the company that fired that worker may end up with a huge lawsuit. When a worker is suspected of using a controlled substance, proceed cautiously; bring the safety director to the site to interview the worker before taking any disciplinary action.

Do Not Put Off Dealing with a Safety Violation

Nothing is more important than acting instead of putting off reporting a violation or immediately stopping a dangerous operation. We have all been guilty of walking briskly through a project under construction on our way to a meeting with the owner or architect—10 minutes late. But along the way we spot a worker using an electrical drill that is sparking and is attached to an extension cord that has seen much better days, with some wires showing without the outer insulation. The potential for a fire awaits. But because we are late, we make note and remind ourselves to come back after the meeting and "red tag" both drill and extension cord. We should instead stop, call the super, and tell him or her to do just that. We can explain away being late, but we cannot explain away a fire on the fourth floor caused by the defective drill that we did not take out of service.

The following procedures should be instituted and distributed to all employees so they fully understand the actions the company will take when safety violations are observed *and reported*.

First Warning

The first time an employee is observed violating any safety rule, the employee shall be given a first warning. The first warning will be an oral one and will be so noted in the employee's personnel file.

Second Warning

The second time the employee is observed violating any safety rule, the employee shall be given a second warning. The second warning will be an oral warning accompanied by a written safety-violation notice. A copy of the written notice will be given to the employee, the employee's union steward (if applicable), and the company's safety director. A copy of the notice will also be placed in the employee's personnel file. The employee will be required to meet with the safety director for counseling.

Third Warning

The third time an employee is observed violating any safety rule, the employee shall be given a third warning. The third warning will be a written safety-violation notice, a copy of which will be given to the employee, the employee's union steward (if applicable), and the safety director. A copy of the notice shall also be placed in the employee's personnel file. A meeting with the employee, the employee's immediate supervisor, the safety director, and a representative of top management will be held to determine why the employee failed to comply with the company's safety program—*for the third time*. Top management must determine what action will be taken at this time. One option open to management, which is included in the disciplinary section of the safety program, is suspension from work. For example, the policy might state, "Employees who accumulate three warnings in a 12-month period may be suspended from work without pay for up to one week."

Fourth Warning

The fourth time an employee is observed violating any safety rule, the employee shall be given a fourth warning. A fourth warning will be a written safety-violation notice. Employees who do not follow safety rules, especially after being warned several times previously, are a threat to themselves and their coworkers. Therefore, employees who receive a fourth warning may, at management's discretion, be terminated from employment or be subject to other disciplinary action deemed appropriate by management.

Consistency and fairness are the hallmarks of an effective safety program. It is essential that the administration of the safety program be consistent and fair. Employees will not take the disciplinary action seriously if they know that safety violations have occurred but management took no notice of them nor issued a violation notice.

A Reward Program

Some companies augment their safety program with incentives to maintain safe working conditions. A short-term incentive can be as simple as presenting a small gift to an employee who exhibits some on-the-spot safety measure. Safety stickers worn on the hardhat, one for every safety measure taken, are a very visible record of an employee's accomplishments.

Long-term incentives for continued safety consciousness can be in the form of a gift certificate at a local department store, a cash bonus, extended paid vacation time, or a weekend for two at a nearby hotel.

Remember, a solid safety program, administered fairly and vigorously, can reduce corporate overhead thereby, making the company more competitive. It also creates a positive public image and makes good business sense. Total construction project management is committed to these goals.

Figure 12.6 is a sample construction-safety checklist. If distributed to the field with a request to use it on a biweekly project walk-through and then sign and submit to the construction or project manager, such a checklist will alert the field supervisor to any unsafe conditions and

CONSTRUCTION SAFETY CHECKLIST

Date inspection conducted: _____ Location:_____

Name(s) of those participating in this inspection: _____

INDICATE EITHER: A = Acceptable/Yes;   U = Unacceptable/No;   N/A = Not Applicable

PERSONAL PROTECTIVE EQUIPMENT		EMERGENCY ITEMS	
Safety glasses and/or goggles available + being used?		Emergency evacuation map posted near work area?	
Protective eyewear use is specified in writing?		Emergency phone numbers posted and known by all?	
Face shield available for bulk liquid tasks? Grinding?		Emergency eyewash and/or shower units accessible?	
Hand protection used/worn as required?		First aid kit available at work site?	
Foot protection worn as required?		First aid trained competent person available?	
Hearing protection worn where required?		BBP kit available/BBP trained individual on site?	
Hard hats worn when falling object hazard is present?		Fire extinguishers readily available (not blocked)?	
Supplies on hand for incidental chemical spills?		Fire extinguishers inspected monthly/yearly as needed?	

ELECTRICAL SAFTEY ISSUES		ELECTRICAL SAFETY ISSUES	
GFCI's used for all portable electrical hand tools?		Strain relief integrity for cords and plugs in tact?	
Extension cords rated for hard or extra hard usage? (2 wire ribbon cord is unacceptable for industrial usage)		For extension cords; hard usage cord includes three wire cord marked = S, ST, SO, STO, SJ, SJO, SJT + SJTO	
Certified or listed equipment is used per manufacturer?		Electrical cords inspected & have all prongs in tact?	
Electrical panels are labeled appropriately?		Strain relief in tact for all flexible cords & plug fittings?	
Electrical panel knockouts are in place?		Portable generators are grounded per NEC requirements?	
Light bulbs for illumination protected from breakage?		Electric power tools are double insulated or grounded?	

CONSTRUCTION SAFETY & HEALTH ISSUES		CONSTRUCTION SAFETY & HEALTH ISSUES	
General housekeeping is neat and orderly?		Flammable liquids are in OSHA/FM metal safety cans?	
MSDS openly available to all employees?		Flammable liquids storage containers labeled properly?	
Concrete work? Silica dust training documented for all?		Fire extinguisher readily available, accessible + inspected?	
All hazardous containers labeled appropriately?		Impact style air tools have safety clips or retainers on them?	
Lockout/Tagout is being used for appropriate tasks?		Pneumatic power tools have hoses secured?	
Hot work permits used for grinding, cutting, welding?		Compressed air used for cleaning limited to 30 psi ?	
Confined space entry work? Check training/permit/etc.		Compressed gas cylinders not in use have caps in place?	

Copy the completed inspection sheet to: _____

If marked"U" for unacceptable or no; list the appropriate corrective action on the reverse side

FIGURE 12.6   Construction safety checklist.

CONSTRUCTION SAFETY & HEALTH ISSUES		CONSTRUCTION SAFETY & HEALTH ISSUES	
Compressed gas cylinders stored secured + upright?		Wall openings + floor holes are covered or guarded?	
Oxygen/acetylene torch units have flash back arrestors?		100% fall protection in place above 6 feet in height?	
Grinders (portable + stationary) have guards in place?		Employees operating lifts are trained on the equipment?	
Stationary grinding wheel tool rest is 1/8 inch or less?		Fall protect. full body harness+ lanyard used at all times?	
Stationary grinding wheel tongue guard is ¼ inch or less?		Excavation? Ladders used > 4 feet deep? Extend 3 feet?	
Grinders are inspected, ring tested + free of defects?		Excavation? Protection from cave-ins for >5 feet deep?	
Safety glasses + face shield used for grinding tasks?		Rebar caps used for protruding reinforced steel posts?	

CONSTRUCTION SAFETY & HEALTH ISSUES		CONSTRUCTION SAFETY & HEALTH ISSUES	
Ladders are safe and inspected as appropriate?		Portable circular saws equipped with protective guards?	
Stair rails = req'd at 30" change in elevation or 4 risers?		Unsafe hand tools are prohibited?	
Stairs or ladder provided for access points > 19" high?		Impact tools, hammers kept free of splinters/mushrooms?	
Extension + straight ladders extend 3' beyond landing?		Wire rope used for lifting? Deterioration is absent?	
Stepladder or commercial stepstool used for high access?		Web slings used for lifting? Deterioration is absent?	
Step ladders are only used in open position?		Crane use? Written lift plan on file listing load capacities?	
Scaffolding = guardrails used? competent person on site?		Hooks used for lifting have safety latch in place?	

CORRECTIVE ACTION PLAN

LIST ITEM, THE PERSON RESPONSIBLE AND EXPECTED COMPLETION DATE !			
ACTION ITEM	PERSON (S) RESPONSIBLE	TO BE DONE BY	STATUS

Status column should be marked – either listed as "open", "in process, or "closed"

Signature of lead inspector: _____

FIGURE 12.6   (Continued)

also act as a strong signal to all workers that jobsite safety is being given the attention it deserves.

Summary

Attention to jobsite safety has many facets: humanitarianism, economics, productivity, and corporate image (an important factor for the sales-development team). Top management must, as a first step, make a strong statement committing the company to the highest safety standards and practices. This is to be followed by a well-thought-out safety program setting forth the company's commitment and the process the company will take to ensure safe practices on the construction site. New hires must become totally familiar with the program as part of their initial hiring program, and experienced workers should attend a reviewing session periodically to remind them of their responsibility to adhere to the program. Diligent and consistent follow-through of the required training, inspections, and enforcement are at the heart of an effective safety program.

CHAPTER 13

Project Communications

Successful construction-project execution is virtually impossible unless you have an effective communications system. Good communication skills are basic to becoming a successful project manager. Effective communication is the lubricant that keeps the machinery of a successful construction project running smoothly.

With today's cell phones, smartphones, and tablet computers, and the ability they provide to communicate verbally or in text or photographs, there should be no excuse for not maintaining communications control throughout the entire project. These electronic devices link subcontractor to contractor to architect to engineer to owner and right back down that line.

Why Do Communications Fail?

There are a myriad of reasons why communications fail in all parts of life. A few of the more common reasons business communications fail are these:

- Not having a clear goal in mind.
- Staying in a negative mode.
- Concentrating on your own thoughts to the exclusion of the other person's ideas.
- Not establishing rapport.
- Assuming that others have the same information on the subject that you have.
- Assuming that others are acting on that information without any prompting from their supervisor.
- Being impatient—not hearing the other party out. Mastering the art of listening instead of thinking of a rebuttal as the other person is talking is a trait that is very important and needs to be worked on.

- Mistaking interpretations for facts.
- Failing to analyze and handle resistance.
- Being unable to admit you do not understand.
- Having an overabundance of ego.
- Thinking that when you make a mistake, somehow it will be corrected, so there is no need to get others alarmed.

Communications is defined as transmitting, giving, or exchanging information, etc., as by talk or writing or electronically. By comparing the common reasons for failure with the definition of communication, we can easily analyze the failures. We are certain that each of us can go down the list and pick out three or four failures that apply to us. These are the ones we need to improve upon to become successful communicators.

One of our techniques is to mentally transpose ourselves into the place of the intended receivers of our communication. Then we read the message while using that person's knowledge of the situation. If we can still understand what we said, the communication might just work. If we cannot understand it, we have to rework the message. If you practice this technique, you can use it even in oral or electronic communications as well as written ones.

If we do not establish rapport in communications, anything we say is likely to fall on fallow ground. The easiest way to affect rapport in your project communications is to build your creditability as a professional manager. People will receive your messages more readily if they respect your opinion. How many times have you heard a colleague say, "Oh no, not another one of those 'critical messages' from Joe Blank that doesn't make any sense—but I guess I'll have to respond." We have wasted everyone's time by sending messages that are redundant, not carefully thought through, or just totally unnecessary. Keep an open mind to build full rapport with your fellow professionals so that when a message is received, they know it is proper, coherent, and in need of a response.

A number-one communications problem is being patient while receiving communications. After many years of trying to improve, some of us still have not perfected the art. If you have that problem, don't give up easily. It can be corrected.

Handling resistance in the mind of the message receiver also requires patience. If putting your idea across is important to you, try to analyze the resistance and anticipate it. You must remember this factor when selling your ideas upward or downward in your work setting.

The last two items on the problems list often go together: an abundance of ego and thinking that if you have made a mistake, it will correct itself, so there is no need to alarm anyone. This usually ends up taking what was initially a small problem and inflating it to a much larger problem. None of us will ever learn all there is to know about managing construction—it is a complex process that varies slightly or significantly from project to project. But showing that we are confident enough to admit that we do not understand something is not a sign of weakness but a sign of strength. Another way to handle a problem is to play the message back to the sender in the way you understood it and ask if this is a correct understanding. That avoids the open admission of not having understood the message.

These are a few basic points for improving your communication skills. Do take advantage of any communication-training courses that your company may offer or that you could attend with your supervisor's permission. Above all, remember to practice, practice, practice.

Communications Systems

Construction managers and project managers are the prime spokespeople for their respective organizations. That duty places them in a position of either creating or monitoring most project communications. To set a good example, CMs and PMs need to make sure that all project communications are handled professionally. They should review all project correspondence generated by their supervisors and project superintendents from time to time, to ensure that

- correct spelling and grammar are being used.
- the facts or questions contained in the communication are concise and understandable.
- the date by which a response is required is included.
- the communication is civil and not accusatory.

Typically, project communications fall into the following key categories:

- Project correspondence
- Audiovisual Presentations
- VIP site visits and resulting reporting
- Project reporting
- Meetings and resulting meeting minutes

As we discuss each of these areas, we will find that the list covers almost every form of project communication one will encounter.

Project Correspondence

The primary project-communications link is that between the client and the construction manager or project manager—that is, unless the client is unhappy with the CM or PM's management of the project, in which case the client will communicate to a higher level. Let's hope that does not happen, but we should recognize that occasionally it does.

One of writers had such an experience, where the owner contacted his boss and said the writer was not doing a good job. My boss, the owner of the company, was concerned and contacted the owner's design team, who related that I was doing a great job. It was a case of oil and water, and my boss told the client he was perfectly happy with my performance and I was remaining on the job. Things got a little dicey at the time, but knowing the owner did not like me helped me to communicate with him in a different (deferent) manner.

CMs and PMs often delegate some specialized correspondence to other key staff members to speed up the communication process. Some examples are procurement, personnel, field engineering, subcontract administration, and issuance of requests for information or clarification. Matters having to do with schedule—particularly if there is a problem, e.g., falling behind—interpretation of the scope of the contract, and matters relating to costs—including the consideration of sending the owner a cost proposal for required work that was inadvertently excluded from the project scope—should fall initially to the CM or PM, who may ask for assistance from a staff member. All correspondence relating to these matters should go out under the CM or PM's signature; if prepared by a staff member, it should be reviewed by the CM or PM.

Project correspondence serves an important function in addition to transmitting messages, instructions, queries, etc. These messages should all find their way into the project's files, whether they be paper or electronic. If and when problems arise later in the job, it often becomes necessary to revisit many documents to build a case to solve the problem or defend against an unwarranted claim. Each document should indicate its recipient, so that if future problems do occur it will be easy to determine whether all interested parties received a copy of the document.

There are numerous ways to produce a document that can be retrieved easily—for instance, by assigning a letter and number to the document based upon a filing system already in place in the company or creating one if one does not already exist. With electronic storage, a series of files can be created so that all documentation pertaining to a file can be retrieved by clicking on the file. There must always be a backup system in case of a computer crash; there are numerous methods, either in-house or via an outside company, for electronic storage.

When these files get too voluminous, they can be transferred to CDs and an index can be created so that one knows where to look. And of course, an index of the CDs can be established, making it easier to select the right one.

Audiovisual Presentations

The old slide show has gone the way of the horse and buggy; Power-Point is one of the most popular audiovisual options of choice nowadays. Compact projectors make the process much easier. But there are also flip charts, overhead-projection transparencies, posters, and 35-mm slide presentations.

Guidelines for PowerPoint presentations include the text and background of the material being presented and the size and types of fonts that are more easily recognized by the audience. Here are some tips on preparing your PowerPoint presentation:

- Know your material. If your presentation flows naturally, audiences will know that you were well prepared.

- Do not try to memorize your material. Using notes is perfectly acceptable, and since your audience is concentrating on the screen, referring to your notes will not detract from the presentation.

- Rehearse your presentation with the slide show. Get some of your people to attend and have them comment on the presentation.

- Pace yourself; avoid the tendency to rush through from slide to slide.

- Assemble the presentation to fit the audience you are anticipating. Look at the presentation from their standpoint. Is it appropriate for that audience?

- Be familiar with the equipment you are using so that if there is a glitch, you can fix it quickly. Use a remote control, not a laser pointer.

Experts say that headings ought to be 32 points or larger, subheadings 30 points or larger, and text 28 points or larger. Bold rather than standard fonts are better.

Text and background should be of high contrast—a dark background should have light-colored text and vice versa. Avoid using combinations of red and green, red and black, dark green and black, or blue and black. Avoid italics and underscoring.

The United States Department of Labor's Occupational Safety and Health Administration (OSHA) has developed procedures for various types of audiovisual presentation. The following subsections cover some of OSHA's suggestions.

Establish the Objectives

Establishing the objectives of your presentation requires you to analyze your own goals, along with the audience's needs and expectations. By considering the nature of your audience, you can more easily determine what you will present and how you will present it. An audience analysis will allow you to

- select the appropriate points of emphasis in your presentation.
- develop a useful level of detail.
- choose and prepare appropriate visual aids.
- create a tone that is sensitive to your audience's circumstances.

Plan and Organize Your Material

After you have determined the characteristics of your audience, you are ready to plan and organize your material. Keep in mind that the use of visual aids will help to produce an effective one-way or two-way communication; as you consider which type of presentation to use, take into account the type of interaction you want to develop with the audience.

Plan Your Material

Be prepared. Do not develop your presentation on the way to the meeting. At a minimum, prepare an outline of your goals, the major issues to be discussed, and the information to be provided to support these main themes. Limit content to five major points and no more than five key supporting points. Analyze your audience and prepare the content with consideration of such things as whether the audience is likely to be friendly or unfriendly, to contain laypeople or people who are technical in their backgrounds, and to want to only listen or instead respond and contribute to your presentation. Select the visual aids most appropriate to the physical setting where you will be giving the presentation.

Organize Your Material

OSHA says to consider an "old chestnut" of public speaking: Tell them what you are going to tell them: tell them; and tell them what you told them. This recommendation:

- recognizes the importance of reinforcement in adult learning.
- completes the communication for the listener.
- informs people who arrive late of what they have missed.
- recognizes the importance of organization, highlighting, and summarizing main points for the audience.
- serves to clarify main themes for the audience at the end of the presentation.

Use Visual Aids

Visual aids help you reach the objective you wish to leave with your audience. Visual aids require a change from one activity to another, from hearing to seeing.

Visual aids add impact and interest to a presentation. They enable you to appeal to more than one sense at the same time, thereby increasing the audience's understanding and retention level. People tend to be "eye minded," and OSHA has found that the retention of information varies considerably according to the mode of communication:

- Oral alone—10 percent
- Visual alone—35 percent
- Visual and oral combine—65 percent

OSHA's studies reveal interesting statistics that support these findings:

- Experimental psychologists and educators have found that retention of information 3 days after a meeting or other event is 6 times greater when information is presented by visual and oral means than when information is presented by the spoken word only.
- Studies by educational researchers have suggested that approximately 83 percent of human learning occurs visually, with the remaining 17 percent occurring through other senses—11 percent through hearing, 3.5 percent through smell, 1.5 percent through touch, and 1 through taste. The use of visual aids is therefore essential to all presentations.

The tips listed here, from OSHA, will help the construction manager or project manager in the selection and preparation of visual aids:

- Start with at least a rough outline of the goal and major points of the presentation before selecting the visual aids. For example, a particular scene or slide may trigger ideas for a presentation, providing the power of images. Do not proceed too far without first determining what you want to accomplish, what your audience wants to gain, and what the physical setting requires.
- Make sure that each element of an audiovisual product—a single slide or a page of a flip-chart presentation, for example—is simple and contains only one message. Placing more than one message on a single element confuses the audience and diminishes the potential impact of the visual medium. Keep visual aids brief.
- Determine the difference between what you will say and what the visual aid will show. Do not read straight from your visuals.
- Ask the audience to read or listen, not both. Visual aids should not provide reading material while you talk. Rather, use them to illustrate or highlight your points.

- Give participants paper copies of various graphic aids used in your presentation. They will be able to write on the paper copies and keep them for future reference.

- Assess your cost constraints. An overhead-transparency presentation can always be used in a formal environment if 35-mm slides are too expensive.

- Account for production time in your planning and selection process. Slides must be developed and video edited. You do not want to back yourself against a wall because the visuals are not ready. You can get production work done in 24 to 48 hours, but it is more expensive than work that is done on an extended schedule.

- Use local photographs and examples when discussing general problems and issues. While a general problem concerning safety or constructability, for example, may elude someone, illustrating with a system in use can bring the issue home.

- Use charts and drawings. They are better for conveying various designs and plans.

- When preparing graphics, make sure they are not too crowded with detail. Do not overuse color. See that line details, letters, and symbols are bold enough to be seen from the back of the room.

- Do not use visual aids for persuasive statements. Qualifying remarks, emotional appeals, and any type of rhetorical statements should be avoided.

- If you have handouts, do not let them become a distraction. They should provide reinforcement after your address—and that is when they should be distributed, unless the audience will use them during the presentation or will need to review them in advance of the presentation.

- Practice presenting the full program. Use the graphical material so you are familiar with its use and order. If you use audiovisual materials, practice working with them and the equipment to get the timing right.

- Seek feedback on the clarity of your visuals. Do this early enough so that if adjustments need to be made, you will have time to do so.

Flip Charts

These are relatively quick and inexpensive visual aids for briefing small groups. The charts, felt-tip markers, and graphical materials are readily available, and with a modest ability at lettering, a presenter can compose the desired visual aid in-house.

Flip charts

- help the presenter proceed through the material.
- convey information.
- provide the audience with something to look at in addition to the speaker.
- can be prepared prior to, as well as during, the presentation.
- demonstrate that the speaker has given thought to his or her remarks.
- can be used to record audience questions and comments.
- can be converted to slides, if required.

But flip charts do have limitations; they

- may require the use of graphic talent.
- are not suitable for a large-audience presentation.
- may be difficult to transport.

There are some useful guidelines to follow in developing flip charts:

- Each sheet should contain one idea, sketch, or theme.
- Words, charts, diagrams, and other symbols must be penned at a large enough size to be seen by the people farthest from the speaker.
- Block letters should be used, since they are easiest to read. Letter should be all capitals, with no slanted or italicized letters.
- Color should be used and varied. A check from a distance is useful to make sure the color works well and is not distracting.

Overhead Transparencies

This type of presentation is useful for audience settings of 20 to 50 people and can be produced quickly, easily, and inexpensively. Any camera-ready artwork, whether word charts, illustrations, or diagrams, can be made into transparencies using standard office paper copiers.

Following are some things to keep in mind when preparing transparencies:

- Most manufacturers of paper copiers offer clear and colored acetate sheets that run through the copying machine like paper but transfer a black image onto the acetate for use as overhead transparencies. (Or perhaps your office has a color copier.)

- The standard transparency size is 8 × 11 inches. The only piece of hardware required is an overhead-transparency projector.
- Overlay transparencies provide a good cumulative presentation.
- The speaker can use an overhead projector with significant light in the room, thereby enabling him or her to maintain eye contact with the audience.

Overhead transparencies also have limitations:

- The projected image size is sometimes too small to be seen from the back of the room.
- Often, the image does not sit square on the screen as the head of the projector is tilted to increase the size of the image.
- It is difficult to write on the transparency while it is on the projector.
- Sometimes the projector head gets in the way of the audience.
- Some speakers feel captive to the machine because they must change each transparency by hand.

Posters

Posters are prepared graphical devices that can be made of a variety of materials and media—photographs, diagrams, graphs, word messages, or any combination thereof. Posters are permanent and portable, can be simple or elaborate, and can be used alone or in a series to tell a story.

There are some limitations on the use of posters:

- They tend to contain too much detail.
- Transporting them can be difficult.
- More elaborate ones require extensive preparation and be quite costly.

When preparing posters, remember these general guidelines:

- Each poster should contain one message or theme.
- Words, charts, diagrams, and other symbols must be penned at a large enough size to be seen by everyone in the room.
- Letters should be all capitals and not slanted or italicized.
- Color should be used and varied, with a check from a distance to make sure the color works.

35-Millimeter Slides

This method has largely been replaced by the use of a digital camera to transfer images, charts, diagrams, etc., onto a computer screen and

PowerPoint presentations as previously noted, but slides are still used in some situations. The same rules that apply to other visual aids would apply to slides: Each picture, diagram, or graph should be easy to read and should deal with a specific topic.

Handling VIP Site Visits

We may all classify VIPs in different ways: In an international construction company, the top management would certainly qualify; high government officials or client owners would also certainly qualify as VIPs, since they are valued customers.

Usually there is a lot of interest on the visitor's part in touring the site and observing the construction activity in progress. This makes housekeeping a prime activity; the site should look as neat as possible, with any trash collected and placed in the dumpster. Safety issues also need to be addressed on the site as well as in the building. Open trenches should have guard rails or other means to protect sightseers. Go through the building and have each subcontractor clean his or her area and dispose of debris. There is generally a clause in the subcontract agreement that housekeeping and debris disposal are to be performed on a daily basis, but sometimes the project superintendent does not strictly enforce this provision. At this time, he or she must.

Sweeping the floors and creating passages through the construction site are both activities that should have been ongoing, but prior to the VIP visit, the CM or PM must tour the building from top to bottom to ensure that these two activities have been accounted for.

Attention to safety is critical at all times, so that a tour of the building before the VIPs arrive should be a no-brainer, but there again, we need to make sure that safety goggles, ear protection, and other personal protective equipment is all in place.

Review the Status of Work and Planned Activities

Be prepared to answer questions about the design, particularly ones posed by the client or a government official, and know the status of each component of construction:

- The site and site utilities
- The structure if it is not complete or the building envelope if the structure is complete
- The type of HVAC system that is to be installed and the status of its installation
- Unique features of the building and the advantages they create
- Planned activities for the coming two weeks

Project Reporting

Another powerful project-communications tool is project reporting. Creating project reports forces the project staff to make a thorough review of their project activities at least once a month. Setting a high standard for quality project reports is an important role for the construction manager or project manager.

Producing concise and well-presented monthly progress reports is the best way for the CM or PM to set the standard for all other reports required of the project. It is also a good time to review the quality of the other project-control reports and to check on how the project is going. The monthly progress report is a summary of how well the construction team is moving toward its project goals.

The main purpose of the progress report is to inform the key interest groups how the work is progressing, including any real or potential problems and the methods by which those problems are to be resolved. An important secondary purpose is to keep a running history of vital project activities. The key groups interested in the progress report are:

- The client
- The design architect and engineers
- The contractor's top management
- Key staff people on the project.

The operative aspects of this discussion are to inform key people of important project activities, progress, and adherence to schedule and budget. A rambling, poorly written progress report will not inform the reader. The progress report must also be factual and results oriented, and it must report any problem areas to present a true picture of project status. It should be written in a positive, direct style to instill confidence in the reader that the construction team is in control of the project. The writing style must be forceful and direct. Stay away from the passive voice and weak verbs. Keep your sentences short and use a new paragraph when the subject changes.

The report format should be consistent from month to month so readers do not waste time hunting for the key indicators each month. Most well-managed companies today have developed standard report formats to suit their type of project. Using a standard format saves the time of having to develop a new format for each project. Make a template that can be called up on the computer to assure a standard format, but remember that even with a standard format, some latitude is possible to meet particular client requirements in the report.

Figure 13.1 shows a typical progress-report format that we have developed over the years. The sample format is intentionally general, but you can readily adapt it to your type of project. Compare it with

0.0 TITLE PAGE OR COVER SHEET

Project name, number, and location
Client and contractor names and logos
Report number, issue date, and period covered

1.0 TABLE OF CONTENTS

2.0 MANAGEMENT SUMMARY

A brief abstract of monthly project activities

3.0 OVERVIEW OF CONSTRUCTION OPERATIONS

Narrative Reports by Construction Activity

3.1 Construction management

3.1.1 Organization and major accomplishments
3.1.2 Site-safety review and report

3.2 Field construction activities by area

3.2.1 Site development and utilities
3.2.2 Foundations and underground
3.2.3 Buildings and structures
3.2.4 Architectural trades

3.3 Major subcontractors

3.3.1 Mechanical contractors
3.3.2 Electrical contractor
3.3.3 Other subcontractors

3.4 Project controls activity

3.4.1 Construction scheduling
3.4.2 Field cost-engineering and -control
3.4.3 Material control
3.4.4 Field engineering and inspection
3.4.5 Design-construction interface

3.5 Procurement Activities

3.5.1 Order placement
3.5.2 Material control

3.6 Administrative activities

3.6.1 Personnel matters
3.6.2 Labor relations
3.6.3 Others

3.7 Design engineering (if applicable)

3.7.1 Status report
3.7.2 Needs list

4.0 SUMMARY OF CONSTRUCTION CONTROL REPORTS

4.1 Cost report summary
4.2 Schedule and progress report summary
4.3 Cashflow report
4.4 Change-order register

FIGURE 13.1 Table of contents for construction progress report.

5.0 PROGRESS CURVES OR CHARTS

    5.1 Construction progress S curves by major activity
    5.2 Procurement commitments
    5.3 Design activities (if applicable)
    5.4 Personnel loading curves
    5.5 Cashflow curve

6.0 WORK TO BE DONE NEXT MONTH

    6.1 Construction work by major area or activity
    6.2 Procurement
    6.3 Design work (if applicable)

7.0 PROBLEM AREAS

    7.1 Information and materials needs list
    7.2 Decisions required
    7.3 Problem areas with proposed solution
    7.4 Review unresolved items from last month

8.0 CONSTRUCTION PHOTOS OR VIDEOTAPES

9.0 DESIGN DOCUMENTATION LISTS (Optional)

FIGURE 13.1   (Continued)

your company's present format to see if it can contribute to better reporting of project performance.

A major report-format decision is whether or not to include complete copies of the project-control reports in the monthly progress report. That may be the way to go on small projects, but it makes a bulky and unwieldy report on large projects. In the latter case, a summary section for the major detailed reports, highlighting any accomplishments and problem areas, produces a more readable progress report. The client's control people are the ones really interested in perusing the detailed project-control reports. It is their duty to pass on any unusual conditions found in the detailed report to their management for action.

The client and contractor management personnel receive and read the monthly progress report, so it should not contain any confidential contractor information. For example, a client should not expect to see the contractor's cost information on a lump-sum contract project. Also, do not discuss any confidential personnel matters in the report. Prepare an addendum covering any confidential company matters as an attachment to your company management's copy.

Construction-project reports should be factual. Painting a rosy picture by ignoring problems only leads to more serious problems later. Bringing problems out in the open as they arise, along with possible solutions, is the best way to handle these situations. This approach shows a positive attitude in the face of an adverse situation. Problem areas require immediate attention from both the owner and contractor to mount a quick and concerted attack to develop the best possible solution.

Do not include a lot of dull and tedious statistics in the report; graphs and charts present data in a more interesting and readable format. Break up long pages with the use of paragraphs. For example, a long list of rebar tonnage set, forms built, and embedded steel items set in the past month may make the reader's eyes glaze over. Detailed information like this may only be of interest and value to a small segment of supervisors—those responsible for those costs; surely management will not express much interest in such details. It is better to list the percentage of each item in place and the balance to complete, or insert a chart showing the amount of foundation in place versus the total amount required.

Preparing the monthly progress report offers an excellent opportunity to look at the report from the owner's perspective. Polish up any areas that many be vague or unclear from the client's viewpoint. Well-presented progress reports can be real point winners toward meeting your career goals. This opportunity comes around every month, so learn from it and improve as the months progress.

Progress-Report Contents

The monthly progress report is an important document, as we have emphasized. Reviewing its contents here is a worthwhile exercise.

Section 1.0: Table of Contents

Using the decimal numbering system makes the table of contents (Section 1.0) and the report itself easier to read.

Section 2.0: Management Summary

This is a key section of the report and perhaps the most difficult to write. This section should give upper management a thorough review of the project in succinct terms. If management is reviewing many projects, the summary section may be the only one that they have time to read, possibly only glancing at the following sections briefly. Ideally, management summaries should be no longer than one page, and never longer than two—then it would not be a summary!

Creating this section for a large project should tax the writer's ability to write clearly and concisely and still cover the ground. It gives the CM or PM a chance to reflect on the whole rather than its parts. It may be difficult to exclude details, but they should be kept out unless they are of such importance that they truly belong in the summary.

Let us say that the square yardage of suspended concrete slabs is ahead of schedule and that the specifications include a flatness provision. Perhaps a note that the average flatness has achieved the required level as set forth in the specifications would be appropriate.

Through the inclusion of some of these figures, the reader is assured that quantity has not sacrificed quality. But generally, if the reader is interested in more detail, he or she can refer to the section of the report where this information can be found.

The summary need not follow the same format as the complete report. Present the higher-priority item up front, in case the reader does not finish the summary. A news-reporting style is perhaps the closest analogy to the writing style needed here.

Section 3.0: Overview of Construction Operations

This portion of the report is a narrative of continuing operations in the various construction activities. Activity leaders should prepare these sections for the CM or PM in a format suitable for use in the final report. As the CM or PM acting as the editor in chief, you should issue the section leaders a standard format for submitting their input to the monthly progress report. Using standardized formats saves a lot of editing time to get the various sections into a consistent format for your report. Once again, create a template in the computer that is to be used for this report.

Collecting data from each construction activity gives you an opportunity to review each group's performance for the month. Go over each report with the activity leader as part of your monthly goals review. Concurrently, you may collect a few news items for the executive summary.

Section 3.0 is where we go into more detail about each discipline's actual progress and what is needed to carry out the mission. Remember not to overload the section with boring statistics; just accent the positive and discuss the problem areas. The tone of this section will evolve over the life cycle of the project, starting with project initiation and site work and then shifting to facility construction. Do not devote a lot of time and space to those areas that have moved into a passive follow-up mode.

Section 4.0: Summary of Construction-Control Reports

This section summarizes the detailed control reports that are too bulky to include in full. The schedule report will be discussed in Section 5.0, so this section should highlight cost-, material-, and quality-control matters. Most of the information in this section comes from the executive-summary section of the control reports. The change-order log, log of requests for information, and project cash-flow projections are other good items to include in this section. Also, this section should relate to how well we are achieving our overall and specific project goals. Are we still likely to meet our overall goal of finishing the project as specified, safely, on time, and within budget? Of course, if this is a lump-sum project, the question of meeting budget is of little importance to the client.

Section 5.0: Progress Curves or Charts

The progress section will discuss progress by monthly updating of the bell and S curves, if such charts are used on your project. For those CMs and PMs using critical-path schedules, the activities on early start/early finish and late start/late finish will be reviewed. For those activities behind schedule, explain the reason for their being behind and the action that has been taken or will be taken to get the project, back on track.

Overall construction progress is covered in this section. If a construction personnel-loading curve like the one displayed in Fig. 13.2 is being used, the sudden increase, leveling off, or decline in personnel—whichever occurred during the time period of the current report—should be explained. Progress can be compared as shown in Fig. 13.3, which is based on a chart created at the beginning of the project as a "goal to be achieved" chart.

Figure 13.4 is the procurement-commitment chart with actual procurement activities plotted against initial goals. Any variations out of the ordinary need to be explained—why procurement is behind, what plans have been made to speed up the process, and what impact, if any, these deviations have on the overall project progress.

Cash flow as shown in Fig. 13.5 may be of interest on a project under a construction-management contract, a cost-plus-a-fee contract, or a cost-plus-a-fee contract with a guaranteed maximum price contract. When a lump-sum contract has been used, only top management in the construction company has any real interest in cash-flow projections.

Section 6.0: Work to be Done Next Month

In this section, each component of construction details its work activities for the coming month. Key milestone dates falling within this period should be discussed—will they be met or not? A brief description of the activities planned for the period is sufficient: possibly new ones to start, others to be completed, and ongoing ones that are under control.

Section 7.0: Problem Areas

This is another key section of the report and requires careful evaluation so as to not underplay a potential problem nor exaggerate the importance of a problem. Summarize the problem areas in the main body of the report. If the problem creates a delay, go into the ramifications of the delay on the construction progress.

Above all, remember to propose some solutions to the problems you have raised. The top management people reading the report are not going to have the answers. They expect the CM or PM to produce the solutions. Leadership, conflict resolution, and problem solving are the strengths of these managers. This is a fertile area in which to practice the fine art of construction management.

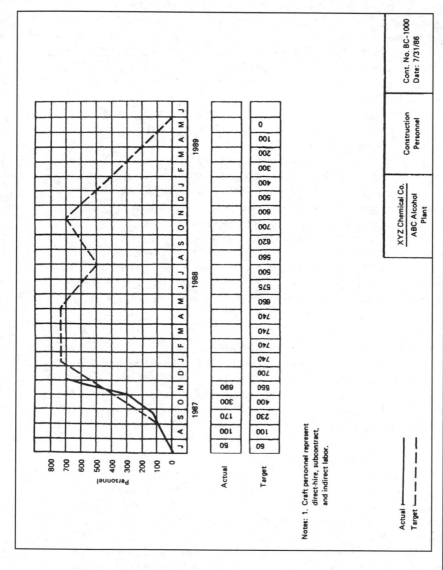

Figure 13.2 Construction personnel-loading curve.

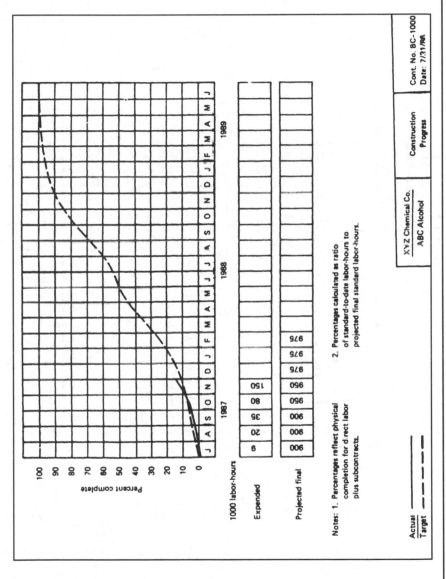

Figure 13.3 "Goals to be achieved" chart.

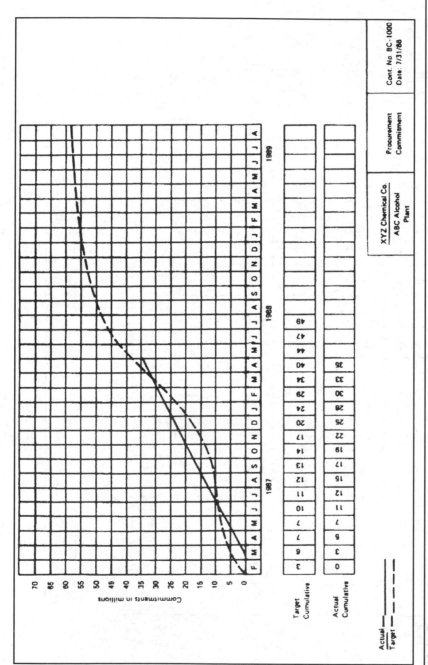

FIGURE 13.4 Procurement-commitment chart.

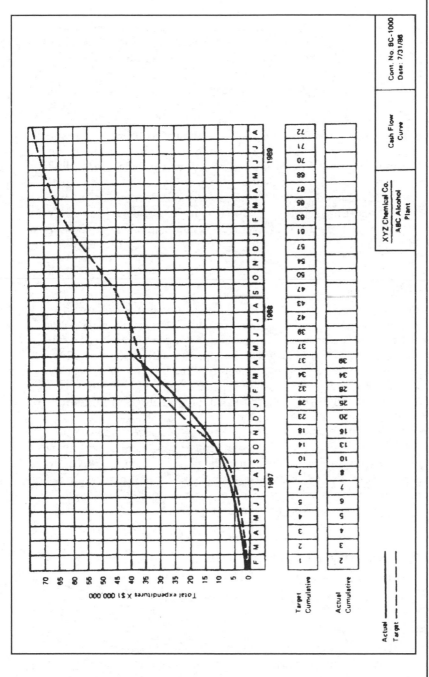

FIGURE **13.5** Cash flow chart for a CM contract, cost-plus-fee contract and GMP contracts.

Section 8.0: Construction Photos or Videotapes

With digital still cameras or video cameras it is quite easy to retain a record of construction progress. When problems occur on the site and it appears that they may lead to a disagreement or dispute with owner, architect, engineers, subcontractors, or vendors, photos can provide powerful documentation. When in doubt, it is best to document with a narrative and a photo.

Site surveillance can be achieved with remote monitoring of the site by the placement of cameras at designated locations. These cameras can record and store the images on a 24/7 basis and allow the construction manager or project manager to monitor the site whenever he or she wishes by accessing the camera in his or her office, at another site, or at home. The system can be set up so that you receive immediate notification of a security breach by phone or e-mail. The record provided by these surveillance cameras can also provide a real-time schedule that may come in handy if a job delay needs to be documented and its impact on the schedule analyzed by the CM or PM.

Project Meetings

Ineffective meetings are major time wasters in project operations. The construction team spends a fair amount of its time in meetings—it is one of the best forms of communication, if it is planned with a meaningful agenda, progresses quickly, and deals with issues raised by resolving them or holding them over until the next meeting, when a resolution is expected. There are working meetings, planning sessions, problem-solving meetings, scheduling meetings, information-gathering meetings, and management-review meetings.

The goal of this section is to make large blocks of time spent in meetings as productive as possible. As construction and project managers are directly responsible for handling or attending most of these meetings—or handing off the responsibility to a staff member or project superintendent—effective planning is an essential ingredient.

The specific goals for improving your meetings are:

- Increasing productivity
- Improving the quality of the results
- Cutting wasted time and money
- Building your personal and company image

Establish the Need

The first consideration is: Is this meeting necessary? Can the objective be obtained in another way, such as a phone call or a conversation with one or two individuals in an informal setting? Unnecessary meetings are a form of overcommunication that is a waste of everyone's time.

When group action is needed to exchange information, resolve conflict, develop ideas, or serve organizational purposes, a meeting is useful. Keep the group as small as possible without excluding someone whose presence may be needed. Meetings with six to eight attendees are preferable; meetings with more than 12 should be avoided.

With smartphones, tablet computers, and personal computers, a lot of information can be exchanged quickly; if a written confirmation is required or desired, a follow-up e-mail will do the trick.

Meeting-Execution Philosophy

Since meetings are similar to a small project, they are the subject of our Golden Rule of construction management: *Plan, organize, control.* These are the key ingredients for a successful meeting. That philosophy applies to all meetings, whether small, medium, or large. Larger meetings are generally more formal and may require a more formalized approach. The meeting checklist shown in the next section applies to all meetings of any size.

It is our recommendation that you follow the meeting checklist for all types of meetings, possibly placing less emphasis on some aspects for smaller, less formal meetings. Although there may be a need for a weekly meeting such as a planning meeting held by the project superintendent with the key subcontractors or vendors scheduled for work that week, in general, meetings should only be held if there is a defined need to do so.

Some construction contracts stipulate that the architect will require weekly meetings and will conduct them, manage them, prepare the meeting minutes, and distribute the minutes to all attendees. These weekly meetings, even though called for per contract might not be necessary as the project kicks-off and settles into a routine. At that point, a suggestion could be made to move them to biweekly, which may be not only more practical but welcomed by the architect and engineers. Although the construction manager or project manager can make the recommendation, its adoption will be contingent upon the architect's obtaining the owner's agreement.

Although it may be rude to cut someone off, if you as CM or PM are conducting a meeting and encounter someone who just likes to ramble on (and we have all experienced that), you need to tell that person that he or she must finish up as soon as possible and move on to the next topic on the agenda. If the person wishes to continue, you can ask him or her to remain after the meeting so as not to tie up the other members. Controlling the meeting means discussing the topic, giving everyone concerned an opportunity to participate if they have something to contribute, and *moving on*!

A Condensed Meeting Checklist

The following checklist can serve as a basic guide for meeting preparation and participation:

Before the meeting

- Explore alternatives to the meeting. Will a phone call or e-mail suffice?
- Set your meeting goals.
- Prepare a detailed meeting plan.
- Make an agenda and prepare any visual aids you plan to include.
- Control attendance.
- Check availability of key attendees.
- Set the meeting's starting time and time limit.
- Distribute the agenda and reference materials before the meeting.
- Select a proper location and facilities.
- Arrange seating and refreshments, if required.
- Assign a recorder to take notes if you are running the meeting.

During the meeting

- Start on time.
- Follow the agenda.
- Control interruptions.
- Keep your meeting goals in mind.
- Record decisions reached and actions needed.
- When action is required, set a date when it is to start and a date by when it is to be finished.
- Do not waste time.
- Be tactful.
- Rotate part-time contributors in and out.
- Adjourn on time.
- If a backup meeting is required, set the date, time, and attendance list.

After the meeting

- Issue minutes of the meeting on time.
- Clear up any unfinished business.
- Follow up on action items.

Reviewing this condensed checklist will give you a good idea of just how involved planning, organizing, and controlling an effective meeting really is.

Planning the Meeting

Thorough planning is essential if you are to meet your goal of a successful, productive meeting. Proper planning is second in importance only to the ability of the meeting manager to steer the meeting.

Meeting plans should be in writing. The amount of writing will be directly proportional to the size and complexity of the meeting. Even small meetings should rate a written outline.

The first step in the meeting plan is to set the goals you wish to achieve. What problems are to be solved? What information is to be exchanged and in how much detail? Setting the goals should go a long way toward getting the desired results and having a productive meeting.

The backbone of any meeting plan is the agenda. As the prime mover for most project meetings, the construction manager or project manager must prepare an agenda for every meeting. It may be necessary to discuss the agenda with some of your staff to see if they have any items they would like to include in the agenda. Failure to prepare a written agenda virtually guarantees that a meeting will be unproductive even before it starts.

Assign a priority and a time allotment to each agenda item to tailor the agenda to the meeting place. The agenda should follow a logical flow. For example, in a problem-solving meeting, state the problem, then give the needed background material before starting on the possible solutions. The agenda logic for project-review meetings can be more flexible in presenting the subject matter. One universal rule for agendas is to place the less-critical items at the end, in case time runs out before you have completed the agenda.

State each agenda item as briefly as possible within the limits of clarity. Each participant on the agenda should have a clear understanding of the matters for discussion under that topic. That is especially applicable to the person who is presenting the subject in the meeting. Actually, the number of items on a properly drawn agenda has no bearing on the time needed for discussion. A well-prepared agenda serves to shorten the meeting time by giving the chairperson an effective tool for controlling the discussion.

After you have properly prepared and approved the agenda, issue it to the meeting attendees before the meeting. Allow attendees enough time to plan and prepare their contributions to the meeting. Be sure to include any written backup material for attendees to read before the meeting. Passing out reading material during the meeting destroys the tempo of the meeting and wastes valuable time.

With the agenda in hand, you can now complete the list of attendees. Limit attendance for working meetings to only those involved in or contributing to the meeting. Attendance is limited because each attendee adds communication channels to the group and increases control problems. Controlling the additional communication channels adds to the difficulty of controlling the meeting effectively.

The meeting plan also must include any visual aids needed to communicate the message you want to deliver at the meeting. These may be flip charts, or any of the other visual aids discussed earlier in this chapter. Use a medium that meets the specific needs of presenting the subject matter to the group. Do not go overboard on visual aids just for show. A chalkboard or whiteboard for simple messages can be just as effective as a PowerPoint presentation, if used properly.

The starting time and duration are important elements in the meeting plan. Calculate the total elapsed time for the meeting so that attendees will be able to plan the balance of their workday after the meeting is finished. Add some contingency time based upon the type of meeting and prior experience, while keeping in mind the productivity goals. Using the elapsed time, back up from a natural cutoff event like lunch or quitting time as a time to start the meeting. A natural time barrier makes it easier for the meeting manager to keep the discussion closer to the meeting goals and agenda. Attendees who may be inclined to wander from the meeting subject seem to be especially sensitive to natural cutoff times.

The window of time for the meeting must be accessible for the major participants if you are to have a quorum. A brief telephone or e-mail poll of key participants can ensure schedule compatibility. Having a couple of options for time and date may be necessary to suit the critical attendees. Of course, qualified alternate attendees are also usually acceptable.

Certain situations may require a dry run for fine-tuning the agenda and presentation; such situations occur regularly in project-review meetings and major management presentations. Include the factor in the meeting plan to allow time for final polishing of the agenda before issuing it.

Organizing the Meeting

Organizing the meeting involves making the necessary physical and administrative arrangements. The key action items are:

- Arranging a suitable meeting site
- Preparing any visual aids you plan to use and assembling any handouts
- Issuing the agenda and venue information
- Assigning someone to record the minutes
- Arranging for refreshments—in the morning, maybe coffee; in the afternoon, perhaps a cool drink
- Making seating arrangements
- Providing pads and pencils
- Holding a dry run if you think one is necessary

Select the meeting site based upon size, creature comforts, visual-aid equipment, proper lighting, and so on. The physical requirements are important to maintain the attendees' undivided attention throughout the meeting. Sometimes for critical meetings, it is even necessary to move the meeting off-site to insulate people from outside interruptions. Perhaps the conference room in your company's office might be a suitable place for some higher-level meetings.

A working meeting can benefit from the use of a portable recording device to record actual proceedings. This also makes it easier on the person assigned to take notes. Publishing the minutes also informs those nonattendees on the project team of the topics discussed at the meeting which may affect them.

It is vital to assign the necessary follow-up actions on matters discussed in the meeting as part of the recording process. This can be accomplished by providing an actions column in the right-hand margin of the minutes of the meeting. This makes it easier for the meeting leader to follow up on the status of the action items in the meeting. For example, if all agree that a design change will be issued by a certain date, then alongside that comment, in the right-hand column, insert the date by which that design change will be issued.

Prearranged seating assignments usually are used only in major management-oriented meetings, but you may be surprised to see that the seats taken by attendees at the first project meeting will also be taken by them at the second and all subsequent meetings. It is a good idea to arrange management people in the normal pecking order to prevent hurt feelings. That factor is especially important when dealing with international groups having strict cultural customs in that regard. Place upper-management attendees near the meeting leader.

Sometimes the meeting may require only part-time attendance by certain contributors. Schedule the part-timers in and out on an as-required basis to save their time and to keep the meeting within reasonable limits. This gives the meeting leader another incentive to keep the meeting on schedule.

Controlling the Meeting

The meeting leader's ability to control the meeting is the most important requirement for the meeting's success. Even a well-planned meeting will fall apart if it is not properly controlled from start to finish. The meeting leader must always keep the meeting goals in mind while controlling the meeting. Writing the meeting goals makes them clearly understandable and available for reference during the meeting.

Loss of meeting control is a prime cause of failed meetings. Although steering the meeting requires a firm hand, it also requires tact and diplomacy to permit valuable input from all attendees. Autocratic control often causes potential contributors to withdraw mentally from the meeting.

As the first act of meeting control, the leader must start the meeting on time. This is a good habit to set up, since it encourages perennial latecomers to improve their schedules. Starting late also places the meeting goals in jeopardy even before the meeting starts.

After a few opening remarks and any necessary introductions, the leader should get right to the agenda and follow it closely. Check a few of the milestone targets as the meeting progresses to ensure that you are maintaining the schedule. Some agenda items will finish early and some late, so the schedule should even out. Make agenda changes only if there is a good chance of improving the schedule.

Occasionally, some part of the meeting plan may break down and force revisions to the agenda. Handle the changes very carefully to protect the overall meeting goals. Quick thinking here can obviate the need to schedule a follow-up meeting.

Meeting leaders spend most of their steering efforts controlling the "problem people" in the meeting. These people fall into these four general groups:

- Those who talk too much
- Those who talk too little
- Those who hold side conversations
- Those who want to usurp the leadership of the meeting

Fortunately most people attending construction-project meetings are technical types who are not overly garrulous by nature. When a participant becomes too wordy, the meeting leader must restate the sense of urgency required to keep the meeting on track. If that does not work, the chair must find another means to cut off the harangue. You may have to double-team the offender with the cooperation of another friendly participant.

Drawing out those who say too little can be even more difficult than controlling those who are too wordy. Quite often, the silent types can make a strong contribution to a successful meeting. Encourage active participation by reticent attendees through positive recognition when they do open up. Build active contact by drawing out their suggestions and opinions. Encouraging the clash of differing opinions also opens the way for those who are careful about expressing an opinion.

Control of side conversations is a must to avoid having two meetings at the same time. Side conversations are upsetting to those who are trying to make a contribution and to the meeting leader. If there are any worthwhile ideas in that side conversation, bring them into the full meeting.

Squelching those who would usurp your meeting is the only way to prevent anarchy. Naturally, you should use extra tact if the usurper happens to be your boss. Occasionally visiting executives do take over meetings, but normally they will back off when gently reminded of their actions.

Meeting control involves many human-relations factors in a live performance. A good way to build your skills in this area is to study the human relations involved and practice various ideas in real situations. Use those that work and discard those that do not. Rate your performance after each meeting to see where you can improve.

Another useful meeting-control device is the tension breaker. Quite often tensions will develop during meetings when conflicts are being resolved or sticky points negotiated. Perhaps the most effective tension breaker is humor. It can relax tensions, but use it carefully: It can backfire if it embarrasses someone.

A prompt adjournment should occur when you have cleared the agenda or the allotted meeting time has expired. If some minor agenda items remain, try to settle them without convening another meeting. If they are important enough, however, schedule a later meeting.

Postmeeting Activities

Issue the meeting minutes promptly to remind people to get started on their action items. Do not give them an excuse to forget about their assignments. Make every effort to issue the minutes no later than 24 hours after the meeting.

Some international meetings require signing a *protocol* before the participants can leave the meeting site. That puts extreme pressure on the actions taken during the meeting. In that case, the production of the minutes must begin before the meeting is over.

In addition to listing the action items, the minutes serve as a record of the actions taken during the meeting. They are major contributors to project documentation. Place a copy of them in the project file as a part of the project record.

Shortly after the meeting, the leader should review the meeting results with key staff members to measure overall performance. Some searching questions to ask and *answer* are the following:

- Did we meet our meeting goals?
- Were we properly prepared?
- Did we control the meeting?
- Did we finish the agenda?
- What were the good points?
- What were the bad points?
- Where can we improve next time?

If the answer to the first four questions is "no", the leader needs to work on those areas to improve meeting performance. Also remember the main purpose of the meeting. Follow up on the action items to be sure that the meeting time was spent effectively.

Approaching meetings with a proactive stance, as we have put forward here, should improve meeting performance. Naturally, these ideas should be compatible with project's operating environment. Effective meetings are too important a communication tools to waste, so the quicker performance in this area is improved, the better!

Be a Good Listener

Most of our discussion to this point has involved outward messages, which comprise only half of the communication equation. Receiving the messages is equally important, but often it does not get the same emphasis.

Much construction-management communication deals with both resolution of both business and personal conflicts. Resolving conflicts effectively requires the construction manager or project manager to listen to both sides of the conflict, especially in the case of personal conflicts. Because field-team morale is a key factor in successful project execution, CMs and PMs often play the project-chaplain role. The most difficult disputes and decisions are the ones that wind up on the CM or PM's desk for final settlement. We do not mean to imply that CMs and PMs have unlimited time to spend on these matters, but they must be good listeners and be able to separate the wheat from the chaff to give timely and sound decisions.

We know of no training given in the area of patience, so this must be self-taught. However, there are several good courses for improving listening ability. The speed of improved listening comprehension from such training courses may pleasantly surprise you. A refresher course every few years is also effective in keeping the skill sharp.

Summary

Effective communication in every form is a potent weapon in the construction manager or project manager's arsenal of skills. This vital ingredient is essential for successful construction-project execution as well as good client, subcontractor, and staff relations. The CM or PM must be an effective speaker, writer, presenter, meeting manager, and listener to present his or her self, ideas, and jobs in the best possible light. CMs and PMs can use the best construction technology and have a spotless site, but poor communication skills may conceal this. Therefore, all of you construction managers and project managers, train and practice to become proficient in all areas of communication.

CHAPTER **14**

Human Factors in Construction Management

I f there were no people involved, construction management would be duck soup! Whenever we talk to construction managers about problem areas, *people problems* always head the list. Perhaps it is because we spend so much time working on construction technology methods and procedures and so little time on human factors.

In-depth investigation of human factors in the business world has taken place only in the past several decades. Frederick Herzberg published a book decades ago, *The Motivation to Work,* based upon interviews conducted with more than 200 employees of engineering and accounting firms. He wished to uncover the factors that cause people to become motivated and, conversely, dissatisfied with their employment. He referred to those factors leading to dissatisfaction as "hygiene factors," in the sense that they are maintenance factors required to avoid dissatisfaction. He listed them in order of importance as follows:

- Company policy
- Supervision
- Relationship with the boss
- Working conditions
- Salary
- Relationship with peers

He also identified the motivators that lead to employee satisfaction:

- Achievement
- Recognition
- The work itself

- Responsibility
- Advancement
- Growth

Herzberg found that one of the important factors in achieving employee satisfaction is job enrichment; the job must fit the ability of the employee and give him or her sufficient challenges, and employees who demonstrate an increasing level of ability need to be given correspondingly increasing levels of responsibility.

Herzberg determined that there were two distinct types of human needs:

1. Physiological needs, such as avoiding unpleasantness, which may be met by acquiring more money.

2. Psychological needs, which are needs for personal development that can be fulfilled by activities that cause one to grow.

The general management community has eagerly accepted Herzberg's work on human factors and has sought to apply it. Today there is an abundance of information, seminars, and lectures on human factors available on the Internet. There are organizations dealing specifically with related project- and construction-management issues as well as management-consulting groups such as FMI (headquartered in Raleigh, North Carolina), which offers courses in team building, training, and development.

But construction managers (CMs) and project managers (PMs) need to improve or hone their human-relations skills, because they continue to rate human factors as a major problem in their job. We believe proper handling of the human factors is the critical third leg of the stool leading to successful projects. It resides right up there with our Golden Rule of construction management and project communication as a requirement for successful projects. Even though we *plan, organize, and control* our project very effectively, the project can still fail if we mishandle the human factors. The human element in construction management involves human relations, personality traits, leadership, and career development. Strength in these areas can help us do a better job of managing projects as well as improve our chances of meeting our personal goals. Years ago, good construction managers with a natural flair for human relations consistently had better projects than those without it. With the research and literature available now, we no longer have to depend so much on chance. All we need is the right attitude, study, training, and practice in human relations.

Obviously, we are not professionals in human relations, but our observations are based upon practical observations made over many decades in managing construction projects. This chapter cannot possibly cover all of the human factors involved in construction-project work, because the field is so broad. The areas we have selected for discussion are the more important ones relating to management performance.

Qualities of a Successful Construction or Project Manager

Let us start by taking inventory of the most important personal traits necessary to manage a construction project successfully. You can then study these lists to see how you shape your own qualifications to meet those requirements. A periodic review of your performance in these key areas can be a valuable tool for keeping necessary skills up to date and improving your performance ratings, if your company holds annual individual performance-review interviews.

A successful manager of construction projects must be

- effective at management and administration.
- people oriented.
- decisive.
- strong in communicating.
- resourceful.
- good at solving problems.
- responsive.
- knowledgeable in the business.
- creative and imaginative.
- patient.

A successful manager of construction must possess

- strong leadership and motivational capability.
- high standards of ethics and integrity.
- personal drive.
- physical stamina and mental toughness.
- multidiscipline capability.

These two lists taken as one present an impressive array of characteristics. It further confirms how broadly based construction management really is. Most of the traits involve human factors in one form or another. We have talked about some of these traits in earlier chapters of the book; this chapter concentrates on the traits not previously covered.

Personality

In looking over the lists, it becomes readily apparent that a person matching the requirements should have a dominant personality. Clearly, a person meeting the qualifications ought to be a self-starter with enough drive to stay the course.

We also know that capital projects are not successfully completed by brute force alone. Because of the people-oriented nature of project work, a lot of persuasion is also necessary. Combing the two personality traits results in the dominant-persuasive personality that is ideally suited to project and construction management.

The largest number of construction-management professionals come from the technical side of the business. Understanding the technical nature of the work is a prerequisite for managing capital projects. The intensity of the technical course work may often exclude study in the humanities, which could broaden a technical person's personality.

Much psychological research shows that human beings are products of our forebears and environment. That research also shows that our personalities are formed by the time we reach age seven. We find it difficult to accept those theories at face value because of our own personal situations, which we find change over the years and are shaped by our experiences in life.

Ethics and Integrity

As in any profession, high standards of ethics and integrity are important to project and construction management. A handshake deal ought to be regarded the same as a signed contract, but only those with proven records of ethics and integrity can make those kinds of deals. And there are times when, perhaps dealing with a subcontractor on a matter requiring urgency, ethics and integrity can make the different between "We've got a deal, I'll follow up on the paperwork" and "I can't proceed without a signed change order."

Capital-project managers often have control over huge sums of money, hundreds of millions of dollars, that does not belong to them. Usually accountants thoroughly audit the funds, but there is still a lot of room for possible conflicts of interest or side deals. A conflict of interest can arise on either side of the client-contractor equation. Both parties must conduct project activities on a businesslike basis and show no favoritism to vendors, suppliers, or subcontractors. That includes refusing favors, kickbacks, or gifts. Now, this does not mean you cannot accept an invitation to a subcontractor's party celebrating 50 years in business—that is just showing good business relations— but giving that subcontractor a peek at other subcontractors' bids after attending that party is a path you do not want to travel, because it will be just the beginning of the road to unemployment.

Today most companies have a statement of policy governing the behavior of those employees who handle company business. Be certain that you stay within those guidelines in all your meetings.

You cannot build a professional reputation in this business if you do not adhere to the rules. There are a lot of insidious ways to circumvent the rules, so be alert for any traps. Some firms vying for project business many not be as scrupulous as you are about not bending the conflict-of-interest rules. As a leader, you are responsible for maintaining the ethical standards of the construction team. Set a good example yourself and make it known to the others where you stand on ethics and personal integrity.

Controlling that situation is fairly easy in the United States, but it can be more difficult in countries with cultures that condone and even promote conflicts of interest. We seldom recommend going against management policy, but if a situation should occur, when working in a foreign country, where your ethics may be compromised, upper management should be made aware of that situation and your reluctance to do so and put it in writing! If your management tries to place you in a conflict-of-interest situation, do not accept it. Trading your professional reputation for a job is not a good bargain; if you find yourself working for that kind of company, you are better off without it.

Personal Drive

Managing construction projects is a demanding profession that often generates long working hours, extensive travel, and perhaps even frequent family relocations, particularly if you work for an international construction company. Conflict resolution and problem solving add more stress to the job. All that is taking place within an environment of change, which further compounds the pressure. Many people will not consider these working conditions to be inducements to enter a field of work, but only people who look on this working environment as a challenge and have a strong personal drive to excel in it can be successful. On the plus side, a certain amount of authority, power, and respect comes with the job. The pay is usually attractive, advancement into upper management is possible, and the personal satisfaction of a job well done is invigorating. Effective construction managers and project managers usually are in short supply, so job security is pretty good. An exception is those occasional dips in the nation's economy, such as were experienced in the 1982–1991 recession and the one that began in 2007 and continues as of 2012.

Physical Stamina and Mental Toughness

The mental side of the job demands a sound physical side to support it. Construction managers and project managers should keep in shape to handle their demanding schedules and the physical demands of just walking a 20-acre site or climbing the temporary stairs to the 12th floor of a high-rise building when the personnel lift is not functioning.

Multidiscipline Capability

Construction people with experience across several disciplines are best suited for the job. Most of us become managers by moving up from the technical ranks, but many of us become managers because of our ability to learn and absorb from our technical managers. Engineering degrees are helpful in attaining management positions, but college graduates in other fields—such as social science, English, and business administration—can become their equals or superiors if they have the will to learn and the desire to acquire skills that are absorbed by osmosis. Listening to a structural engineer discuss problems and solutions at a job meeting can be the beginning of absorbing the rudiments of structural engineering, and if this is repeated on job after job, a nonengineer with the desire to learn can pick up lots of on-the-job structural training. The same is true of other engineering disciplines. By asking questions that add to your knowledge of civil or mechanical engineering, you can gain vast amounts of information that will come in handy when you are faced with a problem requiring consultation with engineers in that field. You begin to understand and speak their language.

The most difficult cross-discipline transition is the one from nonprocess to process projects. The key here is the chemical-engineering expertise required in the process environment. If you are making such a transition, a short course in chemical engineering for people who are not chemical engineers or project managers would be helpful. Obviously, it is not necessary to become a chemical engineer, but the training will give you the necessary basics to better understand the inner workings of the chemical-process equipment and systems. That knowledge is especially useful for reading and understanding flow diagrams and for supervising the final plant checkout and acceptance by the owner. The same advice holds for training in any other discipline in which overcoming a knowledge gap will help you do a better job and further your career.

The major weakness of most construction managers and project managers is a lack of skills on the business side of running projects. Earlier we mentioned the missing humanity subjects in most technical-college curricula. Also missing are business courses. The educational institutions leave it to us to get the business training necessary to supplement our technical-management skills.

We can do that in several ways. The first is through on-the-job training, working as a field or area engineer under a construction manager and mentor for a few years. That approach can be most effective, but it can also be time consuming and costly because of the mistakes made during the learning process. A second way is by taking an MBA degree. Preferably that occurs while you are working as an engineer rather than right after you earn your bachelor's degree. That way you can get a better grasp of the business material and apply it to your technical experience. An MBA is a distinct advantage in gaining higher management positions, but it can be costly. Working during the day and going to college in the evening is a demanding task, and one must be willing to endure the time, expense, and wear and tear on the body.

Community colleges established in or near major metropolitan areas throughout the country offer courses in business administration and may not be so demanding, time- and money-wise, as a four-year college or university. There are also any number of online colleges and universities where you can proceed at a more relaxed pace and learn the important aspects of business administration and how they may relate to your job as construction or project manager. There are also institutions such as the Construction Industry Institute, associated with University of Texas Cockrell School of Engineering, which presents seminars and various types of training opportunities.

Human Relations

The area of human relations in construction management is broad because of the many interpersonal contacts involved in a largely litigation-oriented environment. In addition to external human relations, there are the personal ones concerning the CM or PM's career. Of course, the two are closely related. The rest of this chapter covers those crucial human factors and picks up the traits not covered in the previous lists. We have found the following areas to be the most critical to effective construction management:

- Client-contractor relations
- Contractor-design consultant relations
- Subcontractor relations
- Contract administration
- Project relations (working with fellow supervisors)
- Public relations
- Labor relations
- Leadership
- Common sense
- Coolheadedness
- Negotiating ability
- Patience

Client-Contractor Relations

When contractor's construction and project managers are polled about problem areas, client relations seem to come out on top every time. Interviewing owners will probably result in the same result.

Years ago, the litigious nature of the construction industry seemed to strain the relationship between contractor and owner. But over the past several decades both parties have begun to realize that collaboration, reasonableness, and risk sharing have reduced tensions that often crept into the contractor–owner/client relationship. The whole raft of new contract formats developed by the American Institute of Architects and the Associated General Contractors of America between 2007 and 2010 had several purposes, one of which was to develop ways for owners and contractors to work together harmoniously and achieve their mutual goal: high-quality projects that are produced within the budget and afford a reasonable profit for all involved.

Developing good client relations not only makes the construction process run smoother but makes the life of all participants a little less stressful.

The Associated General Contractors of America's Consensus-Docs stress collaboration, close and frequent communication, and quick and equitable resolution of disputes, as well as recognition of each party's goals. The Integrated Project Delivery contracts from the American Institute of Architects contain similar language. For example, contract C191 2009, Standard Form Multi-Party Agreement for Integrated Project Delivery, includes in its section on risk sharing a blanket statement that the parties waive claims against each other except for seven egregious exceptions that range from willful misconduct to damages arising from liens and claims.

Contractor-client relations involve meting project goals for two very important aspects of the business meeting. They are too valuable an asset to be sacrificed in a breakdown of human relations. Contractor–owner/client relations can make or break

- the success of the current project, or at the very least the satisfaction of the project goals.

- the contractor's opportunities for future work with that client.

These are two powerful incentives for making the contractor-owner/client relationship work throughout the project. The project and construction managers can give up on meeting their personal goals, but they do not have the authority to give away their company's goals. They accepted those goals when they took on their respective project assignments.

It is impossible to write a scenario in which everyone has successful relations with a client or contractor. After all, a lot of personal chemistry is involved. A lot can happen over the complex course of a normal capital project. However, there are some things we can tell you that will improve the odds considerably. The primary necessity in any —contractor-owner/client scenario is for both parties to realize that a good relationship is essential to project success. Neither the project nor the construction manager can afford to have the project branded a failure, regardless of the personalities involved. Poor contractor-client relations can completely wipe out an otherwise outstanding performance. That makes an effective working relationship the number-one priority throughout the project, both for your company and for your personal reputation in the business.

No two people will handle a given situation in the same way. You must handle human relations in a natural, unforced way to make any solution work. You will have to try various techniques with each client to see which one works for the given situation. Some key areas in the contractor-owner/client relationship are discussed in the following subsections.

Ethical Conduct

Any failure in the key area of ethical conduct by either party will seriously undermine the relationship from the start. Neither party can condone unethical practices by the other. Fortunately, unethical conduct does not occur very often. When it does, squelch it immediately. PMs and CMs have a fiduciary responsibility to their respective management organizations to spend the project budget wisely. That does not leave any room for anyone on the project to show favoritism or accepts kickbacks from vendors, suppliers, or subcontractors. Be sure that everyone on your project has a copy of the company's code of conduct and abides by it. If your company does not have a code of ethics, write one for the project to define your own ground rules—and suggest to upper management that they produce a company-wide code of conduct.

Responsiveness

Contractors must be responsive to the client's needs first. This is a typical buyer/seller arrangement that places the customer in a preferred position. It does not, however, give the customer an automatic right to make unreasonable demands on the contractor or supplier. But if you encounter such a client, you will need to find a way to counter those unreasonable demands. Referring to the contract may be one way of pointing out that what the client wants is not part of the contract obligations. If the client is not pleased with the pace of the project, point out some of the difficulties you have experienced and overcome, all as a part of your job.

Responsiveness is also a two-way street: Clients must be equally responsive to the contractor's needs for proper execution of the work. It is vital for clients to make their input and approvals in a professional and timely manner as defined in the contract. A strong effort by both client and contract managers in this area will result in highly successful working conditions. Often the way to the client can be via the architect. Since you are both professionals, you can state your case with respect to a problem—e.g., slow response to a proposed change order—and have the architect advise the owner that prompt review and approval or rejection is required to maintain the costs proposed by the contractor. Delays in approval will only drive the costs higher, since the contractor will have to double back and possibly demolish part of the work before instituting the change.

Mental Toughness

Weakness, just like brashness, is seldom respected in any culture. To keep control of their projects, construction managers and project managers should resist domination by their client counterparts. The buyer/seller relationship is not grounds for an uneven playing field. If there is any doubt in your mind, the contract should define the client-contractor relationship. We recommend that you gently but firmly cut off any one-upmanship activity whenever it occurs on your project. People who are prone to that type of activity will usually get back into line when pressed on the matter.

Subcontractor Relations

Hopefully the subcontract agreement with the various subcontractors was negotiated in such a manner that, although the price may not have been exactly what the subcontractor had in mind, it was a fair and reasonable on. Subcontractors are the core of the construction process and are selected first of all on the basis not only of price but also of performance either on a previous project of your company's or on other contractors' (learned about through investigation). They should be treated much the way you treat your own staff: fairly, professionally, and with reasonableness. We need them as much as they need us.

Contract Administration

Contract administration is an area that sometimes produces friction in the client-contractor relationship. Even the best-crafted contract requires some interpretation from time to time. It is important that both parties take a proactive stance on meeting contractual requirements. This avoids friction over the need for one party to constantly remind the other of the contract requirements.

There are areas of give-and-take in any contract that can make a contractual relationship run more smoothly. However, neither party can give away the store. Likewise, the other party should not ask for it. This area requires a lot of judgment for proper handling, so approach it with caution. Take a few smaller, calculated risks until you get the feel of it.

Relations Among the Project Staffs

Construction managers and project managers are responsible for the contractor-owner/client relations throughout their field staffs. They must resolve any interstaff human-relations problems not resolved at the individual level. The number of personality problems seems to be directly proportional to the size of the staff. The best way to reduce problems in this area is for the client and the construction manager or project manager to set good examples as role models for the rest of the staff. Also, be sure that all members of your staff know the policies and how you expect them to handle their contractor-client relations.

These are a few areas that we feel are critical to ensuring good contractor-owner/client relations. Remember, try different approaches to suit different situations until you find the combination that works for you. Also, nobody bats a thousand in the area of contractor-client relations. There will probably be a time when nothing seems to work, so making a change is inevitable. If it happens more than once or twice, there is cause for concern. You must review your performance and correct the basic cause of the problem.

Internal Project Relations

The next most important human-relations problem area is the field team's working relationships with each other and the rest of the company. Eventually you and your key team leaders will have business contacts with everyone in the company, from the general manager down to the mail-room staff. As the field team's leader, you are responsible for seeing that these contacts are working smoothly.

The most frequent contacts are those with their peers in the firm's home office, as well as subcontractors, vendors, and government agencies. The home-office contacts can include department heads and managers of construction, design, procurement, safety, and estimating. They may be people who will be supplying the members of your project team with assistance in accounting, estimating, or procurement, or giving technical input to the work. To a lesser degree, you will have contact with the business-development people, corporate legal staff, and operating management. Certainly some of the items we covered under—contractor-owner/client relations also apply to these relationships: ethical conduct, responsiveness, mental toughness, and fulfilling commitments.

All this makes it more important for you to establish yourself as a knowledgeable manager with a mature outlook and respect for other people. It is essential to command respect and cooperation from the people who must perform well if your project is to succeed. That reputation is not won by putting on your Superman or Wonderwoman suit every morning before leaving for work. We build our respected reputations gradually, brick by brick, stone by stone, much as we build our projects. Throwing the stones and bricks at others, even when they seem to be asking for it, will not build you a solid reputation with your peers.

Our philosophy is that it takes everybody pulling for you, in addition to a good performance on your part, to have a successful construction project. Having a few people waiting around to pull the rug out from under you increases your chances for failure.

You should also remember that project relations works in all directions—up, down, and sideways. Most managers are attentive to their upward relations as a matter of personal survival. It is from there that both the rewards and retribution flow. We certainly must maintain those relationships if we are to meet our project and personal goals. Relations downward are sometimes another matter. Here we hold the power and authority over the field staff. How we dispense that power and authority is a major factor in maintaining good project morale. Good project morale is the intangible factor that has a strong bearing on a successful performance. You are not likely to meet the project goals of quality, safety, schedule, and budget without it.

Labor Relations

Labor-relations policy and practices are without doubt the most critical human-relations area in the construction arena, because labor relations impinge so heavily on day-to-day site operations and the ultimate success of the project. CMs and PMs have to understand their firm's basic labor-relations know-how to ensure effective choices when making field-level labor decisions.

Any contractor's overall labor-relations policies must be based on their labor posture and the labor laws in effect at the construction site. The basic body of U.S. labor law is enacted by Congress, but there are many state labor laws that further affect the labor practices for construction activity in that state. The most notable of the latter type are so called "right to work" laws, which permit open-shop or merit-shop construction activities along with union shops. The intent of most labor laws is to define and protect the rights of labor and management in an evenhanded manner; the laws have been around for a long time and are well supported by a broad base of case law. Figure 14.1 is a compilation of employment laws, regulations, and technical-assistance services.

In addition to the labor laws directly affecting a project's working rules, there are other laws covering affirmative action and equal opportunity for employment of minority workers. Most contractors have the required plan in effect to comply with these laws—particularly if they are involved in any public-works projects, which generally include provisions for minority hiring and affirmative-action implementation. The project managers and construction managers may need to meet with their purchasing department to insure that these goals are being met and possibly to suggest some subcontractor firms that might be considered. Just as with safety, citations in this area can lead to lawsuits, fines, bad publicity, and termination of contracts.

The main goal of any labor policy is to maintain high productivity and reduce work stoppages which hamper job progress and increase labor costs. Local labor practices and usages develop specific guidelines covering all situations, particularly in a union-shop setting.

The basis for on-site labor relations is the union contract or the open-shop labor conditions that set the working rules for the field labor. Sometimes a supplemental agreement called a *site agreement* is issued to define the special conditions in effect on a specific site for

Employment Law Guide:
Laws, Regulations, and Technical Assistance Services

Prepared by the Office of the Assistant Secretary for Policy

This *Guide* describes the major statutes and regulations administered by the U.S. Department of Labor (DOL) that affect businesses and workers. The *Guide* is designed mainly for those needing "hands-on" information to develop wage, benefit, safety and health, and nondiscrimination policies for businesses.

Statutory and regulatory changes will occur over time, which may affect the information in this *Guide*. For the latest information on all laws check this site periodically.

Table of contents

Updated: September 2009

FIGURE 14.1 Compilation of laws, regulations, and technical assistance services available from the Federal Government.

Employment Law Guide - A Companion to the FirstStep Employment Law Advisor

- Registered Nurses (H-1C Visas) - *Immigration and Nationality Act - H1-C*

VI. Federal Contracts: Working Conditions

- Wages in Supply & Equipment Contracts - *Walsh-Healy Public Contracts Act*
- Prevailing Wages in Service Contracts - *McNamara-O'Hara Service Contract Act*
- Prevailing Wages in Construction Contracts - *Davis-Bacon and Related Acts*
- Hours and Safety Standards in Construction Contracts - *Contract Work Hours and Safety Standards Act*
- Prohibition Against "Kickbacks" in Federally Funded - *Copeland "Anti-Kickback" Act*

VII. Federal Contracts: Equal Opportunity in Employment

- Employment Nondiscrimination and Equal Opportunity in Supply & Service Contracts - *Executive Order 11246 - Supply and Service*
- Employment Nondiscrimination and Equal Opportunity in Construction Contracts - *Executive Order 11246 - Construction Contracts*
- Employment Nondiscrimination and Equal Opportunity for Qualified Individuals with Disabilities - *The Rehabilitation Act of 1973*
- Employment Nondiscrimination and Equal Opportunity for Covered Veterans - *The Vietnam Era Veterans' Readjustment Assistance Act*

INDEX OF LAWS OF PARTICULAR APPLICABILITY TO AN INDUSTRY

Agriculture

- Worker Protections in Agriculture - *Migrant and Seasonal Agricultural Worker Protection Act*
- Child Labor Protections (Agricultural Work) - *Fair Labor Standards Act - Child Labor Provisions*
- Temporary Agricultural Workers (H-2A Visas) - *Immigration and Nationality Act - H-2A*

Note: Under the authority of the Occupational Safety and Health Act, OSHA has issued a number of safety standards that address such matters as field sanitation, overhead protection for operators of agricultural tractors, grain handling facilities, and guarding of farm field equipment and cotton gins. Contact the local OSHA office*(http://www.osha.gov/html/RAmap.html) for more detail (1-800-321-OSHA).*

Mining

- Mine Safety and Health - *Mine Safety and Health Act*
- Black Lung Compensation - *Black Lung Benefits Act*

Construction

- Prevailing Wages in Construction Contracts - *Davis-Bacon and Related Acts*
- Hours and Safety Standards in Construction Contracts - *Contract Work Hours and Safety Standards Act*
- Prohibition Against "Kickbacks" in Federally Funded Construction - *Copeland "Anti-Kickback" Act*
- Employment Non-Discrimination and Equal Opportunity in Construction Contracts - *Executive Order 11246 - Construction Contracts*

Note: Under the Occupational Safety and Health Act, OSHA sets and enforces construction safety and health standards. Contact the *local OSHA office(http://www.osha.gov/html/RAmap.html) for more information (1-800-321-OSHA).*

FIGURE 14.1   *(Continued)*

the duration of the project. The purpose of the contract or site agreement is to lay down a detailed set of rules governing the use of construction labor on the project. The agreement sets out the working hours, pay scales, premium time rates, fringe benefits and their relationship to premium time rates (do all or only some of the associated fringe benefits increase by half?), grievance procedures, and various management and labor policies. The CM or PM uses the site agreement as the basis for administering ongoing labor relations at the site. Those items that cannot be resolved in the field are referred to the home office for intercession and resolution.

The use of nonunion, merit-shop, and union labor varies from state to state and sometimes between major cities within a state. Unions are able to supply contractors with the large numbers of skilled workers often required by contractors working on major construction projects. The nonunion builder's organization Associated Builders and Contractors (ABC) has been operating accredited apprenticeship-training programs to add lots of skilled open-shop trade workers to the industry.

ABC Central Ohio, as an example, offers training in any number of trades—carpentry, concrete finishing, electrical, heavy-equipment operation, HVAC, plumbing, masonry, sheet metal, and fire-protection work. So the truism "If you want skilled labor, you've got to go union" does not hold true in many instances.

It is the nonroutine matters that cause most problems on construction projects. Chief among them are jurisdictional disputes resulting from two or more union crafts claiming authority over a certain operation. We recall working on a high-rise elderly-housing project utilizing union labor. A truckload of refrigerators arrived and our project manager told the labor foreman to unload them and distribute one to each apartment. "Wait a second," the HVAC union representative said, "they have compressors and refrigerant, so we claim the unloading, not the laborers." The electrical union representative chimed in: "They have electrical connections, we claim the unloading." A ridiculous situation, but we had to have a composite crew of workers including plumbers, sheet metal workers, electricians, and laborers unload those refrigerators.

The debilitating effects of craft jurisdictional disputes are reduced by holding a prejob conference. The various craft business agents meet with the construction management to resolve any differences and stake their claim to certain types of work before the job begins. That is the time to establish policy, which can be a tricky process of give-and-take. Discuss each major work activity in detail and agree on who will handle it. The agreements reached when all participants are there and all sign off, in writing, will probably solve many problems, but others will arise, to be sure.

The contract with the owner may include a provision for time lost to labor stoppages as a specific item that allows for excusable delays, or it may place that into the category of force majeure. It is best to read the contract early in the project to know what impact labor-dispute delays may have on the schedule and how the contractor can deal with them without incurring penalties.

A last resort is using the company's legal counsel, but this should be avoided if possible, because it may sever whatever fragile relationship you have established with the various trades.

Public Relations

The construction manager or project manager usually acts as the public-relations (PR) representative for the construction project, since he or she is the one most familiar with the status of construction, scheduling, and highlights of the program. It is natural for you to want your project to be presented in the most favorable light. A proactive approach to project PR is the best way to accomplish that. Bad PR can even get in the way of meeting the project goals. CMs and PMs ought to be alert to newsworthy events in their projects that may be of interest to the local papers or business or technical journals. This is one of those subjects not covered in our technical education, so developing a feel for newsworthy items is another self-taught skill.

It is even important to take a proactive stance on items that are likely to be detrimental to your project or your company's corporate image. Try to look ahead for any negative PR concerning the project that might be developing in your community. Make sure that adverse publicity gets a fair counterpresentation the first time around. If an adverse or erroneous item gets to the media, it is virtually impossible to get a good reaction later.

Establish the initial contacts with local government and community officials before opening the site. Tell them how you plan to deal with traffic, noise, dust, and dirt so they are assured that you are paying attention to some of the problems experienced by the community on previous construction projects in their area. To keep the public on your side, stress the positive effects of the project on the community. The project or construction manager often acts as the project's technical representative in speaking to civil and government groups about environmental matters. These presentations require careful preparation, including clearance from management. Be especially careful in handling the news media. Handing out a well-written and well-checked news release is safer than making an impromptu presentation.

Leadership

Leadership is an area that touches on most of the other human-factor areas in construction management. It is a crucial requirement for effective practice of total construction project management. Building leadership and motivation skills is vital to becoming successful in construction management. The following list of eight areas of importance in management illustrates our thoughts about the nature of leadership in project management:

1. Peer skills—the ability to set up and maintain a network of contacts with equals

2. Leadership skills—the ability to deal with subordinates and the kinds of complications created by power, authority, and dependence

3. Conflict-resolution skills—the ability to mediate conflict and handle disturbances under psychological stress

4. Information-processing skills—the ability to build networks, extract and validate information, and transmit it effectively

5. Skills in unstructured decision making—the ability to find problems and solutions when alternatives, information, and objectives are ambiguous

6. Resource-allocation skills—the ability to decide among alternative uses of time and other scarce resources

7. Entrepreneurial skills—the ability to understand the position of a leader and the leader's impact on the organization

8. Introspection skills—The ability to examine one's own thoughts and feelings

Each of these areas has a direct bearing on proactive construction management. Let us tie them into our discussion.

Peer Skills

We discussed this point earlier in connection with setting up effective working networks with key field staff, home-office department heads, and other operating groups. The contacts with peer groups are vital for the necessary outside support to enable construction managers and project managers to execute their projects. Those networks are built and maintained largely through a mature professional performance and a cooperative attitude with your peer groups.

Leadership Skills

Management delegates to CMs and PMs a lot of authority needed to fulfill the project goals. Knowing how to use that authority to build and motivate an effective project team is essential to successful leadership in the construction area. Wielding authority is the skill most involved with human relations.

Conflict-Resolution Skills

Construction management abounds with stressful conflict-resolution situations. Managers must learn how to cope with those emotionally charged situations quickly, calmly, and fairly without damaging working relationships and project morale. Try to present decisions to the parties involved as win-win situations. That will allow participants in the decision to keep their self-esteem and maintain enthusiasm. There are two keys words to keep in mind that will help you in developing these resolution skills: *give-and-take* (be willing to concede a minor point if that is what is will take to effect a compromise) and reasonableness (is your position reasonable and is the other party's position reasonable?). Attempt to have each party consider the reasonableness of their position; compromising and moving on is better than having the problem advance to a dispute or claim.

Information-Processing Skills

Virtually all project communication passes over the construction manager or project manager's desk. In addition to assimilating that information, the CM or PM must get out into the trenches to find out what is going on. That requires all communication skills: speaking, listening, reading, writing, and presenting information. The information gained is not worth anything until it is analyzed to determine its effect on project performance. This area is crucial to maintaining the good client-contractor and project relations discussed earlier.

Skills in Unstructured Decision Making

Positive leadership leaves no room for shilly-shallying. Decisiveness is one of the character traits we listed for successful construction and project managers. Shooting from the hip is hazardous, and should be avoided. Make use of the time available to make the decision, but do not drag it out unnecessarily.

Resource-Allocation Skills

Project and construction managers oversee the disposition of all project resources. That includes time, money, people, material, equipment, and systems. Each of these makes a contribution toward meeting the project goals, so allocate them wisely. That skill interacts closely with the decision-making just discussed.

Entrepreneurial Skills

This is our favorite. The project is your *business* (profit center) to run. Some people require a small-business approach while others must assume a large-business approach; both are dictated by the physical and monetary size of the project. Managing a $1 million project requires a different business approach from managing a $150 million one. In either case, you should take sound calculated risks when the payout looks good. Through *internalizing* your project numbers, you should know what to expect when the project reports arrive. Most entrepreneurs are creative, so use some imagination running your business.

Introspection Skills

In addition to understanding your position as leader, introspection skills mean periodic self-analysis of your total performance. Is everything possible being done to reach the project goals? Does your performance measure up to the personal standard you have set? Are your leadership skills getting effective results?

Applying introspection to your job environment is one of the best teaching tools available to you. Management schools and training courses can only point you in the right direction. Introspective practice of management theory is the quickest way to learn what really works for you.

Motivational Skills

As good as these eight points of good management are, we also must add motivational skills. Effective leadership in construction management is founded on having a fully motivated supervisory staff and labor force. Success in this key area will ensure a safer, more productive, and smoother-running job that will meet your project goals.

As you reviewed these nine areas of importance to leadership, we are sure you thought of several specific examples of recent project situations applicable to each one. Run through those examples in your mind and rate yourself on how you actually handled them. If the answers are not good in some areas, try to review your leadership skills to improve the outcome the next time. Making a frank appraisal of your performance a couple of times a year can improve your leadership skills dramatically. Improvement in that area also will help you to mold the dominant-persuasive personality necessary for practicing total construction project management.

Common Sense

Common sense is one of those intangible attributes that some of us were lucky enough to be born with; others, not so lucky, must acquire it. The dictionary defines common sense as "sound practical judgment that is independent of specialized knowledge, training or the like; normal native intelligence." To us, the operative words are "sound practical judgment." As a start, we recommend using sound practical judgment in areas where you *do* have specialized knowledge and training, such as construction management. Areas where you do not have specialized training and knowledge, of course, also call for common sense. You need common sense when people seek to promote the use of impractical ideas on your project.

Suppose, for example, a department head is trying to impose an unproven, overdetailed, costly scheduling technique on your small project. That is when common sense should tell you to ask some pertinent questions: Do we really need it on this type of job? Will it really work as well as you say? Can we afford to experiment with it on such a small job? Can we afford the cost?

The construction-management system is really nothing but the application of common sense. The application of sound, practical judgment in practicing total construction project management is what we have been talking about throughout this book. One learns common sense by observation, practice, and experience.

Coolheadedness

Keeping your cool is another trait we can do something about. Construction and project managers cannot survive by pushing the panic button. They have to learn to deal with panic situations calmly to avoid becoming nervous wrecks and instilling a lack of confidence in their staff.

It is better to reserve your energies for clear thinking on how to solve the problem than to be indecisive or avoid the issue at hand. If the project leader is in doubt as to how to proceed, this feeling of, lack of effective leadership will spread to the rest of the construction team. No serious problem ever was solved by displaying lack of direction or leadership.

Panic is a symptom of extreme worry. If you have a tendency to worry, you also are likely to panic in difficult situations. One way we have used to overcome extreme worry is to analyze the problem to see what the worst outcome is that could happen. Many times it turns out to be less catastrophic than first imagined—you can live with it. Starting from that premise, coolly explore ways to improve on the worst-case scenario. Anything salvaged over and above the worst case is an improvement. In many cases, you can turn the panic situation around, canceling out most of the adverse effects.

Furthermore, having successfully handled the panic coolly makes you look more professional in the eyes of your management and peers. That kind of performance builds your desired reputation as a mature, knowledgeable, and respected practitioner of total construction project management.

Negotiating Ability

Construction and project managers routinely find themselves in negotiating situations. Such situations arise with peers, clients, and subcontractors in almost every aspect of executing the project. They include areas such as staffing, estimating, and scheduling, in addition to such normal areas as contract negotiation, purchasing, and change orders.

Negotiation is one of the management arts that you can learn through training and practice. There are several good courses offered in the management-training marketplace and several good books on the subject.

Remember, a negotiation is a special form of meeting, so a detailed plan and agenda are critical to success. That includes setting your short-term and long-term goals and a profit strategy, as well as selecting a negotiating team, if you are not planning to negotiate the deal yourself.

Let us look at the dictionary definition of *negotiate*: "to confer with another so as to arrive at the settlement of some matter." In order to do that, there is usually a process of give-and-take. You may have staked out a position you wish to take, say, on the scope and price of a subcontract agreement after receiving a proposal from that subcontractor. You wish to obtain the most scope at the best price, and the subcontractor may wish to obtain the highest price for less scope. Hence the need to negotiate.

When preparing for a negotiation, you must have full knowledge of the scope of work involved and the added scope you would like to include at, of course, no additional cost. But because negotiation requires a settlement, you should be prepared to arrive at a compromise and should stake out the items you are willing to compromise on and those on which you must stand fast. Starting out the negotiation, you will likely ask for everything and your subcontractor be willing to give up nothing. Knowing when to press and when to back off requires some experience and knowledge of human nature. Body language can tell you a lot. If the person you are negotiating with assumes a "nothing is negotiable" position in either action or speech, it might be wise to merely say, "Well, it looks like we can't arrive at a compromise, so thanks for coming in." The response might well be, "I just can't give into your requests, but let's talk about how we might be able to get together." You have broken the deadlock and through compromise will be able to negotiate a contract. Someone once said that when both parties finish their negotiation and neither party is happy with the results, that is a successful negotiation.

Preparation is key. Knowing what you would like to achieve and what you would settle for is important when conducting any negotiation.

Patience

Patience is a virtue—most of the time. But do not confuse patience with allowing an existing problem area to continue in hopes that it will go away. Patience is something most of us develop naturally as we mature; we base it on common sense and practical judgment. That is why we often see an effective project team made up of mixtures of seasoned hands and young blood, providing both patience and push.

Patience is akin to controlling your temper in a difficult situation. When you lose your temper, you often lose the outcome of the situation. Patience is a must for the client-contractor relationship discussed earlier. This is especially true in international projects where cultural differences are involved.

If you must have an admitted weakness in your construction-management makeup, lack of patience is the most acceptable. We have found this admission to be a useful tool when undergoing annual appraisals. According to good management practice, the reviewer should ask what you think your weaknesses are. The reviewer will probably not dwell on that as a serious weakness and go on to the next subject. Thus you have neatly dodged the minefield of admitting your weaknesses. You are bound to increase your score by a couple of points.

Personal Habits

To wrap up personal traits exhibited by a manager of a construction project, there are a few items that affect the construction manager or project manager personally. The following practices can improve your performance and promote your career:

- Selling the organization
- Knowing the business
- Doing personal public relations
- Bending the rules (very carefully)

Selling the Organization

Selling the organization is an important area that can relate to contractor-client or project relations. Never bad-mouth your company or any part of it in front of client, contractors, vendors, or management. To do so is a common failing, especially after you have developed a close personal relationship with the other person. If you have such a negative feeling about the company you work for, it is better to look for another job than talk disparagingly to your peers about the company.

When you run into a poor performance in the organization, get it corrected and make sure the right people hear about the corrections. But do not reinforce the ill effects by saying, "I always knew that so-and-so was part of a lousy department." If a member of your team makes an obvious blunder, get into the matter and straighten it out. Then go to the client or management and let them know how the matter was corrected and made right.

On the positive side, make sure people hear about a good performance in an unpretentious way. When that is done properly, you build the image of the construction team and improve morale. Proper handling in that area can make an average performance look good and a good one look excellent. It is more difficult to get a poor performance up to average, but it has been done.

Know the Business

You must have a thorough knowledge of the construction business, particularly the fields in which you are practicing. Keep current on general economic conditions and how they may affect your company's business. How are interest rates going? What is the outlook for inflation? What current events are likely to affect your project or the construction industry?

You should take the time to read your industry's trade journals so that you keep current with business trends and industry trends—for example, the green building industry or the trend toward sustainable construction. Keep current on what the competition is doing, what big projects are breaking, and so on. All that background is necessary to participate intelligently in conversations with your management, clients, subcontractors, and peers. Being knowledgeable about your business builds your image as a competent professional in your field.

Do Personal Public Relations

Your personal PR program is something you have to do yourself. One area of effort is getting letters of commendation at the end of a project from either the client or the client's design group. Naturally, you should not expect them after a good or outstanding performance,—you are only doing your job. Do not ask your client or the design team for letters of recommendation; but quite often, when you complete an outstanding project, you could expect such a letter. And if and when you do, keep it in your file for future use.

Another personal PR activity is submitting newsworthy articles about unusual features on a project in progress for local newspapers or trade journals—always clearing it through management first. Such articles are image and company building and may not only help your career but also prompt inquiries from existing or new clients.

Bending the Rules (Very Carefully)

Bending the rules has to do with the art of construction management. We have spent most of the space in this volume laying down strict rules for you to follow in practicing total construction project management. Now we want to talk about *bending* some rules? What is going on here?

Rules are necessary to keep organization in and chaos out. However, they do offer the disadvantage of stifling creativity and *reasonable* risk taking in some instances. Creativity and risk taking can often pay large dividends if they succeed. We hate to miss out on any benefits of that type, so occasionally it may pay to bend the rules. The only qualification we want to put on this point is that you should never consider bending the rules in the areas of safety, risk analysis, ethics, or morals.

The problem is knowing when and how much rule bending is safe. When you spot an opportunity, analyze the situation carefully from all angles, including a sound fallback position. Justify taking the risk by calculating the potential gains and losses to determine if the bending is really worthwhile. List the pros and cons and weigh each one to determine the chances of success. Make sure you have all the information necessary to make the decision and review the plan with your key peers. If the risk looks sound in relation to the reward, you may want to give it a try.

Learning when and where to bend the rules comes only with time and experimentation. We recommend that you start in a small way and try to build on your experience. The cardinal rule is this: *Make sure the potential rewards warrant the risk.*

Summary

Good human-relations skills in construction management are the key to personal and project success. The construction manager or project manager must develop a persona that fosters an image of an ethical, mature, competent, levelheaded professional. Construction and project managers are responsible for the key areas of client, project, labor, and public relations. They must be effective, introspective leaders and role models for everyone on the project as well as their peers.

Job Description

Title: Construction Manager

Whether it be an entry level construction manager/project manager or one with long experience in the construction business, the responsibilities and satisfaction that goes with a job well done is one of the reasons we endure.

1.0 Scope

As we discussed early in the book, we consider the positions of project manager (PM) and construction manager (CM) the same; they are both charged with the responsibility to effectively execute the owner's project they have been entrusted to complete. The only difference remains in the contract format with the owner—the project manager answers to his or her boss, the general contractor, while the construction manager acts as the owner's agent, issuing subcontractor and supplier agreements in the owner's name. So we will treat PM and CM in the same light as we review the concept, duties and responsibilities, authority, working relationships, and leadership qualities of these construction professionals in this appendix.

The construction managers and project managers manage all aspects of construction execution in accordance with the terms and conditions of the contract with the owner. Key areas include field engineering, field procurement, construction activities, facility start-up, and building commissioning and closeout. Other important areas include contractual and financial considerations, personnel matters, safety, and owner and public relations.

The CMs and PMs are the focal point for all construction activities from project initiation to project completion. They are also active in preconstruction bidding and as the construction representative in precontract discussions with the owner.

Company management shall issue a construction-management charter granting the CM or PM sufficient authority to effectively execute the project within the parameters of this job description.

2.0 Concept

Key PM and CM functions are effective planning, organization, control, and monitoring of field operations. PMs and CMs must plan the work and organize the available people and methods, permitting normal field operations to proceed routinely. In addition, they must be able to respond quickly and effectively to unusual or emergency situations as they arise. They shall also establish written goals and priorities for the field effort early in the project and frequently review them with key field supervisors and home-office management to ensure that the planned objectives are being met.

CMs and PMs must be true managers and organizers, not simply doers or supervisors. They must be effective delegates to project their expertise and management philosophy throughout the field organization. Training and developing potential new construction managers is an important consideration in that process.

The CM or PM shall play a strong leadership role in fostering high morale, teamwork, and a motivated workforce geared toward high productivity and meeting the project goals.

3.0 Duties and Responsibilities

Planning, organizing, and maintaining control over field personnel, subcontractors, and vendors in, what is in effect, an outdoor manufacturing facility with potentially disruptive weather conditions, calls on the best talents of the construction manager/project manager.

3.1 Planning Activities

a. Become completely familiar with all the contract documents and special project and client requirements, and prepare and distribute construction information to those with a need to know.

b. Develop the master plan for executing, controlling, and monitoring the construction activities consistent with company operating management or the project director.

c. Prepare and issue written field operating procedures governing all construction activities consistent with approved plans, specifications, company policies, and owner requirements.

d. Oversee preparation and approval of field budgets and establish procedures for controlling and reporting costs for all phases of the construction work.

e. Oversee preparation and approval of the construction schedule and establish procedures for monitoring, controlling, and reporting field progress through project completion.

f. Establish specific project goals and priorities for all facets of the construction work and issue them to the field staff. Incorporate these into an effective delegation and management-by-objectives system to ensure meeting the goals.

g. Participate with management in establishing the labor posture for field operations and preparing the site-labor agreements.

h. Establish a materials-management plan for buying, receiving, storing, distributing, and issuing the project's material resources. Include a field-procurement section for field-procured goods and services if required.

i. Prepare a subcontracting plan for work to be subcontracted to others. The plan shall include the process of soliciting, reviewing, selecting, and negotiating these subcontract agreements and administering those agreements.

j. Review the project insurance requirements and risk-management plan for field operations. Ensure that the correct coverages are in effect throughout the project.

k. Review the heavy-lift requirements and prepare a heavy-equipment schedule and a policy for small tools and light equipment.

l. Set project profitability goals and plans for meeting them in conjunction with supervisory management.

3.2 Organizing Activities

a. Develop a field-organization chart showing the lines of authority and interrelationships of key personnel and their activities.

b. Prepare job descriptions detailing the duties, responsibilities, and objectives for key field personnel. Incorporate the job descriptions into a management-by-objectives system, setting goals with key personnel and a regular performance-monitoring schedule.

c. Initiate and participate in the selection of key field supervisors and establish the requirements for hiring temporary local personnel.

d. Implement a total field-personnel plan in conjunction with key project supervisors and the personnel department in accordance with the project schedule. Review the plan and eliminate any sharp peaks in personnel loading.

e. Finalize policies and procedures for assignment, transfer, and expense reimbursement of field personnel to the job site.

f. Continually review the field organization and adjust to suit actual project needs, especially near the end of the project.

g. Arrange an internal and owner construction-kickoff meeting to initiate the execution of the construction work in an orderly manner.

h. Issue the field-procedure manual within 30 days of notice to proceed (using holds where necessary). Use the manual for field-staff orientation sessions.

i. Keep the owner's field representative informed of changes in the project organization structure and staffing, and secure contractual approvals as required.

j. Set up a jobsite public-relations program to promote the project and company image in the local area.

k. Organize and promote the site safety program, and monitor it closely to ensure a safe site in compliance with company, owner, and government requirements and regulations.

3.3   Controlling Activities

a. Closely monitor field activities for conformance to contract scope requirements and establish a change-order procedure for scope and field revision. Monitor contractual requirements and recommend adjustments when required.

b. Administer and enforce compliance with the terms of the contract, the construction master plan, the field procedures, and management directives, paying particular attention to quality, guarantee, and warranty requirements.

c. Regularly monitor the systems to control field costs, schedule, and quality and to ensure that the systems are effectively meeting project objectives. All control systems should forecast field activities to project completion.

d. Maintain effective communication with client, subcontractor, and key project participants:

- Conducting construction-progress review meetings on a regular basis

- Issuing quality field reports covering the status of physical progress versus schedule, costs versus budget, material status, etc., and explaining off-target conditions

- Discussing any problem areas along with your recommendation for solving them

- Documenting the field work with minutes of meetings, correspondence, telephone, e-mail confirmations, daily diaries, and other such field-generated notes to build working files and a project history

e. Review field-personnel requirements regularly, ensuring that the human resources are properly matched to the workload and schedule.

f. Establish construction quality-control procedures, ensuring quality construction in accordance with the plans and specifications.

g. Monitor the flow of design documentation, vendor data, and project information, ensuring that all parts of the field work can progress smoothly.

h. Monitor all field invoices and payments to ensure adherence to the cash-flow plan. Periodically review the escalation and contingency accounts to see that they are being properly allocated.

i. Promptly inform company management and the project director of any unusual construction events or problems to keep them current on unforeseen field events.

j. Review and approve all outside communications, ensuring that the field and company images are being presented fairly.

k. Practice control by exception and give immediate attention and corrective action to off-target items.

l. Review with company management on a monthly basis the actual project profitability results versus plan.

4.0   Authority

To have strong and effective control of field operations, the CM or PM's authority must be established in writing and supported by top-management policy and actions. The CM or PM shall have authority to:

4.1 Participate in the selection of personnel who will be assigned to the field and to be consulted prior to any proposed changes in assignment of field personnel. (Note: Some owner contracts require owner approval before key personnel changes are made, so check the contract for any such provisions.)

4.2 Act as the company's representative on all matters relating to labor relations in the field. The CM or PM handles all jurisdictional disputes and grievances in accordance with overall company and site-agreement policies.

4.3 Request the presence of any departmental personnel whose services are required to serve the field operations.

4.4 Arbitrate interdepartmental and interdisciplinary differences on matters pertaining to field operations. Upper management may be required to approve the decision.

4.5 Approve all field expenditures and commitments for the project within any limits that may be set by upper management.

5.0   Working Relationships

To successfully complete a construction project, the CM or PM must have the full cooperation of all departments of the company in addition to the authority granted by top management. To gain this cooperation, the CM or PM must maintain good working relationships within and across all organizational lines in the project and the company. At a minimum, the CM or PM must:

5.1 Cooperate with the project director and office staff to meet the overall project goals.

5.2 Cooperate with corporate staff members, department heads, and other management personnel in matters relating to their assigned areas of responsibility.

5.3 Cooperate with other operating units, management centers, or affiliates so that the best interests of the company and the project are served at all times.

5.4 Keep company operating management and department heads current on all project matters that could affect their operations.

5.5 Be responsive to requests for information and services from client, company, and project operating groups.

5.6 Provide routine and special reports required by the company procedures, operating management, or the owner.

6.0   Leadership Qualities

A successful CM or PM must have strong leadership qualities to ensure that the field organization is performing at top efficiency at all times. To develop into a true lead, the CM or PM must:

6.1 Direct all field operations to meet project and company contractual obligations at all times while maintaining high project-team morale.

6.2 Develop and maintain a system of decision making within the field organization whereby decisions are made at the lowest possible level.

6.3 Promote the development and career growth of key field supervisors and encourage them to do likewise with their people.

6.4 Establish written project objectives and performance goals for all key members of the construction team and review them periodically via a management-by-objectives system.

6.5 Promote an atmosphere of team spirit with the field and client staffs.

6.6 Conduct him- or herself at all times in a exemplary manner, setting a good example for all team members to follow.

6.7 Be a good listener and fairly resolve any problems or differences between project personnel, owners, design consultants, department heads, subcontractors, and vendors that may arise.

6.8 Anticipate and minimize potential problems before they arise by maintaining frequent contact with and current knowledge of all field activities, project status, owner and contractor attitudes, and outside factors that might affect the project.

6.9 Maintain a positive attitude toward the field and project staffs, clients, design consultants, subcontractors, vendors, management, and peers at all times.

6.10 Attack problem areas quickly no matter how distasteful they may be. All problems must be brought into the open and resolved as soon as possible.

Do Not Make These Project-Management Mistakes

- Failing to return an owner's voice mail or e-mail because you deem the question or inquiry not relevant to the project

- Not informing the owner when you are going to miss an agreed-upon delivery date

- Changing some technical aspect of the project because you think you have a better way of doing things—but not informing the owner or the owner's design consultants beforehand

- Only contacting your client's office when you are looking for payment of a monthly requisition or a change order

- Advising the owner that you will be late in meeting a delivery date because you had an emergency on another project

- Holding a luncheon meeting containing your owner's competitor's product

Index